Vita Mathematica
Volume 17

Edited by
Martin Mattmüller

Birkhäuser

Johan C.-E. Stén

A Comet of the Enlightenment

Anders Johan Lexell's Life and Discoveries

Birkhäuser

Johan C.-E. Stén
Technical Research Center of Finland
VTT Espoo
Finland

ISBN 978-3-319-34532-1 ISBN 978-3-319-00618-5 (eBook)
DOI 10.1007/978-3-319-00618-5
Springer Cham Basel Heidelberg New York Dordrecht London

Softcover reprint of the hardcover 1st edition 2014

Cover credit: Silhouette of Anders Johan Lexell cut by an engraver in Berlin in 1780. The signature "Haf" suggests that the artist was probably Johann Lorenz Haf (1737–1802). UBB (Basel). Published with permission.

Printed on acid-free paper

Springer International Publisher is part of Springer Science+Business Media
(www.birkhauser-science.com)

Contents

Preface

Multum adhuc restat operis, multumque restabit, nec ulli nato post mille saecula praecludetur occasio aliquid adhuc adjiciendi.
Seneca — Epistulae ad Lucilium

Man sollte auf alles achten, denn man kann alles deuten.
Hermann Hesse — Das Glasperlenspiel, 1943

La dernière démarche de la raison est de reconnaître qu'il y a une infinité de choses qui la surpassent. Elle n'est que faible si elle ne va jusqu'à connaître cela.
Blaise Pascal — *Pensées, fr. 177 (éd. Gallimard)*

Background to This Biography

For the idea of this biography I am indebted first and foremost to Academician Professor Olli Lehto. After delivering a talk at the University of Helsinki on the occasion of the centenary of Lars Ahlfors in January 2007, he told me about the Finnish-born mathematician Anders Johan Lexell and the letters from his European journey in 1780–1781, which are preserved in Saint Petersburg, Russia [15, 106, 108]. Professor Lehto suggested that I should look for these letters and commemorate this little-known scientist and disciple of the great Leonhard Euler. Having made some preliminary inquiries at the Helsinki University Library, I found the material to be more ample and more interesting than I had imagined. I then realised that a biography of Lexell could, and indeed should, be written.

Prima facie it may seem strange that so far there exists no comprehensive biography of Anders Johan Lexell, who, after all, was among the most prominent mathematicians of his age [23]. To quote the Finnish historian of learning, Ivar Heikel (1861–1952), Lexell was [63] "[...] one of the foremost among the several talented scientists, who during the eighteenth century held a chair at the Academy of Åbo [the Finnish university]." According to the history of the Royal Academy of Åbo [80], Lexell was—besides the naturalists Pehr (Peter) Kalm and Peter Forsskål and the chemist Johan Gadolin—the only eighteenth century Finnish scientist to meet with significant international acclaim. As has been noted in [94], these men are also the only eighteenth century Finnish scientists to have their own entries in the *Dictionary of Scientific Biography* [55].

Of course, this is not to say that nobody had ever aspired to write a biography of Lexell. According to Professor Lehto's biography of the mathematicians Lorenz and Ernst Lindelöf [92], father and son, a friend of the former, Professor Wilhelm Lagus at the University of Helsinki, had actually started to collect material connected with Lexell with the intention of writing a biography,[1] but for unknown reasons the project did not materialise. Except for short biographical entries in dictionaries, e.g. [57, 140], the biographical sketches and articles hitherto published in Finland [34, 66, 93, 94, 121, 123, 136, 153, 158], Sweden [62, 99, 101, 122, 130] and Russia [75, 106–109], being based mainly on national source material, are all somewhat biased and narrow in scope. In view of Lexell's cosmopolitan life and his vast scientific correspondence with learned men across Europe, this is neither fair nor adequate. Lexell was certainly Swedish by nationality and language, Finnish by descent, and Russian in that he had chosen to work in Saint Petersburg, but in spirit he was a scientist who knew no narrow national restrictions. Lexell's life and work are therefore clearly worthy of a wider, international contextualisation.

Methodology and Approach

In writing the history of exact sciences, it is natural also to reflect on the nature of the research as such. One of the main differences between historical and mathematical research is the fact that mathematics proceeds analytically, going from the general to the particular, whereas historical research proceeds synthetically, going from the particular to the general [29, 33].

With my background in engineering, historical research inevitably brings to mind a certain class of inverse problems in physics and engineering, where the objective is to determine the source function behind a certain effect. A biography is indeed an attempt to recreate someone's life in given circumstances (scientifically speaking "boundary conditions"), not only this person's actions, but also his thoughts and sentiments, as well as the dreams and ambitions which lay behind them. This is admittedly a difficult task, the more so as the answer will not be unique or exact: the result depends on the interpretation, the point of view and, ultimately, the interpreter himself. Thus, a certain degree of subjectivity is unavoidable.

[1] Lagus's *Lexelliana*, consisting mainly of copies of letters, are deposited at the library of Åbo Akademi, the Swedish university in Turku. Lagus used the material partly for writing the biography of another Finnish-born scientist at the Petersburg Academy of Sciences, the naturalist Erik Laxman (1737–1796) [87].

Another notable difference between the two sciences lies in the question of proof: to know something and to convey that knowledge convincingly to others, you have to be able to prove it. In the exact sciences, proofs are based on logical reasoning and mathematical rigour; in history, they are based on either verbal or written testimony. It stands to reason that in the latter, the credibility of the evidence must be judged from the reliability of the sources and an assessment of likelihoods.

In the historiography of science, explanations of historical events tend to be divided roughly into *external* and *internal*, depending on whether the impetus for the development comes from outside, or is determined by the central subject of the study itself. To achieve a balanced view, the historian of science should therefore follow at least three lines of inquiry simultaneously [147]: (1) the history of a special branch of science, (2) science during a specific period, and (3) the history of science of the relevant country or region. In the study at hand I have considered: (1) mathematics and celestial mechanics, (2) the science of the latter part of the eighteenth century, and (3) the general state of learning and science in Sweden and Russia. Like a set of co-ordinate axes, these three components will lead us to a thorough assessment of the person Anders Johan Lexell.

On the whole, it has been both a privilege and a challenge to study and describe the life of an exceptionally gifted person such as Lexell, living in an intellectually exciting era and surrounded by highly interesting personalities. I have been fortunate to have access to a rich store of letter material. Lexell was indeed prolific not only as a mathematician and astronomer, having been the author of more than a hundred publications during a life that lasted a mere 43 years (see the list of Lexell's publications in the index part of this volume). He was also a regular correspondent: his letters, written principally in Swedish and French, in an elegant style and neat handwriting, range from strictly formal scientific reports to personal thoughts and sheer gossip from the academic world. Lexell's descriptions of his famous contemporaries are precious not only because he was a clear-sighted observer, but because he inevitably brings to his descriptions his own opinions and prejudices. In their sincerity and candour the letters are also quite amusing. Whether or not these descriptions are reliable, Lexell is nevertheless an important witness of the European Enlightenment. Lexell's principal correspondents were the astronomer and Secretary of the Royal Swedish Academy of Sciences in Stockholm Pehr Wilhelm Wargentin (to whom Lexell wrote at least 112 letters), the Secretary of the Petersburg Academy of Sciences Johann Albrecht Euler (27 letters from Lexell have been recovered), the astronomers Johann III Bernoulli in Berlin (16 letters) and Anders Planman in Åbo (8 letters), as well as the naturalist Carl Linnaeus (von Linné) in Uppsala (7 letters). Moreover, he corresponded with the astronomers Joseph Jérôme de Lalande, Charles Messier, Christian Mayer, Nevil Maskelyne and the Secretary of the Royal Society of London, Charles Morton. Unfortunately, almost all letters received by Lexell are now lost. Nevertheless, it is clear that Lexell was well connected. Given the lack of previous biographical studies of Lexell, it is a deliberate choice by this author to let Lexell speak for himself, when appropriate, and in this way make as much use of the letter material as possible.

Formal Issues: Matters of Style

It is suitable at this point to adduce some formal issues that have changed since the eighteenth century. First of all, we note that scientists of old often used different orthography for their name depending on the language used in the context. The names could appear latinised, germanised, and gallicised, i.e. using French equivalents. In the case of Anders Johan Lexell these are, respectively, Andreas Ioannes Lexell, Andreas Johann Lexell and André Jean Lexell. In later Russian literature, his name sometimes appears russified as Андрей Иванович Лексель.[2] However, as a rule in this book, I have tried to preserve as far as possible the original form. In some of the letter citations, where Lexell uses a clearly different spelling of certain names from today, I have changed the spelling. However, especially in the citations, old forms such as de la Lande have been retained (for contemporary Lalande) as there seems no risk of confusion. As to the transliteration from Cyrillic to Roman script, I have followed current English practice. In cases where the transliterated name is not unambiguous, I have given the Russian spelling.

Lexell's main working languages were Swedish, French and Latin. His native language was Swedish, and although Finnish was spoken by most of the inhabitants of Åbo, Lexell's home town, it is not known to what extent he was able to communicate in Finnish. The same applies to Russian. Latin was obligatory at the university in Åbo, as were also German and French; the latter became Lexell's most important working language in Saint Petersburg. In this book, quotations in French are given in their original orthography, translated or summarised in English when necessary and occasionally corrected, with missing diacritical marks added, to facilitate understanding, while the citations in Swedish, German and Latin have been duly translated. In the process, many charming peculiarities of the eighteenth century writing style, such as addressing the recipient of the letter in the third person, have been sacrificed in favour of intelligibility and convenience.

The transition from the Julian (old style = o.s.) calendar to the Gregorian (new style = n.s.) calendar presently in use was still not complete in Lexell's times. When it started in 1582, the Julian calendar lagged 10 days behind the Gregorian. In the eighteenth century, the calendars differed by 11 days; to date, the difference has increased to 13 days. Among the last countries to put the new calendar system into practice were the Protestant countries of northern Europe (e.g. Sweden in 1753) as well as Orthodox Russia (in 1918), a fact which did not fail to cause much confusion in those days, as it was not always clear which calendar was implied. At the Petersburg Academy of Sciences, the minutes of the conferences were written using the old calendar, while in their international scientific correspondence, the new

[2]The Russian custom of using patronyms is unhistorical when applied a posteriori to Western names: Lexell's father was Jonas; thus, the artificial Иванович is misleading. By the same token, Leonhard Euler did not have a second Christian name, although his father's name Paul is sometimes (by confusion with the patronymic Павлович) attached to it in later Russian literature.

calendar was usually applied. In this book, new style dating is mainly used, but in citations and quotes, the original notation is maintained and marked in parenthesis.

Finally, a note is in order concerning the terminology and the names of the scientific disciplines used in this book. In the eighteenth century, the natural or physical sciences in particular—that is to say: biology, geology, geography, chemistry and physics, in the modern sense—were only just beginning to emerge. For those who applied themselves to these sciences, we use the general term *natural philosopher*. We refer to scientists of flora, fauna and minerals as *natural historians* or *naturalists*. Physics was then largely an experimental science—although steadily taking its first steps towards quantification—while rational mechanics and celestial mechanics were considered a part of mathematics.

Acknowledgement

This work would never have been realised without the help and guidance it has been my pleasure to receive. The Saint Petersburg Branch of the Archives of the Russian Academy of Sciences, the Centre for History of Science at the Royal Swedish Academy of Sciences in Stockholm, the Archives of the Académie des Sciences and the Bibliothèque de l'Observatoire in Paris, the library of St. John's College, Cambridge, as well as the University Libraries in Helsinki, Uppsala and Basel, have kindly supplied me with source material.

For all the help, advice and encouragement I have received, I am indebted to Tatiana Artemyeva, Per Pippin Aspaas, Simone Dumont, Emil Fellmann, Tore Frängsmyr, Carl-Fredrik Geust, Tapio Markkanen, Martin Mattmüller, Gleb Mikhailov, Eva Nyström, Vesa Oittinen, Osmo Pekonen, Jean-Nicolas Rieucau, Staffan Rodhe and Andreas Verdun. I am also grateful to Rod Sturdy for improving my English and to Andrei Nikitin for helping me out with Russian texts. My dear wife Päivi Koivisto has supported me through the process, for which I am infinitely grateful.

Financial support from the Finnish Society of Sciences and Letters is gratefully acknowledged.

Espoo, Finland
September 2013

Johan C.-E. Sten

Poem to Lexell

Cum modo disjectis tenebris Aurora rubebat,
    Oreque purpureo jam dabat alma diem;
Floraque per campos suaves spirabat odores,
    Nec procul a nido laeta canebat avis:
Castalidum subeo sylvas & amoena vireta,
    Haec ubi vox Phoebi reddita nube fuit:
Ipse ego, qui fueram Rex inviolabilis orbis,
    Quem coluit semper magna corona Deum:
Ipse ego, qui mitto cunctis mea lumina terris,
    In terris naevos dicor habere meos.
Nempe meis audet radiis praescribere leges,
    Me servire sibi caeca caterva cupit.
Plura locuturo vehemens dolor obserit ora,
    Ergo cum tenebris hunc jubet ire diem.
Continuo at Pallas descendit culmine Pindi,
    Sicque Dei vanas increpat alma minas.
Musarum Praeses, quae Te dementia cepit,
    Cur Te Lexelli cura laborque coquit?
Lexell assiduos inter celebrandus alumnos,
    Qui claret studiis ingenioque suo.
Visere si tentat tremulae penetralia lucis,
    Da veniam placidus namque ea culpa levis.
Arte patent tali sublimia facta Tonantis,
    Ingenuos juvenes hic labor usque decet.
Interea Phoebus commotam temperat iram,
    Atque viam solitam currere mandat equos.

Complimentary poem to Lexell ("Auctori") written by classmate Fredrik Pryss (1741–1767) and published with Lexell's first thesis (1759). For the English translation, I am indebted to Per Pippin Aspaas.

Talibus auguriis ego ovans, perdulcis Amice,
Officium duxi visa referre Tibi.
Quos radiis addit radios Tua docta Thalia
Laudibus hos addet dextera Fama Tuis.
Perge Tuo studio, studiis obstringere Musas,
Sic tandem curis praemia digna feres.

Just as Aurora reddened the darkness-swept sky,
Gracefully bestowing a new day with her purple-red face;
And as the flowers in the fields began breathing out their delicious scents;
While close to her nest a bird began singing her joyful song;
That is when I entered the idyllic, green woods of the Muses
And all of a sudden the voice of Apollo[1] burst forth from a cloud:
"I, who once was the invulnerable King of the World,
Always revered as a God by a huge gathering;
I, who relentlessly bring my light to all lands of the Earth,
Am said to have spots[2] — by the Earthlings!
The blind heap of fools even dares to sanction laws for my rays
And wishes to see me serve as their slave!"[3]
Intending to continue his complaint, he was overwhelmed by the strength of his wrath.
Thus, he gave orders that this day was to end in darkness.
But at that very moment Athena descended from the heights of Pindus,[4]
criticising the meaningless threats of the God in the following manner:
"Head of the Muses, what madness has struck you?
Why are You provoked by Lexell's meticulous efforts?
Lexell should be praised as the one who among assiduous students
shines forth the brightest of all with his studies and intellect.
Should he attempt to venture into the holiest, awe-evoking woods,
Show him some mildness and mercy on account of the crime's insignificance.
The sublime creations of Zeus the Thunderer are exposed through such ingenuity;
Work of this kind is becoming for all youth of distinction".
Meanwhile, Apollo checked his agitated wrath,
Ordering his horses to continue their everyday path.
With omens of such magnitude I deemed it my duty, you Kindest of Friends,
To joyfully share my visions with You.
All the rays that Thalia,[5] Your erudite Muse, has added to the rays of light,
Will a favourable Fame soon add to Your laudable merits.
Persist with Your efforts to bind the Muses with studies,
Which will eventually yield the rewards You deserve for all Your painstaking labour.

---

[1]Phoebus Apollo, the Leader of the Muses in Greco-Roman mythology. He was occasionally identified as the Sun travelling across the sky with his chariot.

[2]Sunspots were discovered early in the seventeenth century.

[3]Apollo seems to consider science (here, the study of optics) as hubris.

[4]Pallas Athena (Minerva) was the goddess of wisdom and knowledge, and Mount Pindus the home of the Muses.

[5]Thalia is the muse associated with idyllic poetry and comedy, sometimes also with geometry and architecture.

Chapter 1
Setting the Scene

1.1 Introduction

Before exploring the life and work of the Swedish astronomer and mathematician Anders Johan Lexell (1740–1784), it is instructive to put him into perspective in history in general and in the history of sciences in particular. Lexell grew up in eighteenth-century Finland, at that time an integrated part of the kingdom of Sweden. Educated in Åbo (Turku in Finnish), in a small provincial university with a relatively short history of natural sciences, Lexell nevertheless managed to reach the summit of international mathematical and astronomical research of his time. Without being in any way radical in his thought, he adopted and virtually embodied the scientific ideals of rationality and logic, to a degree matching the achievements of European colleagues in the intellectual "hot spots" of the Age of Enlightenment. It is the humble intention of this book to give an account of how this was possible.

With this task in mind, we will address and subsequently substantiate the following points: (1) that Lexell, at the height of his career, belonged to the élite of the mathematicians and astronomers of the Enlightenment, and was not a mere epigone of his senior colleague, the leading mathematician of the century Leonhard Euler, (2) that although he came from rather modest circumstances in the province of Sweden, with no evident intellectual example to follow or special family connections to exploit, he managed to establish himself at one of the most prestigious contemporary research centres, the Imperial Academy of Sciences of Saint Petersburg, (3) that he became in a short period of time a name to reckon with, not only as Euler's closest associate but quite in his own right, maintaining frequent correspondence with a number of astronomers, mathematicians and natural historians, (4) that having successfully demonstrated his talents in Saint Petersburg, the Swedish authorities patiently tried to persuade him to return to his home-country, either to occupy a Chair of Mathematics in Åbo, Finland, or some equivalent position in Uppsala, Sweden, and that his return was a concern at the highest national level, and finally (5) that he was very productive, with a number of important scientific publications in Latin, French, German and Swedish (see the list

J.C.-E. Stén, *A Comet of the Enlightenment*, Vita Mathematica 17,
DOI 10.1007/978-3-319-00618-5_1,

of more than 100 publications in the index part of this volume). He was also diligent as a letter writer with a personal style and integrity. Thus he provides valuable insight into the intrigues and relations in the learned world which he frequented.

This biography is organised as follows: In the present chapter, we depict the intellectual landscape of the eighteenth century, in which Lexell lived and to which he contributed. The scholarly and scientific societies and academies emblematic for the Age of Enlightenment are introduced, and their nature, scope and mission are discussed, with a special focus on the Enlightenment philosophies characteristic of the north-eastern "periphery" of Europe, Sweden and Russia in particular. In order to establish the status and development of the exact sciences in the Enlightenment, we focus on the achievements of some principal characters, namely Leonhard Euler, Jean d'Alembert and the members of the Bernoulli family, whose contributions to mathematics and theoretical mechanics were ground-breaking. In the eighteenth century, progress in astronomy was closely linked with that of mathematics and celestial mechanics, and we examine at the end of this chapter how the astronomical profession could be divided into a hierarchy of classes of theoretical and observing astronomers. The challenges of theoretical astronomy and the state of astronomical instruments of the time are summarised.

Following in the main a chronological order of events, in Chap. 2 we introduce the world in which Lexell was born, his education in the small town of Åbo, Finland, where he grew up and made his name as an able mathematician, but where he had only little hope of coming into contact with world class research, let alone of gaining the permanent university position he needed. His unsuccessful attempts to establish himself in Sweden were subsequently followed by a more successful venture in imperial Russia, where, owing to the efforts of Catherine the Great, the Petersburg Academy of Sciences experienced a veritable research boom, and where a talent such as Lexell's was much appreciated. A major factor behind this success was the return of Leonhard Euler to Saint Petersburg. The topic in Chap. 3 is Lexell's invitation and move to Saint Petersburg and his rapid establishment there as an Adjunct[1] of Astronomy. Chapter 4 describes Lexell's remarkable rise to the position of academician in only a few years. Chapters 5 and 8 discuss in some detail Lexell's main contributions to the sciences of astronomy and mathematics, respectively. Chapter 6 is devoted to illustrating his professional relations in the scientific community in Saint Petersburg and, more generally, in Sweden and in Finland. Here, extensive use has been made of Lexell's numerous letters sent to Pehr Wargentin, Anders Planman and Carl Linnaeus, as well as the minutes of the academic conferences in Saint Petersburg [141]. Chapter 7 features some academic occasions, peculiar events and petty scandals at the Petersburg Academy of Sciences, in which Lexell was involved, at least as a witness. Lexell's close and cordial relations with Leonhard Euler and his family are illustrated with several letter citations.

---

[1] *Adjunct* (the corresponding French term is *adjoint*) was the title of an assistant professor at the Petersburg Academy of Sciences.

The most extensive chapter of this book, Chap. 9, is devoted to Lexell's academic journey in 1780–1781 through Germany, France, England and Sweden. This *grand tour* was preceded by a lengthy and agonising decision process, in which Lexell had to make up his mind whether he should return home to Finland to occupy the Chair of Mathematics that had been offered to him five years earlier or to continue on his successful career as an astronomer in Saint Petersburg. The Director of the Petersburg Academy at that time, Sergey Domashnev, although notorious for his despotic leadership, turned out to be helpful in Lexell's case by offering him this scholarly mission to the learned centres of Europe. We will see that Lexell's experiences from the journey and the personal friendships he made there had a decisive impact on his self-esteem and confidence. The events of the journey are illustrated by excerpts from his letters to Wargentin (translated into English) and Johann Albrecht Euler (in the original French), forming an almost continuous travel journal. These excerpts provide interesting and unique observations of Lexell's contemporaries—using a singular kind of physiognomic analysis—as well as his personal reflections on science, religion, the arts and even politics during the Enlightenment.

Chapter 10 describes Lexell's return from his journey, when the Academy of Sciences of Saint Petersburg was in a state of turmoil. Lexell made some headway towards resolving the problem and tried as long as possible to stay on good terms with all parties. With the appointment in 1783 of a new Director, Princess Catherine Dashkova, order was soon restored in the Academy and the leading role and standing of Lexell grew. Chapter 11 outlines what is known about Lexell's last troubled year following the death, in 1783, of Leonhard Euler as well as Lexell's closest associate in Sweden, Pehr Wargentin. In Chaps. 12 and 13 we attempt to synthesise a coherent picture of Lexell the Man, the sensitive but at the same time audacious person he was, his temperament and moral convictions. For the benefit of future studies, the index part of this volume contains a list of Lexell's known printed publications and an English summary of the contents of the Lexell files in Saint Petersburg (Archives of the Russian Acadenmy of Sciences), where, for obvious reasons, most of the *Lexelliana* is preserved. Material nevertheless exists also in Helsinki (HUB), Stockholm (KVAC, KB, Riksarkivet, Bergianska trädgården), Uppsala (UUB), Basel (UBB, Bernoulli-Euler-Zentrum), London (Linnaean Society and Royal Society) and Cambridge (library of St. John's College).

1.2 The Enlightenment

The concept of Enlightenment (*Lumières, Aufklärung, Upplysning*) lies at the heart of every description of eighteenth-century philosophy and science. This biography is not an exception, and thus it seems reasonable to start by giving an idea of its general meaning and character. After Immanuel Kant's sweeping definition of the Enlightenment as "...Man's emergence from his self-incurred immaturity" (1783) [71], historians of science and philosophy generally distinguish between (at least) three ways of characterising the Enlightenment, namely [51]: first, as the emergence of a rationalistic, critical and emancipatory attitude, induced by the rapid progress in certain areas of science; second, through a chronological definition, i.e. by identifying the *Age of Enlightenment* with the eighteenth century (or a part of it); and third, as a phenomenon of the history of philosophy centred around Voltaire, d'Alembert, Diderot and the project of the *Encyclopédie*.[2]

The *Age of Enlightenment*, just like its predecessor, the *Age of Reason*, is a metaphor describing an exceptionally rapid growth of knowledge taking place in a certain historical period. The Age of Reason is usually considered a seventeenth century phenomenon, a successor to Renaissance humanism and a harbinger of the eighteenth century Enlightenment. In the sciences, all the important groundwork made in the Age of Reason started to bear fruit during the Enlightenment. One could also argue that the Age of Reason marks the beginning of the growing conviction that scientific knowledge is the key to social progress [8]. However, as we will see later, the local manifestations of the Enlightenment in different parts of Europe varied substantially, in particular as to the mode and extent of popularisation, i.e. the dissemination of its ideas in a form attractive and amenable to the general public outside the narrow educated élite. Studies of the different aspects of the European Enlightenment, e.g. [8, 51, 71, 73, 116, 131, 148], have shown that this "progressive" or radical form of the Enlightenment was indeed accomplished quite differently, if at all, in the European countries. Most vigorously it was developed in mid-eighteenth century France, Britain, the German-speaking countries and the Netherlands, Austria, Hungary, Switzerland and Italy. Its main ideas were nevertheless formulated earlier in the Age of Reason and thus, the Enlightenment marked an unfolding of these ideas in a somewhat more explicit and palpable form. This progressive spirit of the Enlightenment is epitomised in the goal of the great *Encyclopédie*, as stated in its introduction, namely to fight against prejudices, superstition, intolerance and the authority of religion. In fact, encyclopaedias and encyclopaedism as an epistemological phenomenon were of great importance for the Enlightenment, both for the propagation of its ideals and as a means of improving the standard of general education. Encyclopaedias such as Johann Heinrich Zedler's *Grosses vollständiges Universal-Lexicon aller Wissenschafften und Künste* (Leipzig, 1732–1754) and Ephraim Chambers's *Cyclopaedia, or a Universal Dictionary of Arts and Sciences* (London, 1728) were among the prime sources of inspiration for the French *encyclopédistes* [177].

The Enlightenment was not all epistemology, however. A narrow definition of the Enlightenment, focusing only on scientific and educational aspects, on science as the most valuable creation of the human spirit, and on the belief in progress and utility of knowledge, has an obvious shortcoming: it ignores a significant part of the intellectual, spiritual and aesthetic trends characteristic of eighteenth century culture [45, 90]. Such trends include religious pietism and revivalism

---

[2] *Encyclopédie, ou dictionnaire raisonné des sciences, des arts et des métiers*, edited in 1751–1772 by Denis Diderot and Jean d'Alembert.

(Herrnhut), the spirit of *rococo* art and the *style galant* in music, the passion for nature (Rousseau), neoclassicism idealising especially ancient Greek art and literature, the sentimentalism and proto-romanticism in literature (e.g. the *Sturm und Drang* movement, the *Songs of Ossian*) and the mysticism of the Freemasons and other secret orders. All these expressions of the Enlightenment were historically conditioned by local political, cultural and religious circumstances.

In most European countries during the eighteenth century, tolerance in religious matters was only just emerging and free-thinking was more or less suppressed. In France, known as the pacesetter of the Enlightenment, the authorities, i.e. the Crown and the Catholic Church, were increasingly challenged during the late seventeenth and early eighteenth century by the so-called *philosophes*. Their favourite target was the Jesuit Order (*Societas Iesu*), which during the past centuries had created an efficient and widespread system of education in the Catholic countries and their colonies. Although the Jesuits undoubtedly were important for the advancement of science and learning [64]—after all, a great number of scientists owed their education to the Jesuits—they had acquired a reputation for political plotting and exploitation, which made them unpopular. The decline of the order began in the peripheral and backward kingdom of Portugal in the aftermath of the disastrous earthquake and tsunami which struck Lisbon in 1755. The forceful Prime Minister, the Marquis de Pombal, initiated radical economic and social reforms in his country, involving the expulsion of the Jesuits (1759) and the abolition of slavery. In France, the Jesuits were banned in 1763. Finally, in 1773, the pressured Catholic Church had to yield by dissolving the much criticised order (only to re-establish it in 1814).

In Protestant northern Europe the intellectual atmosphere varied greatly depending on the political circumstances. Germany, in particular, was a conglomeration of states with varying degrees of freedom and independence. In Sweden, a rigid Lutheran orthodoxy had developed during the seventeenth century, which was strictly observed and only gradually liberalised during the following centuries. Since the Catholic tenets of Albertus Magnus and Thomas Aquinas concerning the realms of knowledge and faith were not officially embraced in the Protestant countries, there arose an urgent demand for a new kind of metaphysics to replace the rigid Aristotelianism, which had been challenged by the new physics of Cartesianism. The change was a slow one because of the profound transformation which it implied for both the structure of the sciences and the obligations of the professors. In Sweden, the quarrels between the Aristotelians and Cartesians were formally settled as early as in 1689 when King Charles XI issued a somewhat ambiguous letter that allowed the Swedish universities the freedom to philosophise (*libertas philosophandi*), except when it criticised the Christian belief and dogma [77]. Even if scholastic principles persisted in natural history (in particular, in Linnaeus's systematisation of Nature), the rise of Swedish science during the first half of the eighteenth century took place in a predominantly Cartesian context. Many of the Swedish university professors had either visited or studied in the Netherlands, e.g. at the universities of Leiden or Utrecht, where Descartes had lived and where his influence was strongly felt. The fact that the philosopher had spent his last days in Stockholm as a guest of Queen Christina of Sweden may also have been important.

The empiricist philosophy, which was critical of the rationalist epistemology of Descartes and was represented in England by John Locke (1632–1704) and in the German states by Christian Thomasius (1655–1728), was still regarded with suspicion.

However, in the 1720s and 1730s, a new form of "Leibnizian" philosophy offered by the German rationalist Christian Wolff (1679–1754) successfully contested Cartesianism especially in northern Europe, and after some initial doubt and resistance, Wolffianism was adopted throughout the Swedish universities [50]. In his work, Wolff combined different ideas from Descartes, Leibniz and the scholastics to form an eclectic system of philosophy, which he tried to organise into a systematic science very much like mathematics, operating by definition and syllogistic reasoning. Wolffianism remained, together with Locke's empiricism (although in part incompatible with it), the dominant philosophy in the Swedish universities until the end of the eighteenth century. In contrast to his teacher Leibniz, Wolff maintained (e.g. in *Theologia naturalis*, 1736–1737) that theology should be kept apart from philosophy so as to avoid a conflict between science and religious belief. In consequence, the Swedish Enlightenment was characterised by a very utilitarian and pragmatic approach to science, whose main purpose was to be useful and profitable. Hardly ever was it radical in the sense of challenging the political authorities or religion [51].

In Russia, scientific research and education had literally to be imported with the establishment of the Academy of Sciences in Saint Petersburg in 1724. The country had been isolated from Western influence for a long time and science had played but little part in its development until the reforms initiated by Peter the Great (1672–1725). Accordingly, for Russia the Enlightenment meant essentially a modernisation and "westernisation" of the state. In view of Lexell's life and work in Sweden and Russia, at the end of this chapter we will take a closer look at the special traits of the Age of Enlightenment in these countries.

1.3 Schools and Scientific Societies

Scholarly and scientific societies and academies were crucial for the development of learning and research in the seventeenth and eighteenth century. Whereas the oldest universities had their origin in medieval monastic schools, the learned societies and academies were creations characteristic of this period [116].[3] These new institutions[4] promoted a new culture of scientific research, involving sociability

---

[3]Because of the rapid growth of the number of scientific societies, the Age of Reason has been labelled the "Age of Scholarly (or Scientific) Societies". The term *L'âge des académies* was in fact coined by the French philosophical author Bernard le Bovier de Fontenelle (1657–1757), *Secrétaire perpétuel* of the *Académie Royale des Sciences*.

[4]The difference between societies and academies was principally that of the organisation and membership. In the latter organisations, membership was usually more limited and firmly

(*sociabilité scientifique*) and debate, while the universities remained responsible for basic scholarly education. However, the members of such societies could also make a pedagogical contribution by the engagement of promising young students as their assistants or adjuncts (adjoints), as they were called, who, after gaining experience, could pursue their career and become ordinary members.

The earliest scientific societies established at the beginning of the seventeenth century were in the main local institutions founded by a small number of men of power, sharing an interest in the natural sciences. Notable institutions of that kind were the Italian *Accademia dei Lincei*, founded in Rome in 1603, the *Accademia del Cimento*, founded in Florence in 1657, the French *Académie des Sciences, Arts et Belles-Lettres de Caen*, founded in 1652, as well as the German *Academia Naturae Curiosorum*, also known as *Leopoldina*, likewise founded in 1652. To the second wave of new learned societies devoted expressly to scientific research belongs the *Royal Society* of London, founded in 1660, and the *Académie Royale des Sciences* of Paris, founded in 1666.[5] At the time of their foundation they were local associations, too, but during the eighteenth century they nevertheless developed, together with the *Societas Regia Scientiarum* of Berlin,[6] founded in 1700, into the most influential scientific societies. Their royal or imperial patronage bears witness not only to the national pride and prestige they were associated with, but also to the ever growing importance of the relation between knowledge and power, both political and economic. In addition to the scientific and scholarly societies, from which non-members were strictly excluded, there also developed less formal literary associations convening in salons, coffee houses and the like, according to local circumstances [71].

The new culture of scientific sociability and debate brought about by the scientific academies and societies became successively evident in the use of languages. As a general rule, anything believed to be seminal and of permanent value in the sciences was traditionally authored and published in Latin in order to be understood throughout the European scientific community. Since the scientific terminology was created in Latin, which was universally taught at the universities, this was a natural choice. However, some of the members of the new societies were not formally educated and lacked a sufficient knowledge of Latin. Thus, the scientific societies, besides promoting natural philosophy and research, also took an active

organised than in the former. The Paris *Académie Royale des Sciences*, to take a well-known example, was divided into six sections (Geometry, Astronomy, Mechanics, Anatomy, Chemistry and Botany), each one comprising three *pensionnaires*, two *associés* and two *adjoints* and a number of *surnuméraires*, whereas the Royal Society of London was more informal and broadminded in nominating new *fellows*: besides scientists, physicians, engineers, technicians and industrialists were also enrolled.

[5]The third wave of foundation of scientific societies took place in the nineteenth century. The societies of this period had a national character, such as the Finnish Society of Sciences and Letters—*Societas Scientiarum Fennica*—founded in 1838.

[6]In a reform of 1744, the *Königlich Preußische Sozietät der Wissenschaften* was renamed the *Königliche Akademie der Wissenschaften*, the Royal Prussian Academy of Sciences.

role in fostering national languages. Their transactions were accordingly filled with a growing number of articles and reports in vernacular tongues. Latin was still used in international scientific communication between the learned, in an informal community sometimes termed *Respublica litteraria*, or the Republic of Letters. Nevertheless, Latin was inevitably losing its former position to vernacular languages, especially French [52]. For instance Lexell, in his communication with the Viennese Jesuit astronomer Maximilian Hell, wrote consistently in Latin, whereas in his correspondence with the astronomers Johann III Bernoulli in Berlin and Nevil Maskelyne in London, the language was French.

Scientific journals of the eighteenth century were typically society organs reserved principally for articles written by the society members, but some of them—notably the *Transactions of the Royal Society of London*—were also open to contributions by non-members through communication to some of its members. Hence, a new international forum to render research public was created involving the beginnings of the modern procedure of peer review. The wealthiest scientific societies also organised scientific experiments and equipped costly expeditions in order to promote the resolution of some scientific question of interest, e.g. to determine the exact shape of the Earth [135, 159], or to observe the transit of Venus, to thereby determine the distance between the Earth and the Sun [176]. Additionally, the Paris Academy of Sciences regularly announced prize competitions for the best solution to a challenging scientific problem. Several academies followed the example. In some cases, the prize sums offered were considerable (in the case of the Paris Academy of Sciences, up to 4,000 livres), but usually modest: the honour involved and the mere prospect of a prize was often enough to attract enormous interest.

1.4 The Enlightenment in Northern Europe

1.4.1 Sweden

Sweden was one of the major political and military powers of the seventeenth century, whose influence and wealth was gradually declining during the eighteenth century. To manage the inconvenience of the territorial and economic losses, science and technology seemed to offer a promising pathway back to prosperity and power. In consequence, two learned societies were established in Sweden following the French and British examples, viz. *Kungliga Vetenskaps-Societeten* or the Royal Society of Sciences of Uppsala, founded in 1710 as *Collegium Curiosorum* ("the society of the curious") and centred at the local University, and *Kungliga Vetenskapsakademien* or the Royal (Swedish) Academy of Sciences, founded in 1739 and based in Stockholm [100, 116, 146]. Both societies were—just like their foreign models—commissioned to promote the sciences and to strengthen their influence on society, in particular "... to develop and disseminate knowledge in mathematics, natural science, economy, trade, useful skills and manufacturing". In spite of their comparatively small resources—only a few persons were actually salaried—the two Swedish scientific societies became quite influential on the European scale. The Royal Academy of Sciences was clearly the more dominant of the two and assumed a special role on account of Sweden's northern dimension: the access to Lapland, its exotic flora and fauna and extreme climate conditions; snow, ice, northern lights (*aurora borealis*) and the midnight sun. Scientific contacts and exchange were entertained especially with France [56,138]. The quarterly proceedings of the Royal Swedish Academy of Sciences, *Kongliga Vetenskaps Academiens Handlingar*, were read and appreciated internationally, in spite of the articles being written consistently in the Swedish language (translations into German and French were produced in due course). The use of Swedish in the *Handlingar* fitted in with the utilitarian idea of rendering the fruits of scientific research accessible to the nation as widely as possible.

In spite of the two newly founded scientific academies, the number of universities in Sweden remained unchanged in the period: the three existing universities, those of Uppsala, founded in 1477, Åbo, founded in 1640, and Lund, founded in 1666, were responsible for all higher education in the kingdom.[7] The capital Stockholm had no university, which fact was partly compensated by basing the Royal Academy of Sciences there, perhaps also to counterbalance the more traditional scholarly society of Uppsala. The main mission of the Swedish universities was to provide the government with civil servants and the Lutheran parishes with priests. As it was, Christianity and Classical Antiquity remained the two dominant intellectual traditions for many decades to come. Original research was not particularly encouraged, and that for political and ideological reasons: new and uncontrolled research could have potentially dangerous consequences, such as heresies and revolutions. Aristotelianism as a natural philosophy ceded only gradually to Cartesianism in the early eighteenth century, while the mathematical physics of Newton and Leibniz entered the curricula in Uppsala and Åbo only in the mid-eighteenth century [77,80]. Although this may seem late at a first glance, in a general European perspective it was not exceptional.

The Age of Enlightenment in Sweden (Finland included) coincides with the era of parliamentarian rule in 1718–1772, a period called "the Age of Liberty", when the authority resided nominally in the Four Estates, i.e. the Nobles, Clergy, Burghers and Peasants. In this period living conditions started to improve through the encouragement of entrepreneurship, agriculture, mining and small industry. Foreign trade was still strictly controlled by the Crown. In the same period, the idea that science should serve the economy and be useful to society became dominant. In particular, the success of Linnaeus's taxonomy (*Systema Naturae*, 1735) and

[7] The University of Dorpat (Tartu, Estonia) was founded in 1632 during the period of Swedish sovereignty of the Baltic region. It was closed down in 1710–1802 when under Russian sovereignty. The University of Greifswald (founded in 1456), although belonging to the Swedish Crown at the time, was of German origin and followed a German curriculum.

the numerous scientific expeditions of his "apostles", as his disciples were called, suggested an extensive search for uses for the natural sciences—botany, zoology and mineralogy—to the benefit of man, the more so as this pursuit was not seen to be in conflict with the divine origin and purpose of the world. Characteristically, in every university in Sweden, a Chair of Economy was established in the mid-eighteenth century, focusing on natural history and agriculture. Astronomy was also considered a prime subject for the Swedish universities owing to its importance in cartography, chronology and navigation [127]. The French-Swedish expedition led by Pierre Louis Moreau de Maupertuis to Swedish Lapland to discover the general shape of the Earth was not only a triumph for "Newtonian science", but an important step for Swedish astronomy as well.[8]

In the dissemination of utilitarian ideas among the common people the clergy played an essential role. In fact, in Sweden and Finland, the learned men who were able to grasp at least the rudiments of science were quite often priests, and of them, the most learned ones were bishops. This was possible owing to a broad university education and the fact that the sciences were not as much differentiated as they are today. Such an arrangement enabled effective and rigid supervision of the intellectual climate and the ideas which the government especially wanted to pass on to society.[9] In Lexell's home town of Åbo, for example, three successive bishops in the mid-eighteenth century—Johan Browallius, Carl Fredrik Mennander and Jacob Gadolin—were former professors of physics at the university [68, 80]. Moreover, these men were either friends or students of Linnaeus. Similarly, Samuel Klingenstierna was all-important for the development of mathematical education and research in Uppsala, and through his influence indirectly also in Åbo [144].

1.4.2 Russia

The Imperial Academy of Sciences of Saint Petersburg, founded by Peter the Great in 1724—although he did not live to see its opening in December 1725—was from the outset remarkably elitist in its mission. Its original task, as envisaged by Peter on the advice of Leibniz among others, was to connect Russia firmly with the European network of learning by producing high quality scientific research, and also to reduce

[8]The Swedish physicist and astronomer Anders Celsius (1701–1744) was a key member of Maupertuis' geodetic expedition to Lapland in 1736–1737. Celsius is also (and perhaps better) known for proposing the (inverse) centigrade thermometer scale.

[9]The Finnish-Swedish naturalist and philosopher Peter Forsskål (1732–1763), who had studied in Uppsala and Göttingen, was among the few in eighteenth-century Sweden who for his learning and attitude is close to the spirit of the "French" Enlightenment [51]. In several pamphlets he argued for civil rights and particularly for freedom of expression in speech and print. In consequence, his university career in Sweden was severely hampered. As an able botanist of the Linnaean school, Forsskål participated in the Danish expedition of the geographer and scientist Carsten Niebuhr to the Arabian Peninsula, but fell ill and died there.

the conservative influences of the Orthodox Church in Russian society by means of the education of youth [116, 134, 170]. This task was effected by incorporating the University and the *Gymnasium*[10] into the Academy (which was usually not the case in Western academies). Furthermore, the Academy was commissioned to serve the interests of the imperial court and the government, in particular to improve the defence of the state and the authority of the Czar. Cartography, exploration and naval science consequently enjoyed a high priority in the Academy's curriculum [115]. In addition to its scientific organs, the Academy had an obligation to edit and publish two literary journals and two almanacs, each in Russian and in German (*Sankt Petersburgische Zeitung*), as well as a calendar of the state.

In the absence of a learned middle class, scientific traditions in Russia had to be created almost out of nothing. As to its personnel, the Imperial Academy of Sciences was a very European institution from the start, as its professors were recruited exclusively from Western Europe, often with considerable salaries and benefits. Its first sixteen ordinary members were all Protestants and German-speaking, except one: the French astronomer and cartographer Joseph-Nicolas Delisle. At the same time, Peter sent Russian noblemen to Holland, Italy and the German states to study the sciences and crafts. Thus, a whole new intellectual tradition was taking shape in Russia, very distinct from the previous ones existing within the Orthodox clergy on the one hand, within the nobility on the other [2]. Among the first Russian-born members of the Petersburg Academy of Sciences, the celebrated polymath Mikhail Lomonosov (1711–1765) was appointed, in 1745, Professor of Chemistry.[11]

During the reign of Peter's daughter Elizabeth I (Petrovna) of Russia, i.e. from 1740 to 1762, the Petersburg Academy of Sciences was troubled by several conflicts, both internal and external [69]. While the court culture flourished and large sums were spent on building for example the Winter Palace, the sciences were allowed to wither and one renowned scientist after another left the Academy, notably Leonhard Euler (in 1741), Delisle and Johann Georg Gmelin (both in 1747). The ethnographical collections were destroyed in a conflagration of the Academy's main building (the *Kunstkammer*) in 1747, and the restoration of the museum took several decades. As the resources were lacking to recruit new members from the West, the Academy eventually fell into a decline such that by the time of the death of Lomonosov in 1765, there were only 15 members left out of which a mere seven were scientists. The President of the Academy from 1746 to 1798 was Count Kyrylo Grygorovych Rozumovsky (1728–1803), a Ukrainian Cossack Hetman, whose elder brother was the morganatic husband of Elizabeth I. However, in 1766,

[10]Especially in the German-speaking world and Scandinavia, a *gymnasium* is a secondary school preparing the student for higher education at a university.

[11]Among other things Lomonosov is credited with reforming Russian grammar, and in spite of not being an astronomer, he took part in the observations of the 1761 transit of Venus, at which occasion he discovered Venus's atmosphere. He was also a driving force behind the foundation of the University of Moscow in 1755. Lomonosov vigorously opposed the claims, maintained by the historians G. F. Müller and A. L. Schlözer, that Russian culture was created by Scandinavian Varangians. His main adversary at the Academy of Sciences was F. U. T. Aepinus [69].

Catherine II (the Great) charged her favourite, the young Count Vladimir Orlov, with the practical leadership, while Rozumovsky's leadership of the Academy continued only as a formal presidency.

At the beginning of her long reign in 1762, Catherine was eager to restore the Academy's former splendour and managed, after some negotiations, to call Euler back to Saint Petersburg in 1766, where he at once engaged in the reformation of the vast organisation. This was the beginning of an extremely productive period in the annals of the Academy, which for practical purposes ended with the death of Euler in 1783 and that of Lexell a year later.[12] On the whole it can be said that the quality of research at the Petersburg Academy of Sciences was exceptional during the whole eighteenth century, and its publications, the *Commentarii* and the *Acta*, were indispensable reading for scientists across Europe. Conforming to the aspirations of international prestige of the Petersburg Academy of Sciences, its organs were published mainly in Latin throughout the eighteenth century. In style they resemble the *Histoire* and the *Mémoires* of the Académie Royale des Sciences of Paris. Following the example of the Paris Academy, the Petersburg Academy of Sciences also arranged prize competitions from time to time, but in other respects, such as the authoritarian leadership and heavy bureaucracy, the Petersburg Academy was very unlike its French counterpart.

As to the name of the Academy of Sciences in Saint Petersburg, it is worth pointing out that the Academy changed its official name several times. The Imperial Academy of Sciences of Saint Petersburg, or shortly the *Petersburg Academy of Sciences*, was in the statutes of 1747 renamed the *Imperial Academy of Sciences and Arts in Saint Petersburg* (Императорская Академия наук и художеств в Санкт-Петербурге). However, as Lexell and his colleagues in the eighteenth century consistently used the form "Académie Impériale des Sciences de Saint Pétersbourg" for the institution, we have no reason to avoid this name either. Present-day Russian scholars call it the (Saint) Petersburg Academy of Sciences.[13] The Academy of Sciences is not to be confused with the Russian Academy (Академия Российская), which was established in 1783 as a research institution for the Russian language (its first President was Catherine Dashkova, who from 1783 onwards was Director of the Petersburg Academy of Sciences as well) or with the Academy of Arts (Академия трех знатнейших художеств), i.e. the Academy of the Three Noblest Arts, established in 1757.

---

[12]In an attempt to compensate for the losses of personnel in the mathematical branch of the Petersburg Academy of Sciences, Jakob II Bernoulli (1759–1789), a talented nephew of Daniel Bernoulli, was invited to Saint Petersburg, but he died accidentally by drowning in the Neva river, only 29 years of age [20]. The same year, 1789, Friedrich Theodor Schubert (1758–1825), who was engaged in the Academy since 1785, was appointed to the Chair of Astronomy in Saint Petersburg.

[13]Professor Tatiana V. Artemyeva, private communication

1.5 The Masters of Mathematics

Mathematics, as is so often the case, was at the forefront of the development of natural sciences during the Age of Reason and the Enlightenment. Due to the rigour and the coercive power involved in mathematical demonstrations of natural laws, the mathematical method was generally considered to be an ideal model for all sciences. Mathematical innovations were a prerequisite to the establishment of a solid theoretical basis and structure to a growing number of findings in the empirical sciences, especially physics and astronomy.

However, in the beginning of the eighteenth century, the scientific community was divided into three factions [148], namely:

1. The mainly British followers of Isaac Newton's natural philosophy, including his mathematical style,
2. The mainly continental adherents of the natural philosophy of René Descartes and Christiaan Huygens, and
3. The followers of the mathematics and continuum mechanics of Gottfried Wilhelm Leibniz.

Notable proponents of the mathematical methods of Leibniz were the two rivalling Bernoulli brothers, Jakob I (James or Jacques) and Johann I (John or Jean).

The result of the expedition led by Maupertuis to Torne River Valley in Swedish Lapland in 1736–1737 marked a turning point in the controversy between the Newtonians and Cartesians [135, 148, 159]. The goal of this geodetic expedition was to determine whether the shape of the Earth is flattened at the poles like a mandarin orange, as predicted by Newton's theory of gravitation, or elongated like a lemon, as predicted by the Cartesian theory of vortices.[14] As it subsequently was confirmed that Newton's theory was right, namely that the Earth indeed is somewhat flattened at the poles, the Cartesians were obliged to abandon their position. In spite of this, Cartesianism did not vanish from natural philosophy altogether, in particular to ensure mechanical explanations in physics. In mechanics, conservation principles (e.g. of linear momentum and *vis viva*) were substantially developed during the eighteenth century. In the second half of the eighteenth century, conservation of mass in chemical reactions became the cornerstone of modern chemistry, as did the conservation of the imponderable electric and magnetic substances, or "fluids", as they were called at that time, for electromagnetism [64, 69]. In optics, on the other hand, Leonhard Euler elaborated the theory of the vibrating luminiferous aether first propounded by Huygens into a serious challenger to Newton's corpuscular optics [30, 59, 69].

Towards the mid-eighteenth century, the controversy between the Newtonians and Leibnizians had died down [148]. The leading mathematical scientists of the

[14] Besides Maupertuis' expedition, another mission was sent to Peru in 1735 to measure the length of a meridian arc (i.e. the distance between two points having the same longitude) near the equator [43], but on its return ten years later the question was already considered more or less settled.

Fig. 1.1 Three of the foremost mathematicians of the mid-1700s: Daniel Bernoulli, Leonhard Euler and Jean d'Alembert. Contemporary engravings (Public domain)

time, "the great trio" composed of Daniel Bernoulli, Leonhard Euler and Jean d'Alembert [43], had come to realise that the differences were mainly a matter of notation, style and approach. In essence, the methods were compatible. These men dominated mathematical and physical research in the middle of the eighteenth century, and the progress they achieved in these sciences was unprecedented (Fig. 1.1).

Two of the sons of Johann I Bernoulli, Nikolaus II (1695–1726) and Daniel (1700–1782), were among the first to join the Petersburg Academy of Sciences late in 1725. Moreover, the Bernoullis arranged for Johann I's student Leonhard Euler (1707–1783) to be engaged at Saint Petersburg as well, starting as an Adjunct of Anatomy. Daniel Bernoulli made essential contributions to many areas of physics and mathematics, in particular hydrodynamics, the theory of vibrations and differential equations, as well as to the mechanical aspects of physiology. He returned to Basel in 1733, but as a member of the Petersburg Academy of Sciences, he continued to contribute frequently to its publications [55].

In France, Jean d'Alembert (1717–1783) was among the most prominent mathematical scientists of the mid-eighteenth century. Besides being a first-rate mathematician, he was a co-author and editor with Denis Diderot of the great *Encyclopédie*, a friend of Voltaire and Condillac, and a dominant enlightened character in the literary salons and societies of Paris [61].

With regard to mathematical skill and productivity, Daniel Bernoulli and Jean d'Alembert were nevertheless surpassed by Leonhard Euler. Having attended private lessons with Johann I Bernoulli and taken a degree at the university, Euler left his home town of Basel in 1727 to serve at the Petersburg Academy of Sciences, and subsequently, in 1741–1766, at the Royal Prussian Academy of Sciences in Berlin [43]. Whereas Euler and Daniel Bernoulli knew each other and collaborated harmoniously in their early years, D'Alembert and Euler soon found themselves rivalling each other in the same area of research and thus their relations were often strained, not least due to the harmful intervention of King Frederick (Friedrich)

II of Prussia. Frederick wanted the Prussian Academy of Sciences to be led by a witty French philosopher, such as Maupertuis had been and such as he imagined d'Alembert to be. However, in Frederick's eyes Euler, the most prolific academician and principal mathematician of his Academy, was socially less presentable. Frederick's ungrateful and inimical behaviour eventually led to a complete breakdown in their relations during the 1760s. On the other hand, d'Alembert consistently resisted Frederick's invitations and reconciled with Euler in a personal meeting in Berlin in 1763 [7, 43]. To replace Euler in Berlin, who in the meantime had re-joined the Petersburg Academy of Sciences, d'Alembert persuaded Frederick to employ the young mathematical talent Joseph Louis Lagrange (1736–1813) from Turin. And finally, having returned to Saint Petersburg, Euler attracted the young Anders Johan Lexell into his orbit as well. In such a specific way, a small circle of learned men promoted the conditions and development of mathematical research for several decades.

In the eighteenth century, mathematics was mostly inspired by the recent discoveries in (Leibnizian) infinitesimal calculus as well as its applications to physical phenomena, that is, to the mechanics of rigid bodies and fluids [79]. Even if progress was without doubt remarkable also in geometry and number theory (and to some extent in probability theory), the main topic was still analysis and its applications to physics. Newton's contributions in the area of continuum mechanics were rather limited. Rational mechanics in the early eighteenth century was a heterogeneous collection of partially uncorrelated theories, which the Bernoullis, Euler, d'Alembert and Lagrange were resolutely committed to unify. In fact, the whole discipline of "Newtonian mechanics" was for practical purposes a creation of these men [161, 162]. Their method of unification was to formulate special problems in terms of differential equations, to search for methods for their solution and to investigate their properties. This is, in essence, the modern approach to physics and engineering.

1.6 Astronomy and Celestial Mechanics

Johannes Kepler's laws of planetary motion and Newton's theory of universal gravitation had provided the astronomers with powerful tools to predict celestial phenomena. During the eighteenth century, these laws were applied to various problems of the solar system, that is, to predicting the orbits of the objects which seem to move in the sky relative to the fixed stars. These tasks were accomplished with great skill and an astonishing precision considering the state of the instruments which were used for observation. Astronomy was traditionally important not only as a vehicle to determine the size and structure of the solar system, but increasingly to explore the Earth itself, since geography and navigation at that time depended essentially on astronomy and special astronomical events, such as solar eclipses and conjunctions of the Moon, the Sun and the planets. The discovery of the planet Uranus in 1781—which will be discussed extensively in Chap. 5—had a great impact on the view of the solar system and its extent. The six planets (including the Earth) known since Antiquity were thought to be the only ones in the solar system, but the discovery of Uranus suggested that additional planets might still be found. By observing the relative motion of the nearest stars, the discoverer of Uranus, Friedrich Wilhelm Herschel, also made the first serious attempt to describe the shape of our galaxy, the Milky Way, and the position of the Sun within it.

The main instruments used by astronomers in the eighteenth century were passage instruments, quadrants, sectors, chronometers (pendulum clocks) as well as different kinds of telescopes. These instruments are frequently mentioned in Lexell's correspondence, and as knowledge of them cannot be taken for granted, we include here a brief introduction to them. The *passage instrument* is an apparatus for measuring the time when a fixed star passes through a certain vertical plane, usually the meridian. It consists of a small telescope adjusted in the meridian plane. By measuring the elevation of a star above the horizon and the time of its passage through the meridian, the two astronomical coordinates of the star, the latitude or declination (Dec) and the longitudinal position or right ascension (rectascension, RA) are obtained.[15] Passage instruments were usually found only in the bigger and better equipped observatories, those in Paris, Greenwich and Berlin. The *quadrant*, in French *quart de cercle*, was the most popular instrument for measuring the declination of objects. It was equipped with a telescope and a graduated arc for reading the elevations. The larger the radius of the instrument, the more precise the measurements. The most reliable mural quadrants, i.e. quadrants fixed to a wall, could also serve as passage instruments (the *sextant* is a better known portable version of the instrument).

Telescopes used for the closer study of the celestial objects were of two kinds (and many varieties): *refractors*, where the light from a star is collected by a lens, and *reflectors*, where the light is collected by a concave mirror. At the beginning of the eighteenth century, refractor telescopes suffered from achromatic error, that is, the blurring of the image due to the differing refractive indices for light of different colours. (Starlight is typically not monochromatic; thus, every colour has a slightly different focus). The problem could be alleviated by making the lens very thin, but in consequence the focal distance also grows very large and thus the telescope becomes impractically long. The reflecting telescope invented by Isaac Newton has no objective lens and thus avoids the problem, but for different reasons did not gain much popularity outside England. Other variations of the reflecting telescope are called, after their inventors, Gregorian and Cassegrain telescopes. However, in 1757, the British optical instrument maker John Dollond managed to overcome the chromatic aberration in refracting telescopes by using compound lenses made

[15]For example, the coordinates of the brightest star in the firmament, Sirius (α *Canis Majoris*), may be given approximately in the form Dec: $-161^\circ 42' 58.017''$, RA: $06^h\ 45^m\ 08.9173^s$, the former being measured in degrees, arc minutes and arc seconds, whereas the latter is given in hours, minutes and seconds. This is the equatorial coordinate system. In the eighteenth century longitudes were still often expressed in degrees of the signs of the zodiac (which divides the sky into 12 parts).

of two or several different glass types.[16] His achromatic telescopes or "tubes" became famous during the latter part of the eighteenth century [128, 139]. The newer optical instruments at the time were often equipped with *bifilar micrometers*, which made the measurements of angular distances considerably more precise than by using the traditional *sectors* (of which there were two kinds: equatorial and zenith sectors). Pivotal to the advance of astronomy in the eighteenth century were also the pendulum clocks and chronometers used for timing the celestial events. The mechanical problems of the pendulum clock devised by Christiaan Huygens in the seventeenth century had been mostly surpassed by means of compound pendulums designed to withstand the length variations due to thermal expansion and contraction. Nevertheless, the clocks still had to be synchronised according to the local moment of noon, and their characteristic daily drifts had to be taken into account in correcting the observation moments.

Owing to the improvement of astronomical precision instruments, astronomy gradually changed from an exclusive science which was the domain of a few very mathematical savants, who in their minds could perceive the subtle harmonies of the Universe, into a science where even amateurs and semi-learned enthusiasts could play a significant part. This development was clearly perceived even at that time. The Swiss astronomer Johann III Bernoulli (1744–1807), grandson of the great Johann I Bernoulli and Director of the Royal Observatory in Berlin, designed a rough classification of the members of the "Republic of Astronomy" according to their capacity and importance for the profession [13], [22, pp. 27–39], [149]. It consisted of 14 categories or classes described as follows:

— Ordinary observers, i.e. beginners, *dilettantes* and *voyageurs*, or those who were occupied with determining the terrestrial latitudes and longitudes, belonged to the 1st class; the assistant observers belonged to the 2nd class and astronomers of the Navy and sea captains to the 3rd class.
— The *calculateurs*, who knew how to draw up so-called *éphémérides* (celestial availability charts), to predict astronomical events etc., belonged to the 4th class. Scientists specialised in astronomical instruments and their perfection belonged to the 5th class, those specialised in naval applications to the 6th class.
— Mathematical astronomers developing methods to derive astronomical formulae and tables belonged to the 7th class. The description of this class of astronomers seems to fit among others Edmond Halley, John Flamsteed, Tobias Mayer, Nicolas Louis de La Caille and Pehr Wilhelm Wargentin. The so-called working astronomers of the 8th class were those who rendered services to the mathematical and the learned astronomers.
— The learned astronomers, or *astronomes géomètres*, belonged to the 9th class. They professed theoretical astronomy, including the theory of the sphere

[16] At the same time, Leonhard Euler worked out a theory of achromatic lenses but failed to manufacture such lenses (combining glass and water) in practice, see e.g. "Sur la perfection des verres objectifs des lunettes", *Histoire de l'Académie Royale des Sciences et des Belles-Lettres de Berlin* Vol. 3, 1747 (printed in 1749), pp. 274–296 (OO Ser. III, Vol. 6, pp. 1–21. E118).

and celestial mechanics. The description of this class fits, *inter alios*, Anders Johan Lexell and Johann Heinrich Lambert [149]. Astronomers designing celestial globes and charts belonged to the 10th class, editors of astronomical almanacs and tables to the 11th class.

— The astronomers *par excellence*, who mastered every theoretical aspect of astronomy to a perfection, belonged to the 12th class. This description seems to fit among others Joseph-Nicolas Delisle and Joseph Jérôme de Lalande. The most sublime astronomers of the 13th class were those who, in the footsteps of Kepler and Newton, were able to perceive the "great mysteries" of astronomy and to improve the state of this science by grasping the foundations of analysis. Euler and Lagrange evidently belonged to this exclusive class.

— Finally, the writing astronomers, *astronomes littérateurs*, were those who understood enough of astronomy for the purpose of popularisation, but not to make the science progress in itself. In this special class, the 14th, Johann Bernoulli included philosophical authors such as Rudjer Boscovic and Voltaire.

Obviously, Bernoulli's classification is only indicative, as the qualifications of the different classes are not unambiguous. A different classification has indeed been applied in [149], where the number of living astronomers in 1775 was estimated at approximately 160, but the astronomers of importance (*les grands savants*) at no more than 60. Evidently, the scientists in the latter group were also responsible for the progress in pure mathematics. The professional astronomers of the highest classes were all living in Europe at that time, France being the leading nation with more than a 100 active astronomers, followed by the German-speaking countries and Great Britain. In Sweden, Bernoulli's list mentions 28 astronomers, most of them living in Stockholm, Uppsala and Åbo.

One of the most challenging problems of theoretical astronomy in the mid-eighteenth century was to determine the gravitational interaction between more than two bodies. In the most general case, the problem has no closed analytical solution. Nevertheless, many approximate solutions for the case of three interacting bodies were provided by Euler, d'Alembert, Clairaut and Lagrange (see Chap. 5). Alexis Claude Clairaut (1713–1765) was able to predict, by a method of numerical integration, the perturbations caused by Jupiter and Saturn to the motion of Halley's comet, which reached its perihelion in 1759. Simultaneously with Euler and d'Alembert he also improved the tables of the Moon's motion. In 1764, Lagrange solved the problem of so-called *libration*, a small oscillation in the Moon's motion, while Pierre Simon de Laplace (1749–1827) showed that the perturbations of Jupiter and Saturn are periodic in nature and hence that the solar system is stable, at least to some degree. The problem of determining the solar parallax accurately using the transit of Venus culminated after the observations of the phenomenon in 1769. To this and other related issues we will return in detail in later chapters. We begin with an overview of the state of mathematics and astronomy at the Royal Academy of Åbo, where Lexell received his education.

Chapter 2
Humble Beginnings

2.1 Childhood in Åbo (Turku), Finland

Anders Johan Lexell was born as a "Christmas gift", on 24 December 1740[1] in Åbo, the principal ecclesiastic and administrative centre of Swedish Finland. At the time, the town Åbo, which is the Swedish name of the city today equally known by the Finnish name Turku,[2] had a population of about 5,000 inhabitants. Of them, roughly two thirds were Finnish speaking and one third Swedish speaking. The government officials and the upper echelons of society were mainly Swedish speaking, and Swedish was also Lexell's mother tongue. At the time, the town of Åbo was Finland's largest commercial centre with so-called staple rights, meaning that all exports of the region, such as tar and other forest products, were by decree required to pass through the town. Cultural life in Finland was assembled principally around the university of Åbo, called *Kungliga Akademien* or the *Royal Academy of Åbo* (in Latin: *Regia Academia Aboensis*). Despite its name it had nothing in common with the new scientific societies and academies taking shape in Europe at the time. It was Finland's first university, founded in 1640, and the forerunner of today's University of Helsinki [63, 80]. The cultural and commercial ties between Åbo and the towns of Sweden proper were close. The main examples and influences came from the capital Stockholm as well as the nation's oldest university town and episcopal see, Uppsala. The two major landmarks of Turku—the old cathedral and the majestic stone castle in the harbour—bear witness to the town's importance since medieval times.

For the inhabitants of Finland, the first half of the eighteenth century was a time of particular hardship and distress. First, repeated crop failures in the years

[1]This date is in the Julian calendar (old style), which was used in Sweden until 1753. In the present-day Gregorian calendar (new style), Lexell's date of birth would be 4 January 1741.

[2]Both names—the Finnish 'Turku' and the Swedish 'Åbo' (pronounced /o:bu/)—are used today. In the present book, the names are used synonymously, depending on the context.

J.C.-E. Stén, *A Comet of the Enlightenment*, Vita Mathematica 17,
DOI 10.1007/978-3-319-00618-5_2,

1695–1697 and the subsequent famine had weakened and significantly reduced the population. Next, the expansion of the Russian Empire towards the North-West initiated by Peter the Great in the "Great Northern War" (1700–1721) and continued by his successors led to a gradual retreat of the century-long Swedish governance of the Baltic region. For two periods, 1713–1721 and 1742–1743, Finland was occupied and heavy-handedly governed by the Russians.[3] As the administrative centre of Swedish Finland, Åbo suffered much damage during these years of misfortune, losing both resources and inhabitants. In the aftermath of the wars epidemics were rife, claiming hundreds of victims in the town.

Anders Johan Lexell's father, Jonas Lexell, was a goldsmith and jeweller, born in 1699 in Stockholm as the son of a Lutheran vicar Olaus Lexelius (who died in 1709) and trained by his stepfather Daniel Schultz, a goldsmith and jeweller born in Åbo [53, 86]. In 1725, at a time when Åbo was recovering from a disastrous occupation and in need of skilled labour, Schultz moved back to Finland with his family in the hope of finding work as a goldsmith in Åbo, but apparently without success. The next year, however, Schultz's young and ambitious stepson Jonas Lexell managed to establish himself as a goldsmith and eventually to become a respected and well-liked member of his professional community in Åbo. From his step-father Schultz he inherited the position of Inspector of Pearl Fishing in Finland.[4] He was also engaged in local administration and politics and served three times as the town's parliamentary representative at the Diets (*Riksdag*) in Stockholm. Jonas Lexell was the leader of the Guild of Goldsmiths in Åbo and indisputably its most influential member during the whole eighteenth century (Fig. 2.1).

In March 1740—a year before the outbreak of another disastrous war between Sweden and Russia—Jonas Lexell married the 19 years younger Magdalena Catharina Björkegren. She was the daughter of Anders Björkegren, a Swedish lawyer, who, at the end of his career, had moved to Åbo (where he retired and where he died in 1721), and Margareta Björkegren, née Wittfooth, later to be remarried Heldt. The Wittfooths were an important family of merchants which had arrived from Germany and become naturalised in Åbo in the seventeenth century [24]. A significant proportion of the merchants arriving in Åbo in the seventeenth century were of Dutch or German origin.

The family of Jonas and Magdalena Lexell settled down almost in the centre of the town close to the market place, the cathedral and the Royal Academy.

[3]In Swedish, these military occupations are called, respectively, *Stora ofreden* ("Greater Wrath") and *Lilla ofreden* ("Lesser Wrath"). Both times, the university and its teachers as well as the library were exiled to Stockholm. In 1809, following the "Finnish war" (a part of the Napoleonic wars), Finland was eventually detached from Sweden and the Swedish rule that had prevailed for more than six centuries, to become a formally autonomous Grand Duchy within the Russian Empire, allowed to maintain its old Swedish laws and Lutheran religion. Only in 1917 did Finland gain independence.

[4]To a limited extent, freshwater pearls have been collected in Finnish rivers since medieval times, but the activity has hardly ever been profitable.

Fig. 2.1 Portrait of Jonas Lexell (1699–1768), father of Anders Johan Lexell, painted by Johan Georg Geitel (1683–1771) (Private collection of the Pipping family. The National Museum of Finland. Photo: Per-Olof Welin, published with permission)

Lexell's house was situated about 140 m south-west of the cathedral, at the crossing of Hämeenkatu (in Swedish: Tavastgatan), Uudenmaankatu (in Swedish: Nylandsgatan) and the cathedral park (see Fig. 2.2). All this wooden housing around the cathedral was destroyed in a devastating fire in September 1827. Also the library of the Royal Academy with its valuable documents and holdings was destroyed in the fire, except for the rare material that could be rescued or happened to be on loan outside the town. At the place of Lexell's house there is now a two-storey building in Empire style erected in 1833 (presently the main administrative building of the Swedish University of Turku, Åbo Akademi).

Anders Johan was the first-born child of Jonas and Magdalena Lexell. Next in line was Margareta Elisabeth, born in 1742 during the Russian occupation 1742–1743, when many of the inhabitants of Finland including the Lexell family were evacuated to Sweden. Magdalena Catharina was born in 1745, Anna Lovisa in 1746, Ulrika in 1747, and finally Jonas the younger, during whose birth in 1750 mother Magdalena died. Two of the daughters, Margareta Elisabeth and Anna Lovisa also died at an early age. Magdalena Catharina the younger married a merchant and sea captain named Jost Joachim Pipping, and they had ten children, of whom one, Fredrik Wilhelm Pipping, became Professor of the History of Learning in 1814 and a

Fig. 2.2 Map of the church quarter in Åbo (Turku) showing the cathedral and the house of Lexell (site N° 37). The Royal Academy (university) is situated in the houses on the southern part of the wall surrounding the Cathedral (*C*, *D*, *E*). The Library (*K*) and the Cathedral school (*I*) are on the north side. These buildings were destroyed in the conflagration in 1827. Site N° 48 is the Botanical Garden by the river Aura, the academic bookstore was situated on the bridge; nowadays, the bridge crosses the river closer to the Cathedral. The map was completed in 1756 by County treasurer and land surveyor Daniel Gadolin (1722–1796), whose elder brother, Professor Jacob Gadolin, lived in a house on site N° 8 by the river (HUB. Photo: the author)

prominent figure in the University.[5] Ulrika Lexell, in turn, married a legal councillor named Harald Alfthan and had two children. Anders Johan, as we shall see, did not marry and had no children, neither did his brother, Jonas the younger. Thus, the male line of the Lexells in Finland died out. On the other hand, not all the bearers of the surname in Sweden today are necessarily related.

After the peace treaty between Sweden and Russia had been signed in Åbo in August 1743, whereby Sweden had to cede considerable parts of South-Eastern

[5]F. W. Pipping (1783–1868) was the chief librarian of the University of Helsinki. He was raised to the nobility in 1839. Among his descendants were Nils Pipping (1890–1982), Professor of Mathematics at Åbo Akademi, who worked *inter alia* on Goldbach's conjecture, as well as the world-famous Finnish architect Alvar Aalto (1898–1976). For this information I am indebted to Johanna and Yrjö Alanen, Turku.

Finland to Russia, refugees began to return to their homes, among them the family of Jonas and Magdalena Lexell. Reconstruction of the town started successively and the university returned from its exile and opened its doors to students. But as the inhabitants of Finland were feeling more and more defenceless, the Crown had to invest considerably in the fortification of its eastern territories: as a result, two sea fortresses were built on the islands of the south coast of Finland, viz. *Svartholm* outside Lovisa, and *Sveaborg* (in Finnish: *Viapori*; today *Suomenlinna*) on the islands outside Helsingfors (in Finnish: Helsinki)—at the time a small market town on the south coast of Finland. In 1757–1762, Sweden found itself involved in a war against Prussia (the Pomeranian war, a theatre of the Seven Years War), but this war did not involve Finland directly.

2.2 Intellectual Awakening

Exactly when and how Anders Johan's mathematical talent first became noticeable, and who was his first teacher, is not known. There are no indications of him being a child prodigy, rather he was developing quietly and steadily towards his vocation.[6] Lexell's loss of his mother, at the age of nine, was certainly a severe blow, and it could well have been even more fatal for his development, had the family not been relatively well-off. As it was, Anders Johan could be enrolled in the town's *Gymnasium*, from which he duly graduated in 1755, to enter the University, the Royal Academy of Åbo.

By general European eighteenth-century standards the Royal Academy of Åbo was a small provincial university, admitting around 100 students annually, of which only a fraction left with a degree. There were four faculties, namely those of Theology, Law, Medicine and Philosophy, but only the Theological and Philosophical faculties had several professors. The Head of the Academy was the Chancellor, who was nominated by the King. His deputy, the Pro-Chancellor, was the Bishop of Åbo. Practical matters were dealt with in the Academic Consistory, consisting of the board of professors, presided over by the Rector. The premises of the university were unpretentious, and the lecture rooms of the Academy remained small and incommodious for the whole of the eighteenth century [80]. The lectures normally began at 5 o'clock in the morning on 4 days of the week.[7] Wednesdays and Saturdays were reserved for thesis defences. As there generally was no heating in the classrooms, the students had to endure the lectures in extreme conditions during

[6]The complimentary poem written by Fredrik Pryss (1741–1767) suggests, however, that Lexell was considered the most zealous and talented student of his class (*Lexell assiduos inter celebrandus alumnos, qui claret studiis ingenioque suo*). We note that Pryss himself was the *Primus* at the graduation, i.e. ranked first of his class.

[7]It will be remembered that Queen Christina of Sweden had ordered Descartes to begin his lectures in the Royal Castle of Stockholm at 5 o'clock. Apparently this was not considered inconveniently early, but of course, the French philosopher thought otherwise.

Fig. 2.3 The Academy square, "Akademitorget" in Åbo in 1795. Behind the building on the foreground, the newly built "Anatomy house", is the old main building of the Academy. Above the roofs of these buildings, none of which exist today, are seen the domes and apse of the cathedral. The viewpoint of the artist is approximately site N° 124–125 on Gadolin's map, Fig. 2.2. Tinted drawing by Carl Peter Hällström (1774–1836) (Uppsala University Library. Published with permission)

the cold winter mornings. A good memory came in useful, because the slates were small and writing on them with cold fingers must have been difficult. It goes without saying that books, writing paper and ink were expensive at the time (Fig. 2.3).

The professors usually lectured in Latin, but to a growing extent also in Swedish, when the terminology permitted it. In addition to the public lectures in the Academy's lecture rooms, the professors were entitled to arrange private courses, so-called *collegia*, in their homes for an additional fee. Such courses were rarely organised, however, considering the additional public and practical duties of the professors. For instance, according to the Academy's charter of 1655, the Professor of Mathematics was responsible for the education in arithmetic, algebra and the theory of numbers, theoretical and applied geometry, plane and spherical trigonometry, geodesy, mechanics, music, optics and architecture. Moreover, his responsibility was to teach astronomy, gnomonics (the science of constructing sundials), *computus ecclesiasticus* (computing the Christian festival days), geography and navigation [34, 153]. In reality, every professor organised his subject matter according to his personal interests and inclinations.

In these modest circumstances the young Lexell was fortunate enough to be influenced by two prominent teachers: Jacob Gadolin and Martin Johan Wallenius.

The former was Professor of Physics, the latter of Mathematics, and both of them notable scientists in their own right. Jacob Gadolin (1719–1802) [68, 91, 153, 158] had been trained in Uppsala by two of Sweden's leading mathematical scientists of the time, Samuel Klingenstierna[8] and Anders Celsius. He was an able physicist, astronomer and surveyor, and being also a cleric, he eventually moved to the Chair of Theology, which was better paid at the time, to be finally appointed Bishop of Åbo. It may be noted that Gadolin's father-in-law, Johan Browallius (1707–1755), also was Bishop of Åbo, as well as Professor of Natural Philosophy, and a friend of Linnaeus.[9]

Martin Johan Wallenius (1730–1772), the son of a theologian and a student of Carl Fredrik Mennander (1712–1786), was appointed *docent*[10] in mathematics at the Royal Academy of Åbo in 1755 [81, 123, 153, 158]. Three years later, in 1758, he became Professor of Mathematics: Lexell was one of his first students, and undoubtedly among the most talented. Wallenius's predecessor, the elderly Nils Hasselbom (1690–1764) certainly knew the elements of infinitesimal calculus—the term 'infinitesimal calculus' appears in a thesis by Hasselbom published in 1726 [153, 165]—but his creative period had ended in the 1740s, after which he concentrated on public duties. The main mission of the Royal Academy of Åbo being to educate clergymen, the new mathematical physics was accepted very slowly for philosophical (metaphysical) reasons [77]. For instance, the heliocentric system had been introduced rather tentatively only in the 1690s. But with Wallenius and Jacob Gadolin entering the scene, encouraged by the influential Bishop Mennander, things were gradually to change. In fact, Wallenius was the first to bring infinitesimal calculus (*methodus fluxionum directa et inversa*) as well as modern mathematical physics (theoretical mechanics and geometrical optics) systematically into the university curriculum in Åbo. Thanks to him, Lexell was to learn of the latest developments in these subjects. On the whole, the quality of the dissertations defended by Wallenius's students was remarkable for a provincial university like that in Åbo. In the history of mathematics, Wallenius's fame rests mainly on a thesis published in 1766,[11] where he extended a result by Hippocrates of Chios concerning

---

[8]Samuel Klingenstierna (1698–1765), mathematician and Professor of Physics at Uppsala University [144], had been a student of Christian Wolff in Marburg and Johann I Bernoulli in Basel [144]. Subsequently, he introduced Leibniz's calculus to Sweden. He was also appointed teacher of Crown Prince Gustav.

[9]Jacob Gadolin's son, Johan Gadolin (1760–1852) became a Professor of Chemistry in Åbo. He discovered the element Yttrium, and the element Gadolinium has been named in his honour. Furthermore, Jacob Gadolin's grandson and Johan's nephew, Axel Wilhelm Gadolin (1828–1892), became a military officer in Russia and a Professor of Mineralogy at the Petersburg Academy of Sciences [78].

[10]In Finland and Sweden, the title *docent* is a degree of habilitation and signifies the position of a non-stipendiary lecturer.

[11]*Dissertatio gradualis, lunulas quasdam circulares quadrabiles exhibens*, Aboae, 1766. The thesis was supervised (and most probably written) by Wallenius and defended by Daniel Winquist.

all possible squarable[12] "lunes"—i.e. figures bounded by two circular arcs—the area of which can be determined by strictly Euclidean methods (that is, in principle, by ruler and compass and without the number π). Equally remarkable were his studies of the frictionless descent of a particle along a cycloid, whose axis is vertical, and the solution of the curve along which a particle will move uniformly retarded.[13] However, due to mental disorder, described as mania, Wallenius was obliged to give up work in 1771, when only 40 years old, and 2 years later, in 1773, he died. Lexell, who at that time was already making his career at the Petersburg Academy of Sciences, was appointed Wallenius's successor in 1775, but as we shall see, he never returned to Åbo to take up the post. The reasons for this were quite complicated and also closely linked to the journey he had planned to undertake for a long time. We will return to this issue in later chapters.

Lexell's first printed work was a thesis in Latin, presented *pro exercitio* (for practice), and entitled *Animadversiones subitaneae circa principium universae opticae Leibnitianum, quatenus idem in catoptrica adhibetur* or "Incidental remarks on Leibniz's universal principle of optics, as to its use in catoptrics"[14] defended on 30 June 1759 [Lexell 1]. In this work, which was supervised by Wallenius, Leibniz's principle according to which light moves between two points along the "easiest" path[15] was critically examined. The minimum principle was originally propounded by Pierre de Fermat as the *Principle of least time* and later generalised by Pierre Louis Moreau de Maupertuis and (in the correct form) by Leonhard Euler as the *Principle of least action*. In particular, Wallenius and Lexell pointed out the failure of the principle in the case of concave (hollow) mirrors. Quoting a passage in §1 (p. 3):

> [T]he hypothesis is far from being suitable as a principle for optics on the whole; rather, it can be easily shown by examples, which concern the reflection of light, that it does not hold universally, as there are given not just a few cases in which the path of light is not a minimum but a maximum, or indeed something else, in which neither of these laws prevails.

These cases were examined systematically in the thesis for different types of hollow mirrors: spherical, parabolic, elliptic and hyperbolic. The discovery of this deficiency in Leibniz, albeit already known to the scientific community since the debate on the Principle of least action in the 1740s, was undoubtedly a major stimulus to the young Lexell.

The work is interesting also since it describes the literature available to Wallenius and Lexell. Besides references to Leibniz the work cites Euclid's *Elements*,

---

[12]The process of finding the area of a figure by constructing a square with the same area is called squaring or *quadrature*.

[13]These quite original problems were solved using integral calculus. They appeared along with other mathematical problems in two theses in 1757 and 1758 entitled *Exercitationes miscellaneae mathematico-physicae*.

[14]Catoptrics is the science of mirror optics (reflection of light), while dioptrics deals with the optics of refraction.

[15]*Acta Eruditorum* (Leipzig) for 1682, p. 185, as well as for 1701, p. 20.

Descartes's *Principia philosophiae* and *Dioptrica*, Newton's *Principia* and *Opticks*, the Marquis de l'Hôpital's *Analyse des infiniment petits* (1696) and C.-F. Milliet Dechales's *Mundus mathematicus* (1674). Of more recent literature, Maupertuis' *Essai de cosmologie* (1751), G. W. Krafft's *Praelectiones physicae* (1754), Pieter Musschenbroek's *Elementa physica* (1726), Christian Wolff's *Elementa matheseos universae* (1713–1715) as well as Colin MacLaurin's *Treatise of fluxions* (1742) are cited.[16] It is still not certain that the books were actually available to Wallenius and Lexell; they might merely have been cited from literature.[17] Nor can we say for sure to what extent Lexell himself was the author of the thesis, since it was customary at the time for the professors to write the theses for their students to defend.

Lexell's thesis *pro gradu magisterii* entitled *Aphorismi mathematico-physici* or "Aphorisms in mathematical physics" was supervised by Jacob Gadolin [Lexell 2] and defended in public on 4 June 1760. This work contains 12 short essays (aphorisms) of a rather general character on mechanics (Fig. 2.4). Most of the aphorisms deal with principles governing simple machines, such as the inclined plane, the balance of forces and the centre of oscillation in composite systems etc., with remarks on the contemporary literature in physics, including the popular book *Physices elementa mathematica, experimentis confirmata* (Leiden, 1721) of the Dutch physicist Willem Jacob 's Gravesande. *Aphorismus X* concerns the behaviour of light at boundary surfaces, and it is demonstrated that when light according to the Newtonian corpuscular (particle or emission) hypothesis enters from a medium with a lower refractive index into one with a higher refractive index, the path of light cannot assume a parabolic curve, as some authors seemed to suggest, because this would mean that the attraction of the second medium would be nearly constant in all space (implicitly assuming a force similar to the gravitational attraction near the Earth's surface, where bodies are known to fall in parabolic trajectories). However, as maintained by Gadolin and Lexell in the thesis, the ray of light (or the trajectory of the light corpuscles) is known to interact with the second medium only at extremely short distances from the boundary. Thus, the possibility of a parabolic trajectory could be eliminated.

---

[16]The list contains both new and old, "Newtonian" and "Leibnizian" literature. Accordingly, in Wallenius's and Lexell's work, differentials were called fluxions as in Newton, whereas the notation was due to Leibniz. In general, the attitude in Sweden towards the new calculus, whether of Leibniz or Newton, was pragmatic.

[17]The catalogue of Lexell's books auctioned in Saint Petersburg in 1785 [47] contains almost all the works mentioned (e.g. Newton's *Principia*, the Cologne edition from 1760), but since these books were very valuable, Lexell may have acquired them later. The number of books on mathematics, physics, chemistry and astronomy in the academic library in Åbo was limited: However, the *Opera* of both Jakob and Johann Bernoulli, the principal works of Leibniz, 's Gravesande and Musschenbroek had been acquired before the 1750s [165]. The catalogues of scientific literature compiled by Johan Gadolin and Gustaf Gabriel Hällström in the 1820s (HUB Ms. Coll. 57.68 "Gadolin-Hällström") contain hundreds of titles, although without mentioning the date of their acquisition. Apparently the first copy of Newton's *Principia* in Åbo belonged, not to the library, but to the astronomer Anders Planman [91].

D. F. G.

APHORISMI MATHEMATICO-PHYSICI,

Quos

Consensu Ampl. Facult. Philos. in Regia Academia Aboënsi,

PRÆSIDE,

MAXIME REVERENDO atque CELEBERRIMO

D:NO DOCT. JACOBO GADOLIN,

Scient. Nat. PROFESS. Reg. & Ord. nec non Acad. Scient. Holmiens. MEMBRO.

PRO GRADU MAGISTERII,

Publico examini subjicit

ANDREAS JOH. LEXELL,

ABOA-FENNO.

Die IV. Junii, Anni MDCCLX.

Loco Horisque A. M. Solitis.

ABOÆ, Impressit DIRECT. & TYPOGR. Reg. Magn. Duc. Finland. JACOB MERCKELL.

Fig. 2.4 Title page of Lexell's dissertation *pro gradu magisterii* defended on 4 June 1760 at the usual place (auditorium) and time. 16 pages in 8vo (20×16.5 cm) (The author's own copy)

On 28 August 1760 Lexell graduated in the solemn promotion, only 19 years old. Among his classmates we find Henrik Gabriel Porthan (1739–1804), a renowned Finnish philologist and historian, who was to become chief librarian and Professor of Eloquence at the Academy of Åbo. As a keen proponent of Enlightenment ideas in Finland [60, 80]—in particular of the German *Neuhumanismus* kind— Porthan founded in 1770 a private society called *Aurorasällskapet*, a learned patriotic association organising concerts and publishing the first newspaper in Finland called *Tidningar, utgifne af et Sällskap i Åbo*, i.e. "News, published by a Society in Åbo" [80]. As a close friend of Porthan, Lexell wrote some shorter reports for this newspaper with news from the learned world of Saint Petersburg in 1771–1773. Porthan and Lexell shared a mutual appreciation and friendship, but unfortunately their correspondence has been lost. However, its contents and frequency may be surmised from numerous passages in Porthan's letters to Archbishop Carl Fredrik

Mennander, former Bishop of Åbo, residing in Uppsala [132]. Also considering Lexell's significant book donations in the field of astronomy to the University Library in Åbo [165], Porthan must have been one of his key contacts.

2.3 Development and Frustration

We have very little information indeed as to Lexell's whereabouts and activities during the years after his graduation in Åbo. From the very outset his prospects were not promising, since there were no suitable open positions at the Royal Academy of Åbo, or in the other universities in Sweden, in Uppsala and Lund. In Åbo, Wallenius had only recently been appointed to the Chair of Mathematics, while Gadolin held the Chair of Physics, the young and able astronomer and physicist Anders Planman (1724–1803) being next in line to succeed him. Planman obtained his master's degree in Åbo in 1754 [67, 81, 91, 153, 158] and continued his studies in Uppsala, where he became a *docent* in 1758. He served as Professor of Physics in Åbo from 1763 on and as University Rector for three periods. Being also an ecclesiastic, Planman was later appointed vicar of two parishes near Åbo.

While waiting for a suitable opportunity to present itself, Lexell made his living by private tuition of children of wealthy families. To this end, it was absolutely essential to become known and recommended by key persons in society. Such a person was Bishop Mennander, former Professor of Physics, a friend of Linnaeus and Klingenstierna, and the *primus inter pares* of the learned circles in Åbo [46].[18] In a small town like Åbo, everybody in the upper crust of society knew each other. Thus, Mennander had certainly become acquainted with Jonas Lexell through his public duties and was thereby able to arrange work for his mathematically-inclined son.

During his time as a tutor in Åbo, Lexell diligently pursued his researches in differential and integral calculus. In a few years he prepared an impressive dissertation in Latin entitled *De methodo inveniendi lineas curvas, ex datis radiorum osculi proprietatibus*, that is, "On a method of determining curves from a given property of their radii of curvature" [Lexell 3].[19] In the winter of 1763, with the manuscript in his luggage, Lexell travelled across the frozen Gulf of Bothnia to Sweden to enrol—according to the records on 29 January—as a student at the Academy (or University) of Uppsala. The minutes of the Consistory of the

[18]Mennander was in charge of a new Finnish Bible translation in preparation. Being also professor at the Academy, he established a Chair of Chemistry in Åbo with an appropriate laboratory, a dissecting room, a botanical garden and a greenhouse. Mennander's private library contained 7,000 volumes at his death in 1786 [89].

[19]By chance or not, the title is strikingly similar to Leonhard Euler's pioneering work in variational calculus *Methodus inveniendi lineas curvas maximi minimive proprietate gaudentes*, Berlin, 1744 (OO Ser. I, Vol. 24. E65). On the other hand, the expression they have in common is typical of works in integral calculus of the time.

Philosophical Faculty of Uppsala University for 19 February 1763 state that Lexell, during the absence of Professor Meldercreutz,[20] had quickly applied for permission to preside the dissertation of the thesis as a recommendation for "some position in Stockholm" and, consequently, a student by the name of Erik Östling was assigned to defend the thesis (*pro exercitio*). Which position Lexell had in mind has not been discovered; it might just have been a pretext for getting to know and becoming known by Sweden's most prominent scientists. Quoting the minutes of the consistory meetings[21]:

> *Magister* Lexell, who has recently graduated in Åbo and now arrived in Uppsala, requests from the faculty, for the application of some position in Stockholm, the permission to supervise the defence of a thesis entitled *Methodo inveniendi lineas curvas* ..., by the student Eric Östling from the province of Gästrikland. The faculty has no interest in preventing it, because Lexell has presented his certificate and proved his competence by submitting the thesis. And if Captain Meldercreutz, who as a Professor of Mathematics is responsible, besides the Dean, for verifying the thesis, does not return to the town from his leave of absence to the beginning of the month of March, the responsibility of reviewing the thesis will be given to Professor Duraeus.[22]

The poor student probably only had a faint idea what the thesis was about, as its standard and length were way beyond what was normally required from students.[23] An account of Lexell's thesis will be given in Chap. 8.

We do not know precisely what Lexell was doing or with whom he dealt in Uppsala, nor how his dissertation was received. A fair assumption is that he at least made the acquaintance of the Uppsala mathematicians and astronomers Fredrik Mallet (1728–1797), Daniel Melander (1726–1810; after ennoblement in 1778 Melanderhjelm) [126], Mårten Strömer (1707–1770) and Samuel Duraeus. Whether he also met Linnaeus or Wargentin during this visit is uncertain. Be that as it may,

---

[20]As a young man, Captain Jonas Meldercreutz (1715–1785), had accompanied Maupertuis' expedition to Torne River Valley in 1736. Besides being a Professor of Mathematics in Uppsala, he was a farmer and estate-owner, who spent more and more time at his mine works until he resigned from his Chair in 1772. Carl Linnaeus, in his *Nemesis divina* [102], criticised his colleague Meldercreutz for being greedy.

[21]§1. Anförde Decanus, at Magister Anders Johan Lexell, som i Åbo blifwit promoverad och nyligen hitkommit, ärnar för någon ansökning skul i Stockholm här præsidendo disputera, de Methodo inveniendi lineas curvas, ex datis radiorum osculi proprietatibus, Resp. Studios. Eric. Östling, Gestr. begärande ther til Facultetens tilstånd, hwilket Faculteten icke heller will wägra, enär bemälte Magister upter sitt Magister bref och genom Disputations ingifwande til censur hos Decanus wisar sin skickelighet. Och som Capitaine Meldercreutz, hwilkom, jämte Decano tilkommer pröfwa thetta hans profession tilhörande ämne, nu är borta med Illustrissimi Cancellarii tilstånd, i fall Wälbemälte Capitaine til början af Mars månad, tå permission uphörer, ej kommer i staden, på thet Magister Lexell icke må hindras i the angelägenheter han föregifwer, resolverade Faculteten, at låta Disputation til Profes. Duræus afgå, at öfwerses.
(I am grateful to Dr. Håkan Hallberg at UUB for this information).

[22]Samuel Duraeus (1718–1789), Professor of Physics in Uppsala, was a student of Klingenstierna. His promising career seems to have been cut short by alcoholism [129].

[23]Eric Östling defended his thesis *pro gradu magisterii* on 6 June 1764, presided by Johan Gottschalk Wallerius (1709–1785), Professor of Chemistry. That thesis was only a few pages long.

a few years later he was already engaged in a correspondence with both of them. However, it seems likely that Lexell quickly left Uppsala and returned to Finland, since on 9 September 1763 Lexell was appointed *Matheseos Docens*, i.e. *docent* (lecturer) in mathematics in Åbo. His subsequent application in 1764 for the open position of Adjunct Professor at the Philosophical Faculty of the Academy was at first unsuccessful, but on 24 May 1765 he was eventually appointed [81]. Despite the impressive title, the position was not much valued: the Adjuncts were not entitled to lecture, and their salaries were insignificant. In the long run, such a position could not satisfy a talent like Lexell.

Thus, when the position of lecturer (professor) of mathematics for the marine cadets was declared vacant at the Naval School (*Kongliga Amiralitets Cadett Corpsen*) in Karlskrona, Sweden, the Consistory of the Academy of Åbo issued a strong recommendation for Lexell on 15 February 1766 [81]. In spite of this, the position was not granted to him.[24] However, at the time, Lexell probably had a new and more ambitious plan in mind: he had seen a light in the East.

[24]Lexell's obituary [41] mentions that he had served at the Naval School at Uppsala, but this cannot be correct. First of all, the Swedish marine cadets were trained in Karlskrona, not in Uppsala, and secondly, Lexell's name does not appear in the list of teachers in Karlskrona. Officially, Mårten Strömer was Professor of Mathematics in Karlskrona, but in practice he was deputised for by Bengt Ferner in 1756–1765, while Strömer continued teaching in Uppsala. Ferner was succeeded in 1766–1792 by Carl Gustav Bergström, *Magister* in Uppsala in 1755 [152].

Chapter 3
New Prospects in Saint Petersburg

3.1 The Return of Leonhard Euler

On 1 June 1766, the celebrated Swiss mathematician Leonhard Euler[1] with his family and household left Berlin to return to Saint Petersburg. Euler had been in the service of the Royal Prussian Academy of Sciences in Berlin since 1741, first as Professor of Mathematics, Director of the mathematical section, and eventually as the acting leader of the Academy. The reason behind the move was the long-lasting discord and eventually the complete break-up between Euler and his superior, Frederick II of Prussia. On the other hand, in the Imperial Academy of Sciences of Saint Petersburg, Russia, where Euler had made his early career from his arrival from Basel in 1727 to 1741, he was highly regarded and his return very much appreciated. In fact, in the course of Euler's 25 years of absence from Saint Petersburg, the Imperial Academy of Sciences had declined and lost much of its previous influence and prestige, making it difficult to recruit new members from Western Europe [170]. At the time of his return to Saint Petersburg, Euler was 59 years old, blind in one eye and suffering from a severe cataract in the other, but productive and creative as ever. Euler's departure from Berlin was big news even outside the scientific community, and he was received in a princely manner by both King Stanislaw August Poniatowski of Poland and Catherine II of Russia. Having arrived in Saint Petersburg, Euler was immediately engaged in the reform of the Academy and it was largely due to him that the Petersburg Academy of Sciences achieved its highest glory and reputation ever [37, 43, 170].

In one way or another the news of this remarkable event caught the attention of Lexell, and so the idea of working for this already legendary mathematical genius living scarcely a week's journey from his home town must have occurred to him. He must certainly have known that in 1763 the Professor of Economy in Åbo Pehr

[1]Of the abundant literature on Euler, the biography by Emil Fellmann [43] can be especially recommended, besides the scientific summary in [55].

J.C.-E. Stén, *A Comet of the Enlightenment*, Vita Mathematica 17,
DOI 10.1007/978-3-319-00618-5_3,

Kalm had been offered the Chair of Botany at the Petersburg Academy, and that Kalm had declined the offer. But, then, in his case, how could he obtain such an offer? Although Saint Petersburg was not very far away from Åbo, travelling there without any real prospects of being employed seemed quite risky. Considering the circumstances, there appeared to be only one realistic way: to submit a specimen of his mathematical skill for the approval of the Academy—possibly to Euler himself—and to apply for an academic position in Saint Petersburg. Undoubtedly Lexell must have been aware of the risks involved in such an enterprise. After all, Sweden and Russia had been arch enemies for centuries, where, in particular, the frontier—Finland—was at stake. Although the Cap-party was in power in Sweden at the time (between 1765 and 1769) and relations with Russia were friendly, it may have been regarded as imprudent or even disloyal to enter into Russian service.[2] What would happen in the case of a change of policy or the outbreak of a war? Would he always be accepted back in Sweden? However, in the world of the learned, the "Republic of Letters", national borders are often disregarded, and sometimes the risk is just worth taking. For instance Euler had remained in the service of the Petersburg Academy of Sciences in his Berlin period (1741–1766), even during the Seven Years War, when Prussia and Russia were on different sides (Fig. 3.1).

3.2 Correspondence and Promises

Not every step of Lexell's campaign to win Euler's favour is known in detail, but the first indications of its preparation can be traced to a letter written in Åbo on 14 January 1768 to the Permanent Secretary of the Royal Swedish Academy of Sciences, Pehr Wilhelm Wargentin (1717–1783), Lexell's distant relative[3] living and working in the Stockholm Observatory [125] (see Fig. 9.7). Wargentin was the most influential astronomer in Sweden at the time—for an account of his life, see Chap. 6—and it was probably he who tipped off Lexell about the possibility of finding a position in Saint Petersburg. By correspondence with his fellow astronomer Stepan Rumovski, Wargentin was kept updated on the astronomical expeditions in preparation to observe the transit of Venus in front of the Sun in June 1769 at the Petersburg Academy of Sciences and elsewhere.

For Lexell, his urgent efforts to apply for work abroad happened in a stressful moment due to the fatal illness of his father Jonas Lexell, who died the same spring on 9 May 1768. Whether the deteriorating health of his father actually prompted

---

[2]During the parliamentarian rule of Sweden in 1719–1772, the political scene was dominated by two parties, the "Caps" [*Mössor*] and the "Hats" [*Hattar*], the former being represented mostly by peasants and the clergy. In foreign policy, the Caps advocated friendly relations with Russia, the Hats a strong alliance with France.

[3]Wargentin, whose father had been born in Åbo, was a second cousin of Anders Johan Lexell's mother [24].

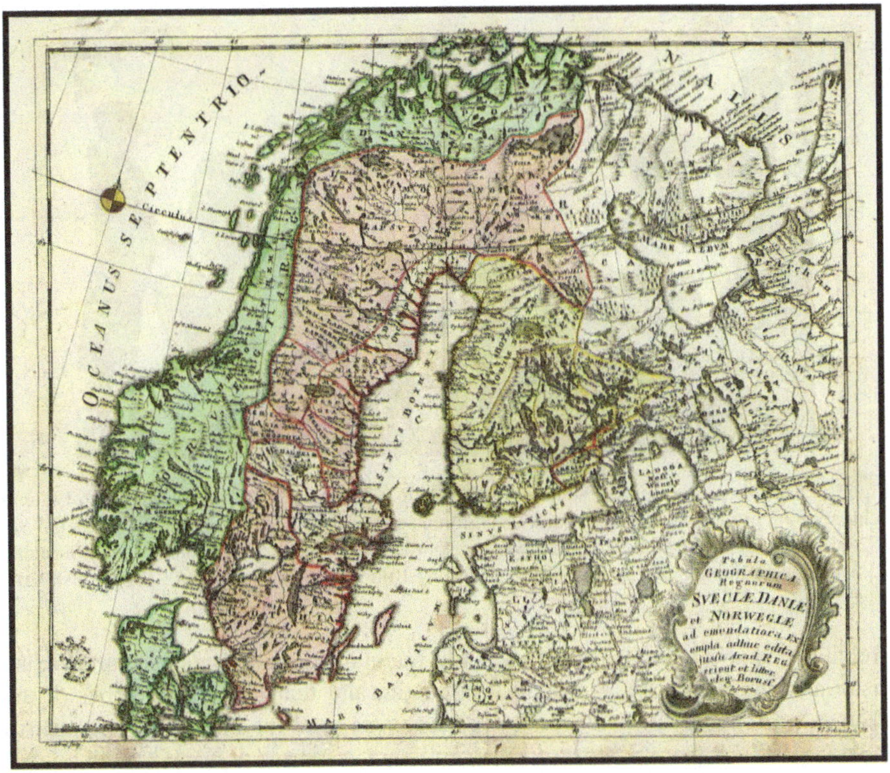

Fig. 3.1 Geographical and political map of northern Europe 1760 in a school atlas, the *Geographischer Atlas*, published by the Royal Prussian Academy of Sciences in Berlin; incidentally, a project of which Leonhard Euler was in charge (Berlin-Brandenburgische Akademie der Wissenschaften. With permission)

Lexell to apply for work abroad is not known. In any case, Lexell's scholarly correspondence with Saint Petersburg was already going on at that time.

According to the minutes of the conferences of the Petersburg Academy of Sciences, on 21 March 1768 a dissertation[4] written in Latin by Lexell was read. The dissertation had been submitted from Göttingen by the German historian August Ludwig Schlözer (1735–1809), a member of the Petersburg Academy of Sciences, who also had close connections with the Royal Swedish Academy of Sciences, of which he had been a foreign member since 1767 [65] (Fig. 3.2). Before entering the Petersburg Academy, Schlözer had been employed in Sweden, and at that time he was writing the *Allgemeine Nordische Geschichte* (which was published in 1771). The only fact that seems to explain why Schlözer—a German Professor of History at Göttingen—recommended a young mathematician from Åbo, whom he most

[4]The German term *Abhandlung* in the minutes of the academic conferences is translated here as dissertation or treatise; in French, the corresponding word is *mémoire* (occasionally translated into English as memoir).

Fig. 3.2 Silhouette portrait of Schlözer's family in Göttingen. To the right August Ludwig Schlözer with his spouse. Among their five children seen in the picture, Dorothea Schlözer (1770–1825), the lady with the globe to the left, was one of the first women to earn a doctorate in Germany (Universitätsbibliothek Göttingen. Schlözer-Stiftung Bilder AL 128. Published with permission)

probably had never met, was the fact that he maintained close contacts with Swedish historians and scientists, especially Wargentin, and that he, as a member of the Petersburg Academy of Sciences, was aware of its need for competent scientists (a number of letters from Schlözer to Wargentin are preserved at KVAC, Stockholm).

Thus, either Wargentin had forwarded the treatise to Schlözer or advised Lexell to write to Schlözer personally for the purpose of contacting the Petersburg Academy. The minutes of the academic conferences, which at the time were written in German, give us the following account of the events on 21 March 1768 [141, Vol. II, pp. 632–633]:

> ...a dissertation was read, entitled *Methodus integrandi aequationes* etc.[5] by a Swedish *Magister* Lexell in Åbo, who asks for employment at the Academy.
> The dissertation was read and Professor Euler pronounced his opinion on it as follows: This dissertation reveals an excellent genius and a great talent of the author in mathematics: therefore, he would deserve to be considered by the Academy according to his wish. It was decided that Professor Schlözer, who sent the dissertation, as well as other correspondents shall be asked to send more information about [Mr Lexell].[6]

[5]This paper appeared in print as *De integratione aequationis differentialis*: $a^n\,d^n y + ba^{n-1}\,d^{n-1}y\,dx + ca^{n-2}d^{n-2}y\,dx^2 + \ldots + ry\,dx^n = X\,dx^n$ in the Academy's *Novi Commentarii*, Vol. XVI, 1769. [Lexell 4]

[6]10 Märtz 1768: "Ferner eine von demselben eingesandte Lateinische Abhandlung Methodus integrandi equationes etc. eines Schwedischen Magisters Lexel zu Abo, der Verlangen trage, bey der Academie in Dienste zu kommen. [...] Diese Abhandlung wurde öffentlich vorgelesen und von dem Hrn. Prof. Euler der Ausspruch gethan: Man entdecke an dieser Abhandlung ein vortrefflich Genie und viele Stärke des Verfassers in der Mathematik: er sey dahero würdig, dass die Academie seinem Verlangen nach Reflexion auf ihn mache. Darauf ward beschlossen, sowohl

Lexell's response came rapidly. His enthusiastic letters to Professor Schlözer and to the Secretary of the Academy, Jakob von Stählin, have been preserved in the foreign correspondence of the Petersburg Academy of Sciences. Both letters were written in elegant French, albeit with somewhat unusual orthography and peculiar verb conjugation.

Lexell to Schlözer (АРАН Фонд 1, опись 3, дело № 51, листы 153, 153об.)

Monsieur,

Je ne sçaurois vous exprimer les sentiments de joje et de la reconnoissance, des quelles je suis pénétrée en récevant votre dernière lettre, car je n'aurois jamais osé, me former un espoir si flat[t]eux que ma dissertation obtiendroit cette grande et parfaite approbation, dont vous m'assurez, vous ne devez pas donc trouver étrange, que votre lettre m'a rendu également jojeux, surpris et confus. De quelle manière aurois je pû mériter, que mon ouvrage seroit comparé, à celui des plus illustres Mathématiciens Messieurs Euler et d'Alembert? Et si je voudrois m'approprier cet éloge, on auroit sans doute raison, de me juger l'homme le plus vain et le plus ambitieux. Quelque fois j'ai pensé que vous m'aviez fla[t]tée, et que mon ouvrage n'aie pas reçu une approbation si honorable, que celle, qui est marquée dans votre lettre. Mais pourquoi aussi douter de votre sincérité, pourquoi soupçonner la fidélité de celui, dont l'amitié pour moi a toujours été sans réserve[?]
Jugez de ma joje, de mon ravissement, d'avoir reçu des éloges si éxquises, par les plus grands Mathématiciens de notre siècle. Je me rappelle sans cesse ces vers de Naevius,

Laetus sum, laudari abs te pater, laudato viro.[7]

Faitez-moi la grâce, je vous en supplie, de déclarer à Monsieur le Professeur Euler la profonde vénération avec laquelle je reconnois ce jugement honorable et gracieux qu'il a donnée de ma dissertation et daignez l'assurer, que je m'éstimerois très heureux, de pouvoir profiter de sa offre généreuse pour m'établir en Pettersbourg, pourvû que j'eusse quelque engagement fixe, car ma fortune est trop mediocre, pour pouvoir subsister quelque tems sans appointement, dans une ville, où il fait extrêmement cher, comme l'on m'a rapportée. Si je puisse obt[eni]r quelque fonction comme Adjoint de l'Académie, je serois content.

Ayez aussi la bonté de présenter la lettre ci-jointe à Mr Le Conseiller d'État von Stählin, qui si je ne me trompe est Secrétaire de l'Académie, et soyez pérsuadée, que je serois avec la plus sincère et parfaite amitié

Votre
très humble serviteur et fidèle ami
André Jean Lexell

Åbo, ce 25 Avril 1768

Lexell to von Stählin (АРАН Фонд 1, опись 3, дело № 52, л. 34–35)

L'accueil favorable et gracieux, qu'il a plut, à la très illustre Académie des Sciences, de faire à cette dissertation, que j'ai eu l'honneur de communiquer avec elle, m'est d'autant plus sensible, que je ne sçais si jamais je pourrois mériter, tous les éloges de quelles les

durch Hrn. Prof. Schlözer, der diese Dissertation eingeschickt hatte, als auch durch anderweitige Correspondenz genauere Kundschaft von demselben einziehen zu lassen".

[7] I am glad, father, to be praised by you, who are a praiseworthy man (Naevius as cited by Cicero, *Tusculanae disputationes* Lib. IV:67).

messieurs les Académiciens ont bien daigné honorer mon ouvrage; c'est pourquoi aussi, je dois les regarder moins comme des récompenses dues à mon mérite, que comme de l'encouragement, pour avoir formé un dessein noble et hardi. Je m'éstime très heureux, d'avoir reçu une preuve si convaincante de la bienveuillance et du faveur de cette illustre Société, quel seroit donc mon bonheur, si je puisse m'en rendre digne?
Encouragée par l'heureux succès de ma première tentative, je prens l'hardiesse, d'adresser le papier ci-joint[8] à l'Académie et quoique convaincu de la médiocrité de son mérite, je me flat[t]e pourtant, qu'il servira comme une épreuve de ce que je sois en vérité l'auteur de la dissertation sur l'équation différentielle

$$a^n d^n y + ba^{n-1} d^{n-1} y dx + \ldots + r y dx^n = X dx^n$$

Souffrez Monsieur que je vous demande la grâce, qu'en présentant cette ébauche à l'Académie, vous voulez bien l'assurer, que je suis penetrée de la plus grande vénération des admirables qualités de ses membres et de la plus sincère reconnoissance de leurs bontés. Daignez aussi être persuadée du profond respect et de l'inviolable attachement avec le quel je serai

Votre

très humble et très obéissant serviteur
André Jean Lexell

Åbo, ce 25 Avril 1768

The entry in the minutes of the conferences for 30 May (o.s.) 1768 [141, Vol. II, p. 640] continues the deliberations about Lexell as follows:

The Secretary of the conference [von Stählin] read a letter from *Magister Legens* Lexsell [sic] in Åbo, where he certifies his great joy and thankfulness for the favourable judgement of the analytical dissertation *Methodus integrandi aequationem differentialem etc.* he had submitted, and encloses for further recommendation another dissertation *Methodus integrandi, nonnullis exemplis illustrata.* As this one, like the previous, has been examined by Professor Euler and turned out to be outstanding proof of his mathematical knowledge, the Director of the Academy, His Excellency the Count [Orlov], ordered the approval of the Academy to be confirmed to [Lexell] and at the same time to tell him that they should keep him in mind and perhaps soon take further decision about him.[9]

More details about the course of events are found in the short obituary notice entitled *Précis de la vie de Mr Lexell* published in the Academy's *Nova Acta* for the

[8]This paper [Lexell 5] was eventually printed with the title *Methodus integrandi nonnullis aequationum differentialium exemplis illustrata* in NCASIP XIV, Part I, 1769. The contents are discussed in Sect. 8.2.

[9]Der Conferenz-Secretair las einen Brief von dem Mag. Legens Hrn. Lexsell aus Abo, darin er der Academie sein grosses Vergnügen und Dank für die gütige Aufnahme seiner jüngst an dieselbe gesandten analytischen Abhandlung, Methodus integrandi aequationem differentialem etc., bezeugt und zu seiner weiteren Empfehlung eine neue Abhandlung, Methodus integrandi, nonnullis exemplis illustrata, [sendet].
Da auch diese nicht weniger als die vorige von der Herren Prof. Euler geprüft, und als ein vortreffliches Specimen seiner mathematischen Wissenschaft erkannt worden, so befahlen des Hrn Directors Hochgräfl. Erlaucht, demselben in der Antwort den Beyfall der Academie zu bezeugen und zugleich zu melden, dass sie seiner eingedenk bleiben und sich vielleicht bald näher gegen ihn erklären mögte.

year 1784 [41]. The author of this obituary—most likely Johann Albrecht Euler[10] —writes:

> In 1768, in order to join the Academy, Mr Lexell addressed to the Academy a dissertation entitled: *Methodus integrandi, nonnullis aequationum exemplis illustrata*, which did not miss its target. The late Mr Euler, responsible for its examination, not only evaluated it favourably, but — and this makes it particularly worthy of praise and special mention — when Count Vladimir Orlov,[11] who at the time was the chief of the Academy, objected that it might have been produced by some skilled mathematician who wanted to favour Mr Lexell, Mr Euler replied with his usual vivacity that in that case there would be only two persons in the world that might have done this, one being Mr d'Alembert and the other himself, Euler. But neither of them knew Mr Lexell at the time.

This was clearly a sensational feat by any standard. Lexell had evidently made a good impression on the Academy, but what is more, he was already compared with the greatest heroes of mathematics, d'Alembert and Euler. No wonder, then, that he was flattered. But how could he know if these kind words were true and sincere? Or were they merely intended to flatter?

3.3 Preparations for the Move

Lexell was by no means unaware of the exceptional circumstances surrounding his employment in Saint Petersburg, since the Imperial Academy of Sciences usually did not enrol its members upon mere expectations. The new members, besides having a strong reputation in science, were expected to manifest a zealous commitment to the pursuits of the Academy. This meant in practice that one had first to become personally known to the academicians and then, little by little, become engaged as an assistant or associate. Such a procedure had been followed e.g. when Lexell's compatriot, the natural historian Erik Laxman [87], was engaged in the Academy (for a biographical sketch of him including his relations with Lexell, see Sect. 6.4). In the eighteenth century, the scientific profession was often passed down within families, or from father to son. In Saint Petersburg, this was the case for the Gmelins, Kraffts, Lowitzs and the Eulers. Outsiders had to work harder to find their place in the community. While the economic and geographical expeditions created a demand for natural historians, only with the rare event of the transit of Venus in front of the Sun in 1769 did there arise a sudden need for precisely the kind of

[10]Johann Albrecht (Jean Albert) Euler (1734–1800), the eldest son of Leonhard Euler, born in Saint Petersburg, became a Professor of Physics and from 1769 on the Permanent Secretary of the Petersburg Academy of Sciences.

[11]Vladimir Orlov (1743–1831), the youngest of four influential brothers from the Russian nobility, had received his education in Germany. Having been appointed Director of the Academy, he changed the working language of the Academy from Latin to German.

mathematical skill Lexell possessed.[12] And as there was no actual vacant position for Lexell to apply for, one was created for him.

All this was the "official" truth, however. Informally, there was another more long-standing problem: managing the vast scientific production of Leonhard Euler. Given that the Petersburg Academy of Sciences during the reign of Catherine II was markedly open to Western influences, it became urgent to find a young and talented assistant to the Master. Nevertheless, according to the minutes of the conferences of the Petersburg Academy of Sciences [141, Vol. II, pp. 639, 662, 665], all through the correspondence with Lexell, negotiations were going on to employ an apprentice of the French astronomer Joseph Jérôme de Lalande.[13] For an unknown reason these attempts were unsuccessful, a fact that possibly delighted Euler, however, who as a devout Christian is likely to have preferred the idea of a skilled mathematician from a Lutheran country to an unknown astronomer from France, known for its widespread free thinking (Euler certainly knew that Lalande was a Mason).

Lexell started the preparations for his move to Saint Petersburg immediately. On 10 June 1768, he addresses his reply to Schlözer via Wargentin,[14] stating two of his main concerns: "first, that [he] is in astronomical observations; second, that [he] would like to know his salary, since he cannot undertake such a long journey on mere expectations of a job". Finally, Lexell mentions an interrupted correspondence with a compatriot, Henrik Lemström (1739–1771), a classmate from the Royal Academy of Åbo and likewise a student of Wallenius and Gadolin [81,88]. On his visit to Saint

---

[12]Catherine II, in a letter to Director Orlov, dated 3 March 1767, quoted in [176], wrote: "[I]f there is not a sufficient number of astronomers in the Academy to make the observations ... competent persons should be searched for ... ". Hence, on the recommendation of Daniel Bernoulli, the astronomer Jacques-André Mallet (1740–1790) and his companion Jean-Louis Pictet (1739–1781) from Geneva were engaged to observe the transit of Venus in Russian Lapland [22]. The astronomer Christian Mayer was invited from Heidelberg to take charge of the Saint Petersburg Observatory.

[13]Joseph Jérôme Lefrançois de Lalande (1732–1807) was the editor of the scientific almanac *Connoissance des temps*, professor at the *Collège Royal* and *École Militaire*, and member of several scientific societies, among others the Petersburg Academy of Sciences [17,38].

[14]Lexell to Wargentin, dated Åbo, 10 June 1768:

"Högädle Herr Secreterare Efter Professor Schlötzers begäran har jag adresserat mit swar på Hans bref til Hr Secreteraren. Det anbudet, som mig göres i Pettersburg, wil jag gärna emottaga, allenast twänne omständigheter göra någon swårighet. Den första är, at jag icke haft tilfälle at wara wid Astronomiska Observationer och således är owan wid et sådant sysslande. Den sednare åter består däri, at jag önskade få weta om, huru stor lön mig skulle lofwas, emedan mina omständigheter äro sådane, at jag ei på blott expectance kan göra någon så långwägad resa. I mit bref til Professor Schlötzer har jag likwäl förklarat mig nögd, allenast jag får weta wilkoren. [...] För någon tid tilbaka hade jag i samma ämne bref af Hr Lemström från Pettersburg och som han wid slutet berättade, at Professor Euler den yngre bedt mig skrifva honom (Lemström) til, så giorde jag det äfven; men har sedermera ei haft något swar. Nu har jag tänkt adressera mig til Swenska Commissions Secreteraren, at af honom blifwa underrättad om Lemström fått brefvet eller icke, ty at det samma måtte komma til rätta är mig angelägit, efter däri war inneslutet et annat bref til Hof rådet Stählin. Jag har äran at med högaktning framhärda..."

Petersburg 1767–1768, Lemström had apparently made the acquaintance of Johann Albrecht Euler and applied for a position at the Petersburg Academy of Sciences, but, as it turned out, unsuccessfully.[15]

Meanwhile, written recommendations for Lexell had started to arrive in Saint Petersburg. According to the minutes for the conference held on 20 June 1768 [141, Vol. II, p. 643], "a very advantageous certificate from the Royal Librarian Mr Gjörwell[16] was read concerning *Magister Legens* Lexell of Åbo", as to his "remarkable talent in mathematics as well as his good manners". On 30 July, according to the same minutes [141, p. 647], a letter from Professor Schlözer in Göttingen was read, citing *verbatim* the Secretary of the Royal Swedish Academy of Sciences Pehr Wargentin's "most splendid testimonial of *Magister* Lexell's mathematical skills and his virtuous conduct". Lastly, on 22 August 1768 [141, p. 648], when Wargentin's laudatory letter of recommendation of Lexell to his colleague astronomer Stepan Rumovski[17] had been read, the Director Count Orlov resolved that the academicians present at the conference would cast a vote for Lexell's employment at the Academy.[18] Thereupon it was decided that a letter of invitation would be sent to Lexell, offering him a post in the Observatory until some other position became available for him. Further, his yearly salary was fixed at 400 Roubles, and 100 Roubles were offered for his journey to Saint Petersburg, with a possibility of an increase dependent on his merits. The letter was written in German on 26 (15) August 1768 by the Secretary Jakob von Stählin; a copy of it has been preserved in Saint Petersburg [108, Letter No. 364]. Lexell's quick reply and acceptance of the proposal was read out at the conference of 23 September 1768 [141, p. 652]. However, Lexell's departure was delayed by a month because of the law which prohibited Swedish subjects to enter into the service of a foreign

[15]Lemström's paper entitled *De tubis capillaribus* was examined by the Petersburg Academy in 1768 [141, Vol. II, p. 626], but seems to have made no impression. Despite the criticism by the Swedish mathematicians Fredrik Mallet and Bengt Ferner [26], Lemström was accepted as an "adjunct" (in Swedish: *ämnessven*) of the Royal Academy of Sciences in 1771, but he died the same year in Stockholm [81].

[16]Carl Christopher Gjörwell (1731–1811), librarian and journal editor, Stockholm. His son, the architect C. C. Gjörwell junior, designed a new main building for the Imperial [earlier: Royal] Academy of Åbo in 1817.

[17]In a letter to Wargentin (KVAC; in Latin), Rumovski confirmed that Lexell's treatise had made an impact at the Academy by its generality.

[18][141, Vol. II, p. 648] (22 August 1768): "Nachdem über die bisherige Recommendation des zu Abo befindlichen Hrn. M. Lexel [sic] auch ein Brief von dem berühmten Astronom und Secretair der Königl. Schwedischen Societät der Wissenschaften, Hrn. Wargentin, an den Hrn. Prof. Rumoffski eingekommen, darin derselbe für den mehr erwähnten M. Lexell ein sehr gutes Zeugniss seiner Stärke in den mathematischen Wissenschaften und seiner guten Sitten ablegt, so wurde auf dem vortrag Sr. Erlaucht des Hrn. Directors von allen anwesenden Mitgliedern auf seinen Beruf in hiesige Kayserl. Dienste bey der Academie gestimmt, und beschlossen, demselben den Antrag mit nächster Post nach Abo zu überschicken, und darin zu melden, dass er für's erste bey dem astronomischen Observatorio wegen Ermangelung einer andern offenen und für ihn schicklichen Stelle gebraucht, einen järlichen Gehalt von Rbl. 400 geniessen, zur Anhero-Reise 100 Rbl. bekommen und mit der Zeit nach Beschaffenheit seiner Verdienste weiter bedacht werden soll".

country without the King's permission. The permission was duly granted, as well as a statement by the University Consistory that Lexell, despite his move abroad, would be notified and taken into consideration in future promotions at the university in Åbo. Thus, once he had ensured his place on his likely return, Lexell arrived in Saint Petersburg at the end of October 1768.[19]

3.4 First Impressions of Saint Petersburg

Lexell's first known written report from Saint Petersburg was addressed to Bishop Carl Fredrik Mennander on 21 November 1768 (in Swedish, KB, Stockholm, in part reprinted in [132]). After a full declaration of esteem and respect to his former benefactor in Åbo, Lexell proceeds with his scholarly account:

> The reorganisation of the Imperial Academy of Sciences which was carried out in 1748[20] is for the most part still observed, with some notable exceptions, namely the History Class, which was reduced at that time, will now be reinstated. According to the same statutes, the number of ordinary members was 20, namely 10 Professors and 10 Adjuncts, but this regulation is no longer observed correctly, because at present there are more than 10 Professors, whereas the Adjuncts are fewer, even if there are some who receive the salary of an Adjunct, without really being one, because the same regulation of 1748 says that all Adjuncts must be of Russian nationality. As to the duties of the Professors, the same regulations state that the objective of the present Academy is twofold; to cultivate the sciences and the instruction of youth, wherefore every Professor is obliged to give lectures for two students; however, as to the latter point this does not apply in the present organisation, for whatever reason I do not know. At present the Professors of the Academy are the mathematicians: two Eulers, Father and Son, Lowitz,[21] Rumovski,[22]

[19]According to the record, this was on 20 October (cf. the list "Source Material in Saint Petersburg" in the index part of this volume). The Journal of the astronomer Jacques-André Mallet of Geneva, who was invited to Russia for observing the transit of Venus [22], mentions on 27 October 1768 having heard from J. A. Euler of Lexell's arrival (p. 254), and that he [Lexell] was entirely inexperienced in astronomy.

[20]The first written charter of the Petersburg Academy of Sciences, granted as late as 1747, divided the institution into the Academy proper and the University. The absence of a formal written charter for the Petersburg Academy of Sciences had created much unrest among the academic personnel and led to the loss of many of its foremost members during the first two decades of its existence [116, 170].

[21]Georg Moritz Lowitz (1722–1774), Professor of Astronomy, was born in Fürth, Bavaria. In 1774, on a geodetic expedition in southern Russia during the Pugachov rebellion, he was captured and murdered by Cossacks in the town of Ilovlya. His son Johann Tobias Lowitz (1757–1804), who accompanied him on his fatal trip, became a Professor of Chemistry and a member of the Petersburg Academy of Sciences in 1793 [137].

[22]Stepan Rumovski (1734–1812) was the principal astronomer of the Observatory. He had been trained by Euler in Berlin.

Kotelnikov,[23] the two last ones are of Russian nationality, the historians: Fischer[24] and Schlözer, the physicians and natural historians: Gmelin, Pallas, Gärtner and Wolf.[25] The Chair of Chemistry is vacant at the moment. The Director of the Academy is Count Vladimir Orlov, the youngest of the brothers.[26] Previously, the supervisor of the Academy was called Chancellor, but nowadays that has also been changed. The present Secretary of the Academy is the State Secretary von Stählin,[27] a rather cheerful and witty fellow. The conferences of the Academy are generally held two or three days a week, mostly on Mondays, Thursdays and Saturdays, sometimes on other days, depending on the circumstances.

The finances of the Academy are managed by a specially appointed Commission, presided by the Director and assisted by some of the Professors of the Academy, chosen by the Director. Previously, there were judges and cabinet ministers in this Commission, who were separately paid for, but with the present reorganisation the government saves a lot of money, and moreover, the issues are more carefully dealt with. This Commission has its own Chancellery and an office belonging to it.

The premises of the Academy are two large stone buildings on Vasilyevsky Island,[28] situated obliquely in front of the Imperial Palace. In the larger one [known as the *Kunstkammer*, Кунсткамера], which has three floors, the Library, the Natural Cabinet and the Observatory are situated, along with two other rooms. I have not yet had the occasion to see the Library and the Natural Cabinet. The Observatory, which forms the dome of this house, has two floors, the upper one being undoubtedly more practical for carrying out observations as regards view; nevertheless, it is surely impracticable during the winter, when the cold and the damp affect all the instruments. On the lower floor there is a hall in the centre and three chambers on the sides, and in one of them the mural quadrant of radius 8 feet was supposed to be installed, but it has not yet been completed. These three rooms seem more practical in winter conditions than the upper one. The instruments at the Observatory are few, since most of them have been removed to be used for the planned astronomical expeditions. Additionally, there are still two quadrants, two telescopes, of which one was unusable, and one Dollond tube of length 6 ft. Also two of the Observatory's Dollond tubes, the larger of which is 18 ft long, are kept in the office of the Chancellery. The Chancellery is located in the other Academy building, where also the hall of the Conferences, the Printing Office, the Bookshop and a number of rooms for artists, which belong to the Academy of Arts, are situated. The Printing Office is large and it has its own foundry for making types.

---

[23]Semyon Kotelnikov (Семён Кириллович Котельников; 1723–1806), mathematician, trained by Euler. He wrote the first textbook of mechanics in Russian.

[24]Johann Eberhard Fischer (1697–1771), historian and antiquary born in Esslingen, Württemberg.

[25]Samuel Gottlieb Gmelin (ca. 1744–1774), natural scientist, born in Tübingen; Peter Simon Pallas (1741–1811), natural scientist and explorer from Berlin [171]; Joseph Gärtner (1732–1791), botanist who returned to Germany in 1770 [119]; Caspar Friedrich Wolff (1734–1794), Professor of Anatomy and Physiology from Berlin.

[26]There were four brothers Orlov in Russian politics at the time. When Count Grigory Orlov (1734–1783), Catherine's favourite, fell into disgrace, while the star of Prince Grigory Potemkin (1739–1791) was rising, Vladimir Orlov was eventually forced to resign from the post of Director of the Academy in 1774.

[27]Jakob von Stählin (Staehlin-Storcksburg) (1709–1785), State Secretary, Professor of Eloquence and Poetry of the Petersburg Academy of Sciences. Born in Memmingen, Württemberg, he arrived in Saint Petersburg in 1735.

[28]"Vassili ostroff" (Васильевский Остров, Vasilyevsky ostrov) was the traditional centre of the German-Baltic community in Saint Petersburg (see the map in Fig. 4.1). The centre of trade, the merchants, the Academies and schools were situated there. However, the church of the Lutheran *St.-Petri Gemeinde* was situated on the *Nevski prospekt* on the mainland.

The above-mentioned Academy of Arts is in some way connected to the Academy of Sciences, but it has its own Director, the Imperial Chamberlain Betskoi.[29] The State Secretary [von Stählin] is likewise Secretary of this Academy, and it is mainly in this subject his greatest strength lies. He owns an important collection of engravings, statues, and paintings, among which there are some really exquisite ones, which he claims cost 1500 Roubles apiece. The Academy of Arts consists of the mentioned workshops for draughtsmen, painters, sculptors and copperplate engravers, which as yet seem to be in their beginnings. All the students in these workshops are native [Russians]. The sculptor has travelled at the old Count Münnich's[30] expense, and I saw at his place the busts of the Grand Duke and Count Münnich, both of them quite well made. The Academy has contracted a geographer from Homann's office[31] and made an agreement with him to deliver 6 maps every year [for the Academy], for which he gets 70 Roubles apiece.

The Professors of the Academy of Sciences have quite different salaries; some of them have 1000 Roubles, others have 800 and some 600, indubitably according to the expectations on their capacity and knowledge. The University once established here is now completely closed down and its members reinstated at the Academy. It could not be otherwise, because a University composed of five Professors, namely one in History, one in Physics, one in Mathematics, one in Languages and one in Medicine, could not cope with its duties. Some of its students are still maintained by the Academy.

The *Gymnasium* here is normally supervised by the Academy. The subjects dealt with are the languages Latin, German and French, History, Geography, Geometry and Arithmetic. It is supervised by Inspector Bacmeister[32] who is renowned for his beautiful version of Botin's "Swedish history"; he is my nearest neighbour, and he has shown me much courtesy.

As to the organisation of the Naval Cadet Corps here and their teaching methods I do not have the honour to say anything, as I have not had the chance to visit them yet. I should not forget to mention that two Fathers from Rome have recently arrived here to be working at the Cadet Corps, one of them mathematician, the other one a teacher of the Italian language. A mathematician from Rome may seem peculiar, but it is nonetheless true. His name is Pezzutti,[33] but as for the name of his companion, I do not know.

The expeditions which are planned for observation of the transit of Venus across the Sun are seven in number, four of which are going to the northern parts of Russia, and three to the southern parts. Professor Rumovski is destined for the Kola Peninsula, Mrs Mallet and Pictet,[34] two Frenchmen from Geneva, are going to Arkhangelsk or thereabout,[35] *Magister*

---

[29]Privy Councillor (*Geheimrat*) Ivan Betskoy (1704–1795) was also the curator (and founder, with Catherine II) of the *Smolny* Institute (Смольный институт) for young girls of the nobility [74].

[30]Burkhard Christoph von Münnich (1683–1767), German-born soldier, who became a Russian field marshal and governor.

[31]Mapmaking company in Nuremberg (Nürnberg) founded by the German geographer Johann Baptist Homann (1664–1724).

[32]Hartwig Ludwig Christian Bacmeister (1730–1806), Adjunct of the Academy, historian and librarian. He wrote contributions to the history of Peter the Great and translated the *Geschichte der Schwedischen Nation im Grundriß* (1767) by the Swedish historian Anders Botin (1724–1790).

[33]Abbé Giovacchino Pessuti (1743–1814) was an Italian mathematician and teacher at the Saint Petersburg Naval school. He returned to Rome in 1769, where he was appointed Professor of Mathematics [22].

[34]See the journals of Jacques-André Mallet and Jean Louis Pictet [22].

[35]Apparently, the destinations had not yet been decided. Finally, Mallet was stationed at Ponoi and Pictet at Umba, two villages on the Kola Peninsula.

Krafft[36] and Lieutenant Euler[37] to Orenburg, although both of these gentlemen have a station of their own, Professor Lowitz is destined for Astrakhan or Guryev [= Atyrau] by the Caspian Sea and the seventh, a certain Lieutenant Islenjef[38] has long ago travelled to Yakutsk in Siberia. For some time now there has been no news from the physical and economical expedition, the last piece of information arrived from Moscow and contained quite remarkable discoveries by Professors Gmelin and Pallas. The first one has become quite famous; he is the nephew of the late Professor Gmelin.[39] The Swedish Professor Falck[40] is there, but he is said to be quite apathetic; he is splenetic[41] as usual.

The large Pharmacy is certainly among the most remarkable institutions in Petersburg, as it provides not only the other pharmacies but also the Army, the Navy and all the cities of the Empire with medicines. The Director of this important establishment is Councillor of the Court Model,[42] a rather competent man in chemistry and pharmaceutics. For a long time he has been in dispute with the local *Collegium Medicum*, and he has defended himself with honour. Presently he has one work in print, but what about, I do not know.[43]

I may mention two remarkable books, which are now in print, namely Professor Fischer's *Historia Sibirica*[44] and Euler's *Calculus integralis*.[45] The latter, that is Professor Euler, I have visited several times, and he is a rather merry and courteous man; in spite of him being a mathematician, he has almost more vivacity than any one of his sons, and yet he is 60 years old. He is not totally blind, but can see only a glimpse of light; in spite of that he works constantly and dictates to *Magister* Krafft, his Amanuensis. He has now the

---

[36]Wolfgang Ludwig Krafft (1743–1814) was Adjunct and from 1771 on Professor of Experimental Physics in Saint Petersburg. He was educated in Tübingen by his father Georg Wolfgang Krafft, former Professor of Physics in Saint Petersburg. In 1767 he travelled to St. Petersburg as an amanuensis (assistant) for Euler. Besides diverse astronomical topics, such as methods to determine longitude, his researches (published in the *Novi Commentarii* and *Nova Acta* of the Academy) concerned electricity, terrestrial magnetism, the theory of heat and applied mechanics. In 1788, he also published a paper concerning Lagrange's theory of attraction between two spheroids [140].

[37]Christoph Euler (1743–1808), military officer, son of Leonhard Euler.

[38]Ivan Ivanovich Isleniev (1738–1784), military officer and Adjunct of Geography.

[39]Samuel Gottlieb Gmelin, introduced earlier in this letter, was the nephew of Johann Georg Gmelin (1709–1755), born in Tübingen, botanist, explorer and academician in Saint Petersburg, renowned for his expeditions to Siberia. J. G. Gmelin later returned to his native town.

[40]Johan Peter Falck (1732–1774), Swedish explorer and naturalist, a disciple of Linnaeus. On his death in Tatarstan, cf. Sect. 6.3.

[41]"Han lär efter vanligheten vara miältsjuk" (the Swedish word 'mjälte' means spleen). In today's medical terminology, Falck's condition is called depression. In the eighteenth century, the ancient humoral pathology was still used for assessing and curing medical condition: According to humorism, a person's health is determined by the appropriate equilibrium between the four basic bodily substances, called humours (Latin for fluid).

[42]Johann Georg Model (1711–1775), physician and chemist.

[43]The second part of Model's *Chymische Neben-Stunden* was published by the Petersburg Academy of Sciences in 1768.

[44]*Sibirische Geschichte von der Entdekkung Sibiriens bis auf die Eroberung dieses Lands durch die Russische Waffen*. St. Petersburg, 1768. At the same time, Fischer's pioneering linguistic study *Vocabularium Sibiricum* (1747) was edited and published by A. L. Schlözer in Göttingen (1768).

[45]*Institutionum calculi integralis*, Vol. I, St. Petersburg, 1768 (reprinted in: OO Ser. I, Vol. 11. E342). Volumes II (E366) and III (E385) of the work were published in 1769 and 1770, respectively.

> *Dioptrica*[46] ready for print and the *Hydrodynamica*[47] ready as a manuscript, and the *Theoria motuum Lunae*[48] under way.
> At present, the common people here seem most interested in inoculation. Ever since both Her Majesty the Empress and His Imperial Highness had taken this treatment with success, it has become *grande mode* here in Petersburg to become inoculated.[49]

The letter continues with Lexell's detailed description of the inoculation process and a chronological account of the symptoms the patients must endure, with the strong recommendation that Swedish physicians should also begin treating their patients against smallpox with similar methods.

Knowing that Lexell had been in Saint Petersburg only for a month when this letter was written, it is of course not to be expected to find in it the most profound insights into the working of the Academy. As Lexell pointed out, however, the Academy of Sciences was a vast organisation, and given the state of development of Russian society, the Academy had to produce itself almost all the services it needed, starting from the manufacturing of instruments, printing, engraving, book-selling and education. The organisation was notorious for its slow and heavy bureaucracy.[50] It was strictly supervised by the Empress by way of a hand-picked director from the Russian aristocracy, who had in his command a force of 70 soldiers of the infantry to watch over the academicians and their premises.

During its whole previous existence, the Academy of Sciences had been a source of irritation for those conservative forces in Russia who were opposed to Western influence in science and culture. Their anti-enlightenment mission, where they temporarily succeeded, was to prohibit the use of Western languages, except Latin, at the Academy. On the other hand, an almost permanent internal conflict was raging between the academicians and the administrators. The directors in charge were often quite unfamiliar with science and education; they were more interested in promoting

---

[46] *Dioptricae pars prima*, St. Petersburg, 1769 (reprinted in: OO Ser. III, Vol. 3. E367). Part II and III of *Dioptrica* were published in 1770 and 1771, respectively [43].

[47] Referring to *Sectio secunda de principiis motus fluidorum*, NCASIP XIV, Part I, pp. 270–386, 1770 (OO Ser. II, Vol. 13, pp. 73–153. E396) and *Sectio tertia de principiis motus fluidorum*, NCASIP XV, pp. 219–360, 1770 (OO Ser. II, Vol. 13, pp. 154–261. E409). Part I of *Principia motus fluidorum* (E258) had been published in NCASIP VI, pp. 271–311, 1761.

[48] *Theoria motuum Lunae, nova methodo pertractata una cum tabulis astronomicis, unde ad quodvis tempus loca Lunae expedite computari possunt incredibili studio atque indefesso labore trium academicorum: Johannis Alberti Euler, Wolffgangi Ludovici Krafft, Johannis Andreae Lexell. Opus dirigente Leonhardo Eulero*, 1772 (OO Ser. II, Vol. 22. E418).

[49] Empress Catherine II and her son, Grand Duke Paul, were inoculated against smallpox by the Scottish physician Thomas Dimsdale in 1768.

[50] In a letter to the Secretary of the Prussian Academy of Sciences Johann Heinrich Samuel Formey (1711–1797), Johann Albrecht Euler writes [37, p. 227]: "Suppose that I submit a demand on Monday, I first have to get it translated into Russian, as every matter is dealt with in Russian. At the next conference session, that is, on Wednesday, the Secretary reads my proposition. We then discuss whether we can accept the proposition, and we do so. The Secretary notes the decision in the minutes of the meeting. The minutes are read for us on Friday, and we confirm them by signing. If the Director or any of the officials of the Commission are absent, the record is sent to him for signing. Next Monday, from an extract of the minutes an order is written to the bookshop authorising me [J.A.E.] to take some books without payment and send them to a given address etc."

their own career and used the Academy as a platform to gain more influence and power. Thus, the academicians had to battle vigorously, and often in vain, for their autonomy [116, 170]. The academicians were divided as to how to cope with the authoritarian leadership, and since they themselves were of different nationalities, there naturally arose tensions between the German and Russian factions in the Academy. In this respect, the long reign of Catherine the Great in 1762–1796 was no exception: while some conditions improved, others deteriorated. Without doubt Catherine's greatest single achievement to the benefit of the sciences was to induce Euler to return to Saint Petersburg in 1766. Euler took charge of the work of reforming the institution in a specially appointed academic commission, and although this work failed, principally because of the conflicting interests and quarrels between the academicians and the directors [37], it was mainly owing to Euler's influence (and, in fact, his leadership) that the Academy experienced its most fruitful period ever.

3.5 Recognition and Approval

Having settled down in Saint Petersburg, Lexell pursued his mathematical investigations and submitted to the conference of the Academy on 12 January (o.s.) 1769 [141, Vol. II, p. 662] a treatise on algebraic number theory entitled *De investigatione numerorum continue proportionalium, quorum datur summa* a *et summa quadratorum* b [Lexell 7] (For details of this paper, see Sect. 8.1). On 13 February (o.s.) 1769[51] [141, p. 668], the paper was read out at the academic conference and Professor Georg Moritz Lowitz promised to study it carefully at home and deliver his opinion of it in the following meeting. Indeed, on 20 February 1769, the question of printing Lexell's treatise in the *Novi Commentarii* of the Academy was already deliberated, but on the following meeting on 22 February the plan fell short on the grounds that Lexell was not yet a proper member of the Academy. Nevertheless, the treatise ended up to be printed in the *Novi Commentarii*, without doubt on the recommendation of Leonhard Euler.

In the academic conference held on 20 March 1769, a letter from Lexell was read requesting permission to take up training for the practical part of astronomy at the Observatory. While the transit of Venus on 3–4 June was coming closer, Lexell, who was supposed to assist in the observations in Saint Petersburg, still had no experience of using astronomical instruments. Having granted him the permission, "...His Excellency the Director proposed to the Conference that Mr Lexell, in regard of his testified diligence and his skill in the mathematical sciences would be appointed Adjunct of the Academy for his further encouragement. When all the members of the Academy had given their consent to the proposal, the Secretary of the Conference was commissioned to announce the news to Lexell and that he will

[51] In quoting the minutes of the conferences of the Petersburg Academy, old style dating is used, unless otherwise indicated.

be introduced in the next meeting of the Academy". In the next conference session held on 23 March (3 April n.s.) 1769, in the magnificent conference hall of the Academy, and in the presence of Leonhard Euler, "Mr Lexell was presented and introduced as an Adjunct of the Academy, and His Excellency the Director assigned him a seat".

Chapter 4
Formation of an Academician

The dream had come true surprisingly quickly. Although Lexell had arrived in Saint Petersburg with modest expectations, he could not possibly have asked for more. He was now an adjunct member of the Academy of Sciences in astronomy, a science which he had never practiced before, but in which all the mathematical faculties he possessed could be brought to bear. His unfailing capacity and willingness to learn had convinced the academicians, and so he was at once engaged in several of the most challenging scientific pursuits initiated by his Master, Leonhard Euler. The ultimate honour of becoming an ordinary member of the illustrious Academy probably had not entered his mind yet. Lexell's new home was now—at least for the time being, as he thought—Saint Petersburg, or Petropolis, in the language of the learned, a growing multi-ethnic imperial capital on the swampy banks of the river Neva,[1] founded by Peter the Great in 1703. Peter's vision for Saint Petersburg was obviously based on Amsterdam with its network of streets and canals. However, like Rome, Saint Petersburg was not built in a day. The original town, built of wood, had been destroyed in repeated fires during the eighteenth century and only gradually replaced by the houses and palaces of stone we know today.[2] Permanent stone bridges across the Neva were built only in the nineteenth century.

At the time of Lexell's arrival, the city had nearly 100,000 inhabitants, a mixture of native Russians and other nationalities from the vast Russian empire serving in the army, the government and the court. In addition, ethnic minorities of Finns—the area being previously populated by Finnish speaking Ingrians—and Swedes were accompanied by mostly German speaking merchants and craftsmen, many of them Baltic Germans. Small communities of French, Dutch, Swiss and Italian immigrants consisting of artists, architects and civil servants, also co-existed in this heterogeneous, artificial assembly, sometimes called the Venice of the North.

---

[1] In Finnish, 'neva' means 'bog'.

[2] For want of suitable bedrock in Russia, the stone material had to be imported from Carelia and Finland. Euler's tombstone is also made of Finnish Pyterlahti granite.

J.C.-E. Stén, *A Comet of the Enlightenment*, Vita Mathematica 17,
DOI 10.1007/978-3-319-00618-5_4,

4.1 Life in Saint Petersburg

Although the Imperial Academy of Sciences and its illustrious scientists were in the centre of Lexell's thoughts, his abode both culturally and spiritually was without doubt in the Swedish community assembled around the Swedish Lutheran Saint Catherine's congregation in Saint Petersburg[3] [74]. The parish church as well as the Swedish legation were then, as they are even today,[4] situated at the northern end of the *Malaya konyushennaya* (Малая конюшенная)[5] on the mainland (east from the tip of the Vasilyevsky Island; see the map in Fig. 4.1), a quiet side-street to the well-known *Nevski prospekt* (Невский проспект) in the vicinity to the Imperial Palace (the "Winter Palace") and the Admiralty.

Due to long-standing internal disputes between the Finnish and Swedish speaking members, the congregation was split in 1745. A new church building for the Swedish congregation (to which Lexell belonged) was eventually consecrated on the neighbouring "Swedish lane" (in Swedish: *Svenska tvärgatan*, in Russian: Шведский переулок) on Ascension day 1769—an occasion in which Lexell may have been present. Five years later, in 1774, the parish commissioned Lexell to procure to the church its first pipe organ, which a Swedish master Olof Schwan was invited to build in 1775–1776. Lexell's commission was without doubt suggested by his intensive correspondence with Wargentin at the Royal Academy of Sciences in Stockholm, which at the time supervised the building of church organs in Sweden [100].

As to where Lexell's living quarters were situated, nothing can be said for sure. However, he is likely to have lived in an apartment in the *Gymnasium* belonging to the Academy, as he mentions in the letter quoted in the previous chapter to have the Inspector of the *Gymnasium* Ludvig Bacmeister for a neighbour.[6] The *Gymnasium* was situated in the second large building of the Academy next to the *Kunstkammer* (Fig. 4.2).

The Petersburg Academy of Sciences was working the usual six days a week, except Sundays and holidays. Working hours were not strictly supervised, as they had been during the previous decades [170]. Nevertheless, every academician was expected to always do his utmost to achieve his goals to the honour and glory of the Academy and its High Patron, the Sovereign of Russia. For natural historians, geographers and explorers this could mean long journeys on foot and years of

[3]Although Peter placed the Orthodox Church under state control, he was generally tolerant in matters of religion [115]. He wanted the Western influences to show and permitted the establishment of parishes and churches for the foreign congregations in Saint Petersburg. Among the oldest were the Swedish and German Lutheran congregations.

[4]The buildings are not the same as in the mid-eighteenth century.

[5]In Swedish: *Lilla stallhovsgatan*, or the "Small imperial stable street".

[6]This information is confirmed in the Journal of Johann III Bernoulli (1744–1804), who visited St. Petersburg and its Academy in 1777. Bernoulli says [12, Vol. 4] that Lexell and Rumovski were living in the *Gymnasium*.

Fig. 4.1 Detail of a map of Saint Petersburg in 1765 by the mapmaking company Reinier & Josua Ottens, Amsterdam: *Nova ac verissima urbis St. Petersburg Delineatio*. The river Neva flows into the Gulf of Finland on both sides of the Vasilyevsky Island (Helsinki University Library. Photo: the author)

field work in difficult, sometimes extremely dangerous conditions. For astronomers, night-time was also working time. Naturally, it was the duty of the professors of the Academy to instruct their adjuncts and students at the Academic University. Only those professors who were ordinary members of the Academy were entitled to give lectures. However, the Academy had very few students at the time[7]; the most popular lectures were actually attended to by members of the court and the nobility (Fig. 4.3).

As a new Adjunct of the Academy, Lexell would receive numerous tasks and commissions. He took up studies in astronomy, both the theoretical part and the practice under the instruction of Jesuit Father Christian Mayer, who stayed in Saint Petersburg from May 1769 to June 1770 [120]. Born in Moravia, in today's Czech Republic, Mayer had studied theology and philosophy in several Jesuit colleges

[7]Princess Catherine Dashkova, Director of the Academy of Sciences in 1783–1796, mentions in her *Memoirs* [31] that at the time she took over the leadership, the Academy had only 17 students and 21 adjuncts, but that owing to her efforts their numbers rose to 50 and 40, respectively. In a report to Catherine II in 1786, she mentions that in 1783 there were 27 students for the *Gymnasium*, but 3 years later 89 [104].

Fig. 4.2 The *Kunstkammer* by the river Neva, the home of the Imperial Academy of Sciences. On top of the building is the Observatory. In the background is seen the Peter and Paul Fortress and Cathedral on the Hare Island. In the eighteenth century, small boats were used to traverse the Neva; during the winter, the frozen river could usually be crossed on foot. Lithography in *Collection des vues de S:t Pétersbourg & des environs*. Vol 1. S:t Pétersbourg: la Société d'encouragement, 1822 (Kungliga biblioteket, Stockholm. Published with permission)

and at the University of Würzburg. In 1745, he entered the Society of Jesus as a novice and was ordained as a priest in 1750. At the same time, he lectured at the Jesuit College in Aschaffenburg until 1751, when he was appointed to the Chair of Philosophy, later on to that of Experimental Physics at the University of Heidelberg. On a mission in Paris in 1757 he acquainted himself with French astronomers and ordered the construction of his first astronomical instruments. In 1759, Mayer observed the comet named after Halley in Heidelberg and in 1761 the Venus transit in the yard of the Schwetzingen castle. In the 1760s he was already widely known as a diligent cartographer and experienced astronomer of the court of Prince Elector Karl Theodor. Thus, when the Petersburg Academy of Sciences was looking for reinforcements for the observations of the forthcoming Venus transit, he was an ideal candidate. In the following chapters we will return in more detail to the cooperation between Mayer and Lexell relative to the project of observing the transit of Venus in 1769.

In addition to the astronomical work, Lexell assisted Leonhard Euler by reading and writing letters and papers for him. Moreover, in the conferences he presented to the Academy the mathematical work of Euler and Daniel Bernoulli.[8] In the minutes

[8]As a foreign member of the Petersburg Academy of Sciences, Daniel Bernoulli was entitled to have his papers printed in its organs.

Fig. 4.3 Lexell's place of work, the Kunstkammer, by the Neva in February 2008 (Photo: the author)

of the conferences we find the following entries (All dates quoted here are in the original old style calendar used in the records of the Academy) [141, Vol. II]:

—18 August 1769 [141, p. 698] The paper recently submitted to the Academy by Mr Daniel Bernoulli: *Commentationes physico-mechanicae de frictionibus, variis illustratae exemplis*, was read to the assembly by Mr Lexell.
—4 September 1769 [141, p. 699] Lexell absent due to illness.
—19 October 1769 [141, p. 710] A letter has been received from the illustrious mathematician Mr Daniel Bernoulli of Basel of the 13 September new style [...] and he submits to the Academy a very important dissertation: *Mensura sortis ad fortuitam successionem rerum naturaliter contingentium applicata*,[9] which was read by Adjunct Lexell.
—19 March 1770 [141, p. 737] Adjunct Lexell read the paper submitted by Mr Leonhard Euler: *Solutio problematis, quo duo quaeruntur numeri, quorum productum tam summa quam differentia eorum, sive auctum sive minutum, fiat quadratum.*
—11 October 1770 [141, p. 768] Adjunct Lexell read: *Continuatio argumenti de mensura sortis ad fortuitam succesionem rerum naturaliter contingentium applicata* by Daniel Bernoulli.
—15 October 1770 [141, p. 768] Adjunct Lexell read: *Examen physico-mechanicum de motu mixto, qui laminis elasticis a percussione simul imprimitur* by Daniel Bernoulli.[10]

[9]This paper, whose title translates as "Chance estimation applied to the accidental succession of naturally occurring events", was concerned with calculations of probability assuming a binomial distribution of natural events, say, that there be born a certain number of boys, given the total number of births annually.

[10]The papers by Daniel Bernoulli mentioned in this quote refer to St.60, St.59, St.59b and St.61, respectively, in Straub's catalogue [55]. The paper by Euler which translates as "Solution of the problem concerning finding two numbers, the product of which increased or decreased by

One can hardly imagine a better apprenticeship and a greater honour for a young man like Lexell than to present the works of the greatest mathematicians of his time.

A glimpse of Lexell's daily activities during his first years in Saint Petersburg is given in a letter written to his astronomer colleague in Åbo, Professor Anders Planman (HUB Ms. Coll. 171) dated 25 June 1770. There, Lexell interrupts his writing by saying that "it is already 11 o'clock and I have to go to the conference. In the afternoon, I am going to Euler to calculate". On the next page he nevertheless continues by writing that "my plan was changed and I am now free for the afternoon", whereupon he starts an account of his analysis of the eclipse of the Sun observed in 1769. Towards the end of the letter, Lexell passes over to discuss the events in the Academy, telling a somewhat malicious anecdote about Father Christian Mayer, who had recently left Saint Petersburg to visit Wargentin in Stockholm:

> Almost nobody was happy with Mayer at his departure. Those who had supported him previously in every way were now no more favourable to him than those whom he considered to be his worst enemies. In sum, he let it show too much that he was a Jesuit. We two separated as very good friends; whether his friendship was only for his self-interest, I cannot say. It is true that in spite of all his faults and weaknesses, I cannot hate him, maybe because I have my own share of faults. However, I am neither keen nor self-interested. Before his departure he had made a remarkable and incomparable discovery. You might guess, a *perpetuum mobile* or some such thing. No, it was to draw a map of the whole of Russia by means of Harrison's clock.[11] A simple enough proposal, but also very chimerical, as nobody can procure such a clock. However, he himself thought it was a real discovery comparable to Newton's discovery of the law of attraction, and there are some people who admire this proposition, and some of them even think they have not praised it enough. *Sed mundus vult decipi.* [But the world wants to be deceived.] A golden *tabatière* [snuffbox] was his prize. Here *ratio inversa* has taken place manifoldly. His great book was awarded with 100 Ducats and the last short piece[12] with a golden tabatière.

Usually Lexell told nearly the same news to Wargentin and Planman, but apparently he found this anecdote too imprudent and spiteful to be presented to the respectable Wargentin. In the same letter to Planman, Lexell continues about his difficulties to get along with his elder colleague, astronomer Stepan Rumovski (Fig. 4.4):

> Since he [Mayer] left the Observatory to Rumovski I have not been there a single time, and so it will probably be. Rumovski loves to be alone when he observes, and I do not wish to make a fuss in order to get to work. But I should not blame myself, since the Count [Director Orlov] knows that Rumovski does not tolerate anybody else at the Observatory. I will just have to wait until the new observatory is built, which is a main point in our new Charter. Whether it will be accepted is uncertain.[13]

their sum or difference is a square" is item E405 in Eneström's catalogue (OO Ser. I, Vol. 3, pp. 148–171).

[11] John Harrison (1693–1776), English clockmaker and inventor of the marine chronometer.

[12] Mayer's proposal was published as a book entitled *Nouvelle méthode pour lever en peu de tems et à peu de frais une carte générale exacte de toute la Russie* in 1770.

[13] Only in 1839 was a new observatory built in Pulkovo, 19 km from the centre of the town.

Fig. 4.4 Lexell's astronomer colleague Stepan Rumovski [108]

On the other hand, the letter to Wargentin (at KVAC, Stockholm) dated 11 June 1770, is more colourful in this matter:

> Professor Rumovski is as we say "like a dog in a manger", he does not make observations himself, thus he will not accord to anybody else the same favour. The Count has asked if I could not observe together with Rumovski, but Rumovski has said no, because when one of us is away the other could bring the instruments out of order, such as the quadrant, which is in the meridian plane. But the fact is that he is lazy and presumptuous. It suits him better to permit a helmsman from the Admiralty to observe from time to time some corresponding elevations, when he himself for his convenience may observe some more particular phenomena. On the other hand it would vex him if somebody else would observe with him and be more studious than he, although as for me, I would never aspire to steal from him an honour on his account. I thus have to wait until a new observatory is built, and before that I hope to be so lucky as to get a small employment in Sweden. Otherwise I am not discontented with my situation; I can live here with my 400 Roubles in salary but not save anything. I nevertheless think it is so much money, that I do not know if I am worth it. If I was to be promoted in the future, the expenses will increase much like the salary. In general, it is hardly noticed at all that one is regarded as living on charity, and that one should take it graciously, whatever one receives.
> When the new Charter is finally realised I imagine the body of professors like the gods of the Romans, *Dii maiorum et minorum gentium*, when I compare myself and other of my kind with such a respectable man as old Professor Euler. And so it will probably be in the other sciences, too.

In the meantime, extracts of Lexell's letters of the academic events in Saint Petersburg began to appear in Porthan's newspaper *Tidningar* in Åbo. In the issue of 15 April 1771, Lexell writes about the scientific expeditions of the natural historians of the Petersburg Academy of Sciences, Pallas, Gmelin and Güldenstaedt to various parts of Caucasus, Persia and Siberia, respectively. In the issue of 15 November he reports among other things that

> ...[T]he famous Professor Euler the Elder (who has been blind for some years, but who continues to work industriously in his discipline) had his cataract operated on, with the happy outcome that he can now see well. The operation was performed by the skilled oculist B[aron] Wentzel.[14]

As we now know, the happiness ended but too soon, when following an infection in the operated eye Euler once more lost his sight (for details of Euler's sight, see [43] and references therein). He was nevertheless able to distinguish light from shadow and continued to write on a blackboard in his study.

On 8 July 1771, Lexell reports to Wargentin of the fire in Saint Petersburg, in which Euler's house was damaged, and in which Euler all but perished (KVAC, Stockholm):

> The great fire, which raged here in Petersburg in the end of May, has probably already become known from the news, as well as the hard fate of the decent old man Euler and his family. Later it has been said that Her Majesty the Empress has made him an offer of 2500 Roubles to repair the house, nonetheless he must account for damages up to 3000 Roubles.[15]

In the same letter, he remembers to recommend his friend Porthan for a vacant position at the Academy of Åbo:

> From Åbo I have recently been informed that a proposal for the position of Adjunct of Philosophy has been put forward, and that my especially good friend *Amanuensis Bibliothecae, Magister Docens* Henric Porthan is suggested for the second place; therefore, if you, Herr Secretary, could show me and him the grace to recommend him to His Excellency the Councillor Count Ekeblad,[16] it would mark a favour which I would consider as if it was to myself.

In our times this kind of lobbying may seem peculiar and even unfair, but in the era of patronage and enlightened absolutism this was quite a normal procedure, as the whole administration was essentially built on personal networks of favours and returns.

Under Euler's leadership, Lexell participated in several challenging astronomical pursuits (for an overview, see Chap. 5): he was simultaneously engaged in the determination of the orbit of the comet of 1769 [Lexell 6] and, together with J. A. Euler and W. L. Krafft, in the laborious calculations for Euler's lunar theory [Lexell 19]. He also continued his own mathematical investigations and submitted

---

[14]Michael Johann Baptist de Wenzel (1724–1790).

[15]Euler's house was situated ca. 1.3 km from the Academy on the same embankment of the Neva as the *Kunstkammer* (at the present address: 15 Lieutenant Schmidt Embankment, Набережная Лейтенента Шмидта).

[16]Count Claes Ekeblad (1708–1771), Privy Councillor of Sweden and Chancellor of the Royal Academy of Åbo. His wife, Countess Eva Ekeblad, née De La Gardie, was the first woman to be elected a member of the Swedish Academy of Sciences.

his dissertations for publication in the Academy's *Novi Commentarii*. In addition, the transit of Venus on 3–4 June 1769, the eclipse of the Sun on 4 June, and the collection and compilation of data from the numerous expeditions was an enormous enterprise for the Academy, and Lexell under the supervision of Euler was occupied at the very heart of the matter.[17] As to Lexell's role in this work, Christian Mayer wrote to Wargentin that in the application of Euler's special method to all the observations made around Russia, "[i]n all these calculations L. Euler has made use of the help of that excellent man, Mr Anders Lexell. From him you will easily learn all that has been done . . . ".[18] In the next chapter we will discuss more specifically the efforts and challenges of eighteenth century astronomy, especially from the point of view of Lexell's contributions.

4.2 Euler's Praise

Lexell's participation in the scientific team of Leonhard Euler was clearly well received and much appreciated. Still, it must have come as a surprise, not least for Lexell himself, when Euler at the conference session on 13 December (o.s.) 1770, that is, on Christmas Eve new style, held a panegyric speech with an exuberant commendation for him ([141, Vol. II, pp. 792–793])[19] which cannot but have made a deep impact on the participants of the meeting:

> Mr Leonhard Euler reported that having now calculated with the help of Adjunct Lexell all the observations of the latest transit of Venus and by this means found the correct value of the parallax of the Sun, and thereafter worked out a complete theory of the comets and thereby the orbit of the wonderful comet of 1769, he has also worked out a theory of the irregularities of the Moon with the happiest outcome, so that it now by way of his efforts has acquired the greatest degree of perfection; that they now will be compared with real observations and afterwards, under his supervision, precise Moon tables will be computed, so that by means of them the position of the Moon will be known at any time to a precision of some few seconds, but also they may be calculated with far less effort than that required by the best currently available Moon tables. Now, since Adjunct Lexell does not have enough time to pursue this important work alone, especially since he has committed himself to His

---

[17]On 21 July 1769, according to the minutes [141, Vol. II, pp. 695–696]: "In Ansehung der von den Herren Observatoren eingeschickten Journäle wurde festgesetzt, dass erstlich aus denselben, wie bisher geschehen, nur dasjenige, was den Durchgang der Venus vorbey der Sonnenscheibe betrifft, ausgezogen, durch den Druck bekannt gemacht und an die fremden Academien, sowie auch an die fürnehmsten der auswärtigen Gelehrten geschickt werden soll. Hernach wird der Hr. Adjunct Lexell unter der Anführung des Hrn. L. Eulers alle eingelaufenen Beobachtungen dieses Durchgangs der Venus berechnen, und daraus die Parallaxe der Sonne als den Hauptzweck aller diesen Verschickungen zu bestimmen suchen, und darüber eine Abhandlung ausarbeiten, welche hernach sowohl besonders als in den Commentarien gedruckt werden könne".

[18]Letter in Latin, dated 22 October 1769 (KVAC, Stockholm): "... In his omnibus usus est opera Cl. Viri D. Andrea Lexell, ex quo, si velis, ea, quae facta sunt, facile cognosces".

[19]In the index of the *Procès verbaux* of the Academy [141], this is mentioned as "Valde laudatus" = "Great praise" by L. Euler.

Excellency the Director to perform astronomical observations, Mr Leonhard Euler has also asked Adjunct Krafft to participate in this particularly glorious work and thereby assist in bringing it to a speedy conclusion, in particular since this work is much more urgent than anything else in astronomy can be. Adjunct Krafft took this call with as much joy as desire, and promised to exert all his strength in the quick completion of this useful and invaluable work. Mr Leonhard Euler added on the occasion, that the world owes it solely to the untiring zeal of Adjunct Lexell, that the enormous expenses that were invested for the latest transit of Venus did not end up utterly wasted, in the same manner as those spent on the previous transit of 1761; that without him, perhaps no one would have been able to determine from the observations made of the last transit of Venus the true parallax of the Sun, since the methods of this calculation known hitherto are entirely inadequate, as was also learned from the previous experience, and since to this day not a single scholar has proven himself to be so brilliant as to deduce only one, certain conclusion from all the observations. In general, works such as that which Adjunct Lexell has done till now, and still does jointly with Adjunct Krafft, would bring astronomy much more profit than a series of numerous astronomical observations would ever make.[20]

[20]Der Herr Leonhard Euler meldete, dass, nachdem er mit Beyhülfe des Hrn. Adj. Lexell alle Beobachtungen des letzten Durchgangs der Venus der Sonnen vorbey berechnet und daraus die wahre Parallaxe der Sonnen auf das Genauste gefunden, hernach aber eine vollständige Theorie der Cometen ausgearbeitet und nach derselben die Bahn des im Jahr 1769 erschienenen merkwürdigen Cometen bestimmet, er anjetzo auch eine ganz neue Theorie der Ungleichheiten des Mondes mit dem glücklichsten Erfolge zu Stande gebracht, so dass dieselbe nunmehro durch seine Bemühungen den grössten Grad der Vollkommenheit erreichet habe; dass er dieselben anjetzo mit wirklichen Observationen vergleichen und hernach solche genaue Monds-Tafeln unter seiner Aufsicht wolle berechnen lassen, vermittelst welchen man den Ort des Mondes zu einer jeglichen Zeit nicht nur bis auf etliche wenige Secunden, sondern auch noch darzu mit weit weniger Mühe wird ausrechnen können, als man es bishero auch durch die besten Tafeln in Stande gewesen ist. Da nun der Adjunct Lexell dieses wichtige Werk allein auszuführen nicht Zeit genug haben möchte, zumalen er sich gegen Se. Erlaucht den Herrn Chef verbindlich gemacht hat, fürnemlich denen astronomischen Observationen obzuliegen, so bot der Hr. Leonhard Euler den Herrn Adj. Krafft auf, an dieser vorzüglich ruhmbringenden Arbeit gleichfalls Theil zu nehmen und zu helfen, dass dieselbe je eher je besser zu Stande komme, zumalen an der Vollendung dieser Arbeit weit mehr gelegen sey, als an allem, was in der Astronomie sonst Wichtiges gethan werden könnte. Der Herr Adjunct Krafft nahm diese Aufforderung mit eben so grosser Freude als Begierde an, und versprach, alle seine Kräfte zur geschwinden Vollendung dieses nützlichen und unschätzbaren Werks anzustrengen. Der Hr. Leonhard Euler fügte bey dieser Gelegenheit hinzu, die Welt habe es einzig und allein dem unermüdeten Fleiss des Hrn. Adjuncti Lexell zu verdanken, dass die auf die Beobachtung des letzten Durchgangs der Venus angewandten grossen Unkosten nicht gänzlich und eben so unfruchtbar geblieben, als diejenigen gewesen, welche auf den vorhergehenden Durchgang der Venus im Jahr 1761 gewendet worden; dass ohne ihn vielleicht Keiner in Stande gewesen wäre, aus den angestellten Beobachtungen des letzten Durchgang der Venus die wahre Parallaxe der Sonnen zu bestimmen, zumahlen die bishero bekannten Methoden zu dieser Bestimmung gänzlich unzureichend sind, wie denn auch dieses die bisherige Erfahrung gelehret, und sich bis zu dieser Stunde Keiner unter den Gelehrten hervorgethan hat, der nur einen gewissen Schluss aus allen Observationen gezogen hätte. Überhaupt aber brächten dergleichen Arbeiten, wie der Hr. Adj. Lexell bishero unternommen, und noch inskünftig gemeinschaftlich mit dem Hrn. Adj. Krafft unternehmen werde, der Astronomie weit mehr Nutzen, als eine Reihe von unzähligen astronomischen Observationen es jemals zu schaffen in Stande sind.

After such a praise—a real seal of approval on Lexell's work—we can only imagine the emotions he experienced this Christmas Eve, which also happened to be his 30th birthday. Without in any way belittling his own contribution, he may even have asked himself: "Can this be true?"

The reward did not have to wait for long: according to the minutes of the conference held on 8 April (o.s.) 1771 [141, Tom III, p. 12], Lexell and three other Adjuncts were proposed as candidates for ordinary membership of the Academy:

> [...] In another letter sent by His Excellency the Director to the Secretary of the conference, it is suggested to the assembly that
> Adjunct Lexell be appointed Academician of Astronomy,
> Adjunct Krafft be appointed Academician of Experimental Physics,
> Adjunct Lepechin[21] be appointed Academician of Natural History,
> and Adjunct Güldenstaedt[22] be appointed Academician of Natural History, as well.

4.3 Growing Responsibilities

Lexell's nomination did not have immediate consequences for his responsibilities. Even two months after the appointment, he still had to write a special request, which was read out at the conference held on 4 July 1771 [141, Vol. III, p. 23], to establish his working office on the lower floor of the Observatory cabinet. The permission was duly granted and the necessary instruments were promised for his disposal. He was further granted a well-earned leave of absence, which he spent by travelling home to Finland in January 1772.

As a Professor, Lexell also had a duty to teach at the academic university. For students, he was appointed some young Kalmyks from the Court, whose education Catherine II supplied for personally.[23] Some extracts of the *Procès verbaux* [141, Vol. III] of the conferences illustrate the nature of Lexell's varying assignments and activities at the Academy in 1772–1773:

> —10 January 1771 [141, Vol. III, p. 2] Adjunct Lexell presented: *De criteriis integrabilitatis formularum differentialium*, Author A. J. Lexell, a paper which by its outstanding quality will be inserted in the next volume of the *Novi Commentarii*.
> —14 May 1772 [141, p. 56] Academician Lexell delivered to the Physico-Mathematical Section of the XVIth volume of the *Novi Commentarii* a paper written by him under

[21]Ivan Ivanovich Lepyokhin (Лепёхин; 1740–1802), Russian natural historian and explorer, obtained his doctorate in medicine in Strasbourg [55]. In Latin texts, his name is often transliterated Lepechin.

[22]Johann Anton Güldenstädt (1745–1781), Baltic-German naturalist and explorer born in Riga.

[23]This is mentioned in a letter from Bishop Mennander to his son Carl Fredrik Mennander junior (Fredenheim after ennoblement in 1772) dated in Åbo on 19 March 1773 [132]. The Kalmyks are a Mongolian people living in southern Russia.

the direction of Mr Euler the Elder, *De perturbatione motus Terrae ab actione Veneris oriunda*.[24]
—25 May 1772 [141, p. 58] Academicians Rumovski and Lexell informed the assembly of the contents of a letter they have received from Chevalier Wargentin of Stockholm: the letter concerned mainly the longitudes of the places Ust-Kamenogorsk, Barnaul, and Astrakhan, which Mr Wargentin has taken the pains to determine by comparing with observations of the same time at these places. [...] Wargentin further asks Mr Lexell to procure for him the first two parts of the *Flora sibirica*,[25] because the Academy had sent him only the last two parts. It was decided that these parts of the *Flora sibirica* would be donated to him.
—25 June 1772 [141, pp. 61–62] A letter from Mr de la Lande in Paris dated 12 June, and another from Mr La Place[26] dated 30 May were read. The latter submits for the Academy's consideration two papers *Disquisitiones de calculo integrali*, and *Disquisitiones de maximis et minimis fluentium indefinitarum* and requests that they be published in the Commentaries. However, it was argued against it that it was not the practice to publish papers in the Commentaries by someone who is neither a member, nor has been requested by the Academy to write. In the meantime, Academician Lexell took these papers with him to read through.
—17 September 1772 [141, pp. 67–68] Academician Lexell delivered *Experimenta thermometrica sive de medicamentorum vera refrigerandi et calefaciendi facultate, Thermometri ope declarata &c.* Author Anth: Rol: Martin.[27] Cand: med:. The author of this writing wishes the Academy to publish it in its Commentaries, and promises a continuation for it. Academician Lexell will write in the name of the Academy a due thanks for the communication of this laborious work, but will then say that the Academy cannot accept the proposal, as it would contradict its charter to print dissertations in its Commentaries, which are not written by its members.

According to the charter, the Academy's publications were reserved solely for works written by its members or associates. The unfortunate duty of the Academy was thus to inform the non-members—in this case through Lexell—of the refusal to publish their manuscripts. Nevertheless, it was precisely this way Pierre Simon de Laplace, still young and unknown to the scientific community, became known to the Saint Petersburg academicians.

—21 September 1772 [141, p. 68] Academician Lexell delivered for the Commentaries the following two papers by Mr Euler the Elder: I. *Novae demonstrationes circa resolutionem*

[24]"Perturbations of the motion of the Earth provoked by the action of Venus", NCASIP XVI, pp. 426–467, 1771 (OO Ser. II, Vol. 26. E425). For a discussion of this article, see Chap. 8.

[25]Referring to J. G. Gmelin's and S. G. Gmelin's: *Flora sibirica sive historia plantarum Sibiriae*, 4 volumes. St. Petersburg, 1747–1769.

[26]Pierre Simon de Laplace (1749–1827), mathematician and physicist, author of the influential work *Traité de mécanique céleste* (1799–1825) and known for his ground-breaking work in theoretical astronomy and physics, a member of the *Académie française* and many scientific societies.

[27]Anton Rolandsson Martin (1729–1785) was a botanist and medical doctor, born near Reval (Tallinn), Estonia. He had studied in Åbo and defended a thesis in natural history for Linnaeus in Uppsala, after which he made scientific expeditions to Spitsbergen. In 1761, as a result of a gangrene, Martin's right leg was amputated. He moved to Finland and continued his medical experiments on a grant from the Royal Swedish Academy of Sciences [129, 138]. The work in question, which he submitted to the Petersburg Academy, is a thermometric study of medicaments.

numerorum in quadrata[28] and II. *De motu vibratorio laminarum elasticarum, ubi plures novae vibrationum species hactenus non pertractatae evolvuntur.*[29] He also delivered his own work *Disquisitio de investiganda parallaxi Solis ex transitu Veneris per Solem A. 1769* [Lexell 22] and *Animadversiones in tractatum Rev. P. Hell de Parallaxi Solis* [Lexell 23]. At the author's request it was further decided that both of these treatises be quickly published separately: the number of copies, the format and letter types were fixed, as in the Commentaries, to 400 copies *in quarto*.

—19 November 1772 [141, p. 74] Academician Lexell delivered to the Commentaries a paper by Mr Euler the Elder *Resolutio aequationis* $Ax^2 + 2Bxy + Cy^2 + 2Dx + 2Ey + F = 0$ *per numeros tam rationales quam integros.*[30]

—23 November 1772 [*loc. cit.*] Academician Lexell delivered I. Supplement for his treatise in press on his remarks against Father Hell on the parallax of the Sun, and II. *Solutio problematis analytici* [Lexell 21], Author A. J. Lexell, for the Commentaries.[31]

—3 December 1772 [*loc. cit.*] Academician Lexell delivered for the Commentaries L. Euler's paper *De resolutione irrationalium per fractiones continuas, ubi simul nova quaedam et singularis species minimi exponitur*[32] and the same academician presented his own paper delivered in the last conference: *Solutio problematis analytici.*

—7 December 1772 [p. 75] Academician Lexell delivered for the Commentaries the paper by Mr Euler the Elder *De criteriis aequationis* $fxx + gyy = hzz$ *utrum ea resolutionem admittat nec ne.*[33] [...] Doctor Bergius[34] requests that his dissertation, which was sent to the Academy early in 1768 and read out at the conference on 17 March, entitled *Triplaris Americana descriptione atque icone illustrata*, which cannot be printed in the Commentaries, would be returned. The dissertation was next given to Academician Lexell, who will take care of the reply.

—29 November 1773 [p. 109] Professor Lexell presented his treatise *Observationes variae circa series quae ex sinibus vel cosinibus arcuum arithmetice progredientium formantur.* As this treatise[35] relates to those of Mrs Daniel Bernoulli & Euler the Elder, which are to enter

---

[28]"New proofs regarding the resolution of numbers into squares". The paper was later withdrawn from NCASIP and instead first published in the *Acta Eruditorum* in 1780 (OO Ser. I, Vol. 3, pp. 218–238. E445).

[29]"On the vibratory motion of elastic laminates, where many new types of vibrations not previously treated are solved", NCASIP XVII, pp. 449–487, 1772 (OO Ser. II, Vol. 11, pp. 112–141. E443).

[30]"Solution of the equation $Ax^2 + 2Bxy + Cy^2 + 2Dx + 2Ey + F = 0$ by rational as well as integer numbers", NCASIP XVIII, pp. 185–197, 1774 (OO Ser. I, Vol. 3, pp. 297–309. E452).

[31]Paper [Lexell 21], entitled in English "Solution of an analytical problem", concerns a co-ordinate transformation introduced by Euler in the paper *De solidis quorum superficiem in planum explicare licet* ("Concerning solids, the surface of which can be unfolded onto a plane"), NCASIP XVI, 1772, pp. 3–34 (OO Ser. I, Vol. 28, pp. 161–186. E419). In particular, Euler discovers a new family of developable surfaces besides general cylinders and cones, namely those generated by the tangents to a space curve [79, p. 564] and [142]. Lexell indicates another description of these surfaces by spherical geometry, conjecturally arriving at a restriction to Euler's formulae.

[32]"On the solution of irrationals by continued fractions, where also a certain new singular and small kind is presented", NCASIP XVIII, pp. 218–244, 1774 (OO Ser. I, Vol. 3, pp. 310–334. E454).

[33]"On the criteria of whether the equation $fxx + gyy = hxx$ can be resolved or not", concerning a diophantine equation of the second degree, was first published in *Opuscula analytica* I, 1783, pp. 211–241 (OO Ser. I, Vol. 4, pp. 1–24. E556).

[34]Peter Jonas Bergius (1730–1790), physician and Professor of Natural History, Stockholm.

[35]Various observations concerning series formed by an arithmetic progression of sines or cosines of an arc [Lexell 28].

> in the mathematical part of the 18th volume of the Commentaries, it was judged appropriate to insert it there directly after Mr Euler's paper entitled *Summatio progressionum* ....[36]

Between the two latter entries, in 1773, the language of the minutes of the academic conferences changed without notice from German to French, as if it were to announce the beginning of a new era.

In the following chapter, the contents of the astronomical subjects mentioned in these entries will be discussed more thoroughly.

[36]"Sum of the progressions $\sin\varphi^\lambda + \sin 2\varphi^\lambda + \sin 3\varphi^\lambda + \ldots + \sin n\varphi^\lambda$, $\cos\varphi^\lambda + \cos 2\varphi^\lambda + \cos 3\varphi^\lambda + \ldots + \cos n\varphi^\lambda$", NCASIP XVIII, pp. 24–36, 1774 (OO Ser. I, Vol. 15, pp. 168–184. E447).

Chapter 5
Professor of Astronomy

Although Lexell during his apprenticeship with Christian Mayer had learned the art of observing by means of astronomical instruments, it was clear that his talent and inclination lay principally in the field of theoretical astronomy. As a mathematical subject, it suited him perfectly, and by adopting Euler's approach and rigour to the problems at hand, he learned to master the science to a great perfection. Furthermore, his research in spherical astronomy suggested to him related geometrical problems, which eventually led him into a deeper understanding of spherical geometry as a special kind of non-Euclidean geometry; a subject we will return to in Chap. 8 dealing with Lexell's contributions to mathematics, where also the contributions to celestial mechanics are discussed.

Following Maupertuis' expedition to measure the length of a meridian arc, astronomical methods were increasingly used in the eighteenth and nineteenth century to determine the general shape of the Earth. Also navigators, explorers, geographers and mapmakers had to rely on astronomical methods to determine the latitude and longitude of a place. In fact, the latitude of a location on the Earth's surface can be determined by appropriately observing the elevation of the North Star, but the determination of longitude or the east-west distance to a reference meridian (for instance that of Paris or Greenwich) posed a considerable scientific problem. Anyone familiar with the geographical maps of the time knows how inaccurately the contours of the continents were generally drawn (the map in Fig. 3.1 is much more accurate than many 17th century maps). Crucial to determining longitude were those phenomena in the firmament which provided an accurate time reference by which all observations made around the globe could be synchronised and compared. Such phenomena include the various conjunctions between the Moon, the Sun, the stars and the planets and their satellites—generally called eclipses or occultations—which could be observed at different places simultaneously. Furthermore, from the small geometrical and temporal differences of the observed phenomena, differences in the locations of the observers could be deduced by geometrical and dynamical reasoning.

J.C.-E. Stén, *A Comet of the Enlightenment*, Vita Mathematica 17,
DOI 10.1007/978-3-319-00618-5_5,

In the eighteenth century, the most popular method of determining longitude was by means of the Moon's angular distance to nearby planets or stars (the so-called lunar distance). However, due to the relatively short distance between the Moon and the Earth,[1] the prediction of the Moon's position posed one of the most intriguing problems for the astronomers of the eighteenth century. Moreover, since the lunar tables were prepared for the Earth's centre point, the coordinates ultimately had to be evaluated by laborious computations. Additionally, the distorting effect of refraction of the atmosphere had to be taken into account when the Moon was observed at a low altitude. Another reliable method for determining longitude was by observing the moments of the eclipses of the satellites of Jupiter. Also eclipses of the Sun (by the Moon) could be used for determining distances on the Earth: exact observations of the progress of the Sun's *umbra* (shadow) across the surface of the Earth provided an accurate terrestrial measuring rod.[2]

During his career as an astronomer, Lexell was to use practically all available astronomical means to resolve the longitude problem. Among other things, taking advantage of the eclipses of the Sun in 1764 and 1769, he managed to determine more precisely the longitude of Saint Petersburg [Lexell 10], Åbo [Lexell 20], and many other towns of interest [Lexell 25]. His results for the meridian difference (in time) between Swedish towns and sites were given as a table [Lexell 25]:

Difference in time	Stockholm	Uppsala
Lund	$19'29''$	$17'46''$
Uranienburg (Ven Island)	$21'11''$	$19'28''$
Åbo	$16'51''$	$18'34''$
Cajaneborg (Kajaani)	$38'43''$	$40'27''$
Pello	$23'59''$	$25'43''$

Hence, the longitude difference, expressed in time, between the observatories of Stockholm and Uppsala was $1'43'' \ldots 1'44''$; these meridians were subsequently compared to the Greenwich, Paris and other meridians in Europe.

Lexell's best known contributions to astronomy are nevertheless related to the determination of the orbit of the comet bearing his name, as well as with the discovery of the planet Uranus. Lesser known, but no less important, was his effort to determine the solar parallax and thereby the distance between the Earth and the

[1]The size of the Moon is not negligible in comparison with its distance to the Earth, and as Kepler's laws are formulated for point-like bodies, the distance of the centre of mass of the Earth-Moon system to the Moon's centre had to be determined.

[2]An alternative to the purely astronomical methods to determine longitude was to measure, by means of an accurate chronometer, the difference between the moment of noon and the simultaneous reading of a timepiece. The method is beautifully expounded in Euler's *Lettres à une princesse d'Allemagne sur divers sujets de physique et de philosophie*, Vol. III, 1772, Lettres CLXIII–CLXIV (OO Ser. III, Vol. 11, pp. 1–312. E417). However, the great distances on the Earth, the slow means of transportation and the unreliable timepieces made it technically very difficult to determine longitude with sufficient accuracy.

Sun by means of observations of the transit of Venus in 1769. As the latter also happened to be Lexell's first scientific assignment at the Petersburg Academy of Sciences, the following section is devoted to these investigations in some detail.

5.1 The 1769 Venus Transit and the Solar Parallax

The position of any object in the sky relative to the most distant ("fixed") stars depends on the distance of the object from the Earth and the position of the observer on the Earth. In order to quantify this so-called daily parallax effect, the distance to the object must be known, as well as the diameter of the Earth, which has been known rather accurately since Classical Antiquity.[3] However, in the eighteenth century the main astronomical distances were known only approximately. Not even the mean distance between the centres of the Earth and the Sun, which is a fundamental measure of length in astronomy (today known as the astronomical unit and abbreviated as AU, ca. $149.6 \cdot 10^6$ km),[4] was known with satisfactory precision. The best method for finding the distance to the Sun in the eighteenth century was to observe the innermost planets of the solar system, i.e. Venus and, to a more limited extent, Mercury, during their so-called transits in front of the Sun's disc. Before the eighteenth century Venus transits, the main distances in the solar system had been deduced approximately by observing the motion of the planet Mars when in opposition (for Wargentin's early attempts, cf. Sect. 6.1).[5] In our times, with the benefit of radio astronomical methods, the transits of Venus or Mercury no longer have the particular scientific value they had before.

Venus is the second planet of our solar system counting from the Sun. It is also known as the Evening Star or the Morning Star, depending on which side of the Sun it happens to be seen from the Earth. Conjunctions between Venus, the Earth and the Sun occur at a regular interval of 1.6 years, but only in those rare conjunctions where both planets are located simultaneously close to the point of intersection of the two orbital planes does Venus appear as a black dot against the bright disc of the Sun. In fact, transits of Venus are very rare events: after occurring twice in 8 years, there is typically a pause of alternately 121.5 and 105.5 years. The phenomenon has been observed only seven times since the seventeenth century, namely in 1639, 1761, 1769, 1874, 1882, 2004 and 2012 [136, 176]. The peculiar (double) periodicity of

[3] An ingenious method of deducing the diameter of the Earth was invented by Eratosthenes of Alexandria (ca. 276–195 BC).

[4] The astronomical unit is essential for determining the distance to the stars using the so-called annual parallax effect.

[5] Observations of the transit of Mercury in 1631 gave the first quantitative evidence of the size of the planets and the dimensions of the planetary disc.

the Venus transits is due to the inclination angle (of 3.39°) between the orbital plane of Venus and the ecliptic.[6]

Just as the transit of Venus in 1761 had done, the 1769 transit caused a considerable mobilisation of scientists globally,[7] but given that the results from the transit of 1761 had not been satisfactory, the preparations were even more careful and systematic in 1769. At the forefront of these campaigns were the scientific societies of the great sea powers, that is to say the Royal Society of London and the Académie Royale des Sciences of Paris, but also the Stockholm and Petersburg Academies took an active part in the project. Around 150 scientists were observing the phenomenon in 1769 to ensure a sufficient amount of data, knowing that adverse weather conditions and other difficulties were bound to interfere with a part of the observations. On the other hand, the large mass of numerical data collected by the observers posed a new mathematical problem, namely that of processing the data appropriately so as to determine the solar parallax with the highest possible precision.

The preliminary calculations of visibility zones for the transit using astronomical tables began many years in advance. With the best possible observation sites being selected, precise tables were computed for appropriate estimation of the observables—e.g. the duration of the transit and the moments (time instants) of internal and external contacts at ingress and egress—valid for the chosen locations on the Earth and for the Earth's centre, to which the observations had to be reduced. This calculation also required an a priori estimate of the solar parallax.

Next followed the practical organisation of the expeditions, the determination of the precise geographical longitudes of the observation stations, the calibrations and finally the measurements of the contact moments. These tasks presented enormous challenges for those participating in the project. Some of the expeditions became legends: Among the most celebrated Venus transit expeditions was Captain James Cook's voyage to Tahiti in the Pacific Ocean. Less fortunate, but equally famous, was the French astronomer Guillaume Le Gentil, who had sailed to India to observe the 1761 transit, but arrived too late at his destination. Instead of returning to France, he stayed in the region to await the transit of 1769, but on that occasion the weather was cloudy [136, 176]. Some of the expeditions have remained less known until very recently, see e.g. [4] on the journey of the Hungarian astronomer Maximilian Hell (1720–1792) to northern Norway and [22] on the journeys of the two young Genevan scientists Jacques-André Mallet and Jean Louis Pictet to the Russian Kola Peninsula. Although the transit of 1769 was best visible on the Pacific Ocean, when it was night time in Europe, the phenomenon could also be observed in the northern

[6]The virtual path of the Sun around the Earth, or the plane in which the Earth orbits the Sun, is called the ecliptic.

[7]Precise astronomical knowledge meant better maps, which in the long run offered a strategic benefit in the command of the overseas territories. For the significance of the eighteenth century Venus transits in general, see e.g. [4, 136, 156, 167, 176]. An analysis of the methods involved has been given by Verdun [167].

parts of Scandinavia and Russia, where the Sun does not set in summer, or sets below the horizon only for a few hours.

To see how a phenomenon like the transit of Venus can be used in order to determine the distances in our solar system, let us first recall that for a body orbiting around the Sun, Kepler's third law implies that

$$\frac{T^2}{r^3} = \text{constant},$$

where T is the period of one revolution of a celestial body and r the semi-major axis of its orbit. The constant is specific for our Sun as the principal attractor; thus, it applies to all the planets (and comets) orbiting around it. By observations we know that the period of Venus $T_{♀}$ is approximately 5/8 of that of the Earth ($T_{♁} = 1$ year)[8]; hence, the distance from Venus to the Sun is about 0.72 times the Earth's distance to the Sun. But since the distances between the Sun and the planets (including the Earth) are unknown, this gives only the *relative* dimensions of our solar system.

In the book *Optica promota* (1663), the Scottish mathematician James Gregory (1638–1675) considered the problem of determination of a planet's parallax by measuring the duration of its conjunction with another celestial body. The idea was proposed anew in 1716[9] by Edmond Halley (1656–1742), the future Astronomer Royal of Great Britain, who had recognised that the forthcoming transits of Venus of 1761 and 1769 could be ideal for accurate determination of the distance between the Sun and the Earth (and thereby the size of the solar system). Halley's idea was thus to observe the transit of Venus in front of the Sun concurrently at different places on the Earth and to determine the duration of the transit at these places. Because of the parallax, Venus appears to be at a somewhat different place on the Sun's disc and to move at a slightly different speed simultaneously when seen from different places on the Earth. The further north the place of observation, the further south Venus appears on the disc of the Sun, and the closer to the equator the observer is located, the faster Venus appears to move due to the greater local velocity of the observer.

The geometry can be illustrated schematically as in Fig. 5.1: at the top is seen a cross-section of a transit, with Venus in front of the disc of the Sun as seen from two locations A and B. For clarity, the dimensions have been exaggerated. The ingress (entry) and egress (exit) of Venus and the four critical moments are represented in the middle; at the bottom are some of the angles pertaining to the simplified calculations given below. Let A and B represent schematically the two observation stations and let D_{AB} be a chord joining perpendicularly two different virtual paths

---

[8]The fact that $T_{♀}$ is very close to $\frac{5}{8}T_{♁}$ is a remarkable numerical coincidence, which has puzzled many an astronomer.

[9]Halley's influential paper "Methodus singularis qua Solis Parallaxis sive distantia a Terra, ope Veneris intra Solem conspiciendae, tuto determinari poterit" appeared in the *Philosophical Transactions of the Royal Society*, Vol. XXIX, pp. 454–464. Considering that the following Venus transit would occur in 1761, Halley was indeed looking very much ahead of his time.

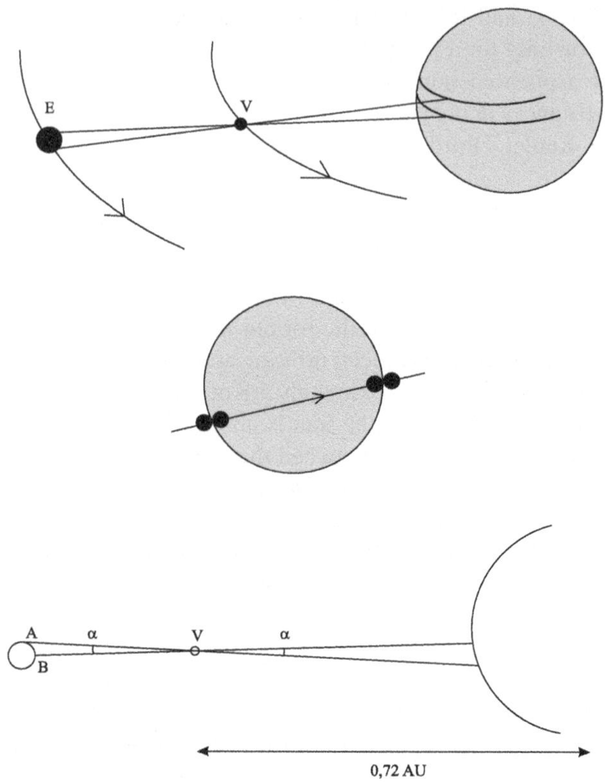

Fig. 5.1 Three illustrations relating to the transit of Venus (NB: Not to scale. The dimensions have been exaggerated for clarity): The different paths followed by Venus on the Sun's disc as viewed from different places on the Earth (*top*), the four crucial contact moments (*middle*), and the geometry for determination of the solar parallax by Halley's method (*bottom*)

of Venus seen against the surface of the Sun. If β is the angle subtended by D_{AB} as viewed from the Earth's centre, then

$$\tan\frac{\beta}{2} = \frac{D_{\mathrm{AB}}}{2\ \mathrm{AU}},$$

where the unit AU is unknown. Applying Kepler's third law, the angle α subtended by the same line as seen from the distance of Venus is related to β via

$$\tan\alpha = \frac{\tan\beta}{0.72}.$$

Clearly, α is also the angle subtended by the chord between the points A and B of the known length d_{AB} perpendicularly to the ecliptic as seen from Venus. Thus,

$$\tan\frac{\alpha}{2} = \frac{\frac{1}{2}d_{\mathrm{AB}}}{(1-0.72)\ \mathrm{AU}},$$

whence AU can be determined. Hence, the solar parallax is

$$\pi = \arctan\frac{d_{♁}}{2\,\mathrm{AU}},$$

where $d_{♁}$ is the diameter of the Earth. The mean value for the angle π is about $8.79''$ (seconds of arc).

The methods actually used in 1769 to determine the solar parallax were more elaborate than Halley had envisaged. The French astronomer Joseph-Nicolas Delisle (1688–1768) had devised a method which was based on the moments of ingress and egress, without the need to determine the duration of the transit. Many variations of the method existed, but typically they involved the following principal steps [167, 168]: (1) the measured contact moments were corrected from errors due to clock drift, giving the primary observables; (2) the epoch or the exact duration of transit were calculated as the secondary observables; (3) these observables were next reduced to the Greenwich or Paris meridian or to the Earth's centre and subsequently compared to theoretical values obtained for the same location; (4) by averaging the differences between each pair of reduced observables (the observed and theoretically estimated ones) a set of averaged observed differences Δ_{obs} and theoretical differences Δ_{theory} were obtained; (5) for each observation pair, the observed parallax was finally obtained as the product of $\Delta_{\mathrm{obs}}/\Delta_{\mathrm{theory}}$ and the theoretical a priori parallax estimate π_{theory}. All the parallax values obtained in such a way were subsequently averaged and scaled according to the mean distance between the Sun and the Earth. However, for this procedure to be correct and valid, the modelling equation must be linear, which was only assumed at the time.

Many variations of the above method of averaging and comparison were used in 1769, and as can be expected, they did not produce identical results, which caused much confusion and suspicions of deliberate manipulation. The solar parallax being an absolute constant, every pair of observations of Venus on the Sun's disc should of course produce the same value for π. The fact that the results nevertheless differed from each other was principally due to measurement errors, including imprecise instruments, individual reading errors, the effect of atmospheric refraction and so on. This posed a new problem of modelling: how to describe, by means of physical laws, the true observables, taking the different sources of error into account. The problem was rarely understood at the time; among the few who did were Euler (and Lexell as his associate), as well as the French astronomer and mathematician Achille-Pierre Dionis du Séjour (1734–1794) [167].

Euler's method, which Lexell also adopted (in its essentials) in his subsequent studies, was presented in a large volume entitled in English "Presentation of the methods used, both for the determination of the solar parallax on the basis of the observation of a transit of Venus in front of the Sun, and for finding longitudes of places on the surface of the Earth on the basis of observations of solar eclipses, along with the calculations and the conclusions drawn upon these".[10] In this method the solar parallax was determined by fitting as many reliably measured observables into the observation equations concurrently—instead of comparison and averaging by means of pairs, which had been used previously, and was still used by most of the astronomers involved in the observations of the 1769 Venus transit. As indicated in the title, the method was also applicable to the determination of longitude by means of solar eclipses.

In Euler's method the observables were modelled by mathematical relationships: all physical laws involved in the process and the quantities in these equations, the so-called model parameters, which are known only approximately because of the errors inherent in the observations, were described. The goal of the process was to adjust the parameters so as to minimise the sum of all estimation errors. The so-called observation equations were derived in three steps: 1° the geocentric angular distances between the centre of the Sun's and Venus's discs at conjunction are determined from astronomical tables, 2° these elements are reduced to the pole of the equator and from there to the zenith of any place, and 3° the apparent distance between the centres of the Sun's and Venus's discs is expressed in terms of the solar parallax π. The derived observation equation contained several variables and constants to be determined. The ensuing adjustment process involved the following stages: minimising the number of parameters by linear combinations of equations, grouping the equations according to the four contact moments, averaging the equations to arrive at approximations of the unknown parameters, computing more accurate theoretical elements and setting up new equations with correction terms as new unknowns. Finally, the corrections are determined such that the sum of the estimation errors is a minimum. The idea of the process resembles in broad terms that of the method of least squares, but is technically not the same.

Euler's method thus involved at least two novel ingredients, namely statistical data processing and the minimisation of estimation errors. The calculations were obviously very time-consuming in those days—for today's personal computers they would be the work of a few seconds—but considering that the theory of measurements, linear algebra and the method of least squares had not yet been developed, the results were astonishingly accurate, as will be seen below.

As an Adjunct of Astronomy, Lexell was instrumental in the process of editing the second part of the 14th volume of the *Novi Commentarii* containing all the results of the Russian observations and the relevant calculations. Although he was not mentioned as an author in the articles (such as those mentioned below), he was clearly the one who performed the calculations. The final and complete account of the observations of the transit of Venus and the ensuing solar eclipse (on 4

[10] *Expositio methodorum, cum pro deteminanda parallaxi Solis ex observato transitu Veneris per Solem, tum pro invendiendis longitudinibus locorum super Terra, ex observationibus eclipsium Solis, una cum calculis et conclusionibus inde deductis*, NCASIP XIV Part II, pp. 321–554, 1769 (OO Ser. II, Vol. 30, pp. 153–231. E397) [28, pp. 342–574]. For an analysis of the work, see e.g. [167, 168].

July) made in Russia in 1769 was published in 1770 as a separate publication entitled "Collection of all the observations made on imperial command in the Russian Empire on the occasion of the Venus transit in front of the disc of the Sun in 1769" [28]. The articles include Christian Mayer's *Expositio utriusque observationis et Veneris et eclipsis solaris factae Petropoli in specula astronomica,* where the independent observations made at the Saint Petersburg Observatory by Mayer, Johann Albrecht Euler, Lexell and Mayer's assistant Father Gottfried Stahl SJ were given. Both the beginning and the end of the transit could theoretically be observed in Saint Petersburg, where the Sun went down in between the ingress and egress of Venus, but only the final two contacts were actually perceived distinctly.[11]

In a letter dated 18 August 1769, Lexell reports to Wargentin that all the observation journals completed in Russia of the transit of Venus had already reached the Academy, except the one from Yakutsk, and that five of these journals had already been printed as reports, namely those from Saint Petersburg, Kola, Ponoi, Umba and Orenburg. Furthermore, Johann Albrecht Euler had commissioned Lexell to ask Wargentin to compare the data obtained by Christoph Euler and Krafft in Orsk and Orenburg respectively on the occultations (or eclipses) of the satellites of Jupiter with those simultaneously observed in Stockholm or Uppsala, or by means of Wargentin's own semi-empirical model. In this way, the geographical coordinates of Orsk and Orenburg were to be established.

From the minutes of the conferences we can see how Lexell's work progressed [141, Vol. II]:

> —28 May 1770 [141, Vol. II, p. 747] The Secretary of the Conference presented a publication printed in Åbo: *Expositio observationum transitus Veneris per Solem, Cajaneburgi A:o 1769 d. 3 Junii factarum, quam praeside Mag. Andreas Planman publice ventilandam sistit Car. Gebh. Widquist,*[12] which Professor Planman has sent to him from Åbo by way of Adjunct Lexell.
> —5 July 1770 [141, p. 753] Adjunct Lexell delivered: *Observationes circa transitum Veneris per discum Solis d. 24 Mai / 4 Junii a. 1769 facta in oppido Gurief*[13] *a Georgio Mauritio Lowitz,* which he has arranged to be published in the Second Part of the 14th volume of the *Novi Commentarii.*

[11]Even Catherine II, assisted by F. U. T. Aepinus, observed the rare phenomenon at the Oranienbaum palace some 40 km from St. Petersburg [120, p. 205]. These observations apparently did not suffice for scientific purposes, however.

[12]This thesis (presented *pro exercitio*) contained the results of Anders Planman's observations of the transit of Venus in Kajaani (Cajaneborg), Finland. Both in 1761 and 1769, the Royal Swedish Academy of Sciences had sent Planman on an astronomical mission to north-eastern Finland to observe the transit of Venus. In 1761, due to the late arrival of spring, Planman had to stop at the town of Kajaani, which nevertheless was at a sufficiently high latitude for observing the phenomenon. Also in 1769, Planman observed in Kajaani using a 21 foot telescope. In 1761, Planman managed to observe all four contacts, but in 1769, because of bad weather and smoke, only two. He also witnessed an optical phenomenon called the "black drop" (*gutta nigra*). On the way back to Åbo, Planman also determined (for the first time) the latitude and longitude of six locations in Finland [67, 91].

[13]Guryev, present-day Atyrau in the Republic of Kazakhstan.

—13 August 1770 [141, p. 754] Adjunct Lexell presented: *Observationes transitum Veneris per Solem d. 24 Mai / 4 Junii 1769 spectantes, in castello Orsk institutae a Christophoro Euler*, which will be joined with the other reports of the same volume presently in press.
—14 January 1771 [141, Vol. III, p. 2] Adjunct Lexell presented: *Additamentum continens calculum observationum in California et nonnullis aliis Americae Septentrionalis locis institutarum*, Author Lexell.
—28 February 1771 [141, p. 8] Adjunct Lexell delivered: *Expositio observationum astronomicarum A. 1770 in urbe Zarizin*[14] *institutarum ab Adjuncto Inochodsov*,[15] and at the same time *Urbis Zarizin latitudo et longitudo*.

One year after the transit of Venus, Lexell wrote to Planman about his work with the method of determining the solar parallax (letter dated 25 June 1770):

> Having finally completed the calculations of the solar eclipse which I have performed under the supervision of Professor Euler, as well as on my own, I have the honour to communicate to you a little draft of both the methods I have used, the former is Professor Euler's and the latter I may call my own, as long as I do not know if anybody has already discovered it. You may judge yourself what they are worth, I only fear that my draft may lack some necessary precision, which as to the former method is helped by the example, but as to the latter, I do not have the time to work out anything the like. If you would be so kind as to communicate to Secretary Wargentin the values I have found for the corrections as well as for the moments of conjunction, it would be a great favour for me [...]
> [...] Professor Euler's new method of calculating the observations of Venus is exactly similar to that he used for the Moon. With these calculations I have been able to determine the parallax and the corrections for the longitude and latitude of Venus. The parallax is on the average 8.61″, the longitude correction 16″ and that of the latitude 9″. The observations from California and South America should give the right results. I cannot see how la Lande has deduced the parallax of 9″ for Cajaneborg [Kajaani, Finland]. His correction for the diameter of the Sun of 7″ is also contrived and groundless. As to the observations of Father Hell I do not know what to say, he may have made them up according to those from Petersburg, and then he has not been very lucky, as our observations are surely not the most accurate. The observation of the eclipse of the Sun in Wardohus [Vardø] is quite incorrect, as to the beginning, unless the moment for the beginning in his printed treatise is incorrect due to a mere misprint.
> I have also thought of applying my method to compute the observations of Venus, but I will have to postpone that for other more urgent things. There are two volumes of the Commentaries this year. The first contains the classes *Math:* and *Physicas* and the latter only the section *Astronomica*. All observations will be there, with their necessary reductions, and finally Professor Euler's methods and calculations of the [solar] eclipse and [the transit of] Venus. I am commissioned to calculate Lowitz's and [Christopher] Euler's observations, thus the biggest task in that section will be my duty. As to the latter work I wish to wait for the observations from California and Zudsee [South Pacific]. For the rest that I can accomplish is for my friends, since I have not yet planned to publish my calculations of the eclipse or those I plan to undertake on Venus.

[14]Tsaritsyn, or present-day Volgograd, has also been known as Stalingrad.

[15]Petr Inokhodtsev (Петр Борисович Иноходцев) (1742–1806), astronomer and Lowitz's Adjunct [119]. He assisted Lowitz in the observations of the 1769 transit of Venus in Guryev and determined the latitude and longitude of numerous places in Russia. In 1779, he was appointed academician. Besides being an active astronomer and meteorologist, he was a historian of astronomy.

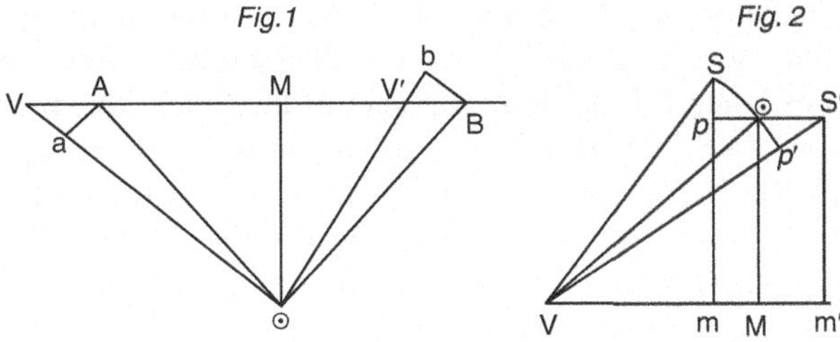

Fig. 5.2 Referring to the geometry of the transit of Venus [Lexell 13]. On the *left*, the effect of parallax on the observation of Venus at ingress and egress. On the *right*, the corrections due to uncertainties in the observation of Venus's position

In his first published results in the (anonymously edited, but de facto Euler's) report of the Russian observations [28, p. 538–539], as well as in the anonymous article in the *Novi Commentarii* (NCASIP XIV, Part II, pp. 518–519, 1769) Lexell had determined using Euler's method the parallax corresponding to the mean solar distance to be 8.80 seconds of arc. The anonymous author concludes:

> Parall[ax]is Solis nobis erit $\pi = 8,67$ quae respondeat distantia Solis a Terra, quae hoc tempore erat 1,0154. Pro distantia media, quae unitate exprimi solet, haec parallaxis aliquanto fiet maior scilicet 8, 80 quae quum referatur ad semiaxem Telluris, distantia media inter centra Solis et Terrae censenda erit aequalis 23436 semiaxibus Terrae, hincque pro Perigeo parallaxis = 8, 95 et pro Apogeo 8, 65.[16]

In 1771, having taken the results of the observations made in Tahiti into account, Lexell published in [Lexell 17] the mean parallax of 8.68 arcsec. Publications [Lexell 13] and [Lexell 14] in the proceedings (*Handlingar*) of the Royal Swedish Academy of Sciences contain somewhat different deductions. In 1772, in his public reply to the criticism of Father Maximilian Hell [Lexell 23], he further adjusted his value for the solar parallax down to 8.63 seconds of arc. Lexell's posterior parallax values differ from the earlier and more correct value presumably because of the rather arbitrary importance he gives to the observations of the inner contacts, which tends to make the parallax value slightly too small [168].

The method used by Lexell in his own contributions contains essentially the same elements as Euler's method described above. In its simplest form, it is given in publication [Lexell 13]. Let us consider Fig. 1 on the left in Fig. 5.2 [Lexell 13]: Let VMV′ be Venus's orbit, ⊙ the centre of the Sun, and ⊙M the smallest distance between the Sun and the orbit of Venus. Further, let A be the position of Venus at ingress (entry) and B its position at egress (exit), for either inner or outer contact, as seen from the centre of the Earth. For the outer contacts, ⊙A and ⊙B equal the

[16]The value 8.67″, obtained at the moment of the transit (when the distance ⊙♁ was 1.0154 times the mean distance) yields the mean parallax of 8.80″. The distance between the Earth and the Sun is stated as 23,436 times the mean radius of the Earth, while with today's values, it is on average 23,480 times the Earth's mean radius, corresponding to a parallax of approximately 8.794″.

sum of the semi-diameters of the Sun and Venus. Correspondingly, for the inner contacts, ⊙A and ⊙B equal the difference of the semi-diameters of the Sun and Venus. Lexell's estimate for the semi-diameter of the Sun at the moment of transit was 947″ and for Venus 29″. Thus, for external contacts ⊙A = ⊙B = 976″ and for internal contacts 918″. The geocentric latitude of Venus (degrees above or below the ecliptic) was 10′13.4″ and hence, the smallest distance ⊙M = 606.7″. Thus, for the external contact AM = 764.52″, the corresponding duration for AM = $3^h11^m8^s$ and hence, for the inclination of the ecliptic, the angle will be A⊙M = 51°33′56″. Similarly, for the internal contact AM = 688.94″, the duration of AM = $2^h52^m14^s$ for the elements of the Sun on 3 June 1769. Hence, the angle will be A⊙M = 48°37′55″. Now, V and V′ are the positions of Venus when, at some point on the Earth, a contact (either external or internal) is observed at ingress or egress, respectively. If ⊙a = ⊙A and ⊙b = ⊙B, then Va and V′b represent the parallax effects in the directions V⊙ = V′⊙. In the direction of the orbit VA, the effect is obtained approximately by the relationship VA = Va sin(VAa) csc(VAa) when the a priori estimate 8.5″ for the solar parallax is used.

Next, Lexell considered the consequences of the uncertainty of ⊙A and ⊙M (see Fig. 2 to the right of Fig. 5.2). To this end, let ⊙V first be constant and let ⊙M suffer a slight augmentation. From V, a circular arc CS is drawn, and the line Sm parallel to ⊙M. Sm denotes the true distance between the centres of the Sun and Venus. Further, ⊙p is drawn parallel to VM. Sp is the small correction needed for ⊙M. Then, in the triangle S⊙p, p⊙ = Mm = Sp, which is the amount by which VM has diminished.

Second, let ⊙M be constant while ⊙V undergoes an augmentation S′$p′$. If S′$m′$ is parallel to ⊙M, then M$m′$ = ⊙S′ = $p′$S′. Hence, the total correction is VM = S′$p′$ sec(⊙VM) − Sp tan(⊙VM). When the parallax effect in the direction ⊙V is denoted $\alpha\pi$, where π stands for the horizontal parallax, and setting Sp = y, and S′$p′$ = μ for external contacts and ν for internal contacts, one obtains the equations

$$\mathrm{VM} = 764.52 \pm \alpha\pi \sin \mathrm{V}a\mathrm{A} \csc \mathrm{VA}a - y \tan \odot\mathrm{VM} + \mu \sec \odot\mathrm{VM},$$

$$\mathrm{VM} = 688.94 \pm \alpha\pi \sin \mathrm{V}a\mathrm{A} \csc \mathrm{VA}a - y \tan \odot\mathrm{VM} + \nu \sec \odot\mathrm{VM},$$

for external and internal contacts, respectively. Correspondingly, from the knowledge of the speed of Venus with respect to the Sun, the durations for the respective distances VM are

$$\mathrm{T} = 3^h11^m8^s \pm 15(\alpha\pi \sin \mathrm{V}a\mathrm{A} \csc \mathrm{VA}a - y \tan \odot\mathrm{VM} + \mu \sec \odot\mathrm{VM}),$$

$$\mathrm{T} = 2^h52^m14^s \pm 15(\alpha\pi \sin \mathrm{V}a\mathrm{A} \csc \mathrm{VA}a - y \tan \odot\mathrm{VM} + \nu \sec \odot\mathrm{VM}).$$

Finally, for every observation point, expressions are formed for the external contact at ingress T, the internal contact at ingress T′, the internal contact at egress T″ and the external contact at egress T‴. For each site there will arise a series of equations, which, when they are compared to the actually measured times at respective sites, lead to an overdetermined system of equations to which a best possible estimate is sought (with respect to some norm), allowing a certain correction in longitude (μ and ν) and latitude y. The outcome of Lexell's calculations in [Lexell 13] involving the measurements from Vardø, Kola [near today's Murmansk], Cajaneborg (Kajaani), Prince of Wales Fort in Hudson Bay and at the tip of Baja California was $\pi = 8.54''$, the margin of error being estimated to "no larger than one fifth of a second". In the subsequent paper [Lexell 14], where the Tahitian measurements were included, the estimate was $8.68''$.

To Lexell, the transparency of the calculations was all-important. His agenda was not to prove some observer's outstanding qualities or to question the credibility of any observer in particular. Neither did he imagine himself capable of determining the solar parallax to a tiny fraction of an arc second; rather, he was careful to state the limits of doubt in all conclusions. Not all astronomers shared his ideals, however.

As might well have been expected, disagreements soon arose between the astronomers concerning the correct method of processing the data as well as the reliability of the individual measurements. The chief players in the ensuing affair were Anders Planman, Joseph Jérôme de Lalande and the Hungarian Jesuit and astronomer Maximilian Hell [3, 4]. Father Hell had been observing the transit in Vardø, on the north-eastern coast of Norway. However, for many reasons the report containing his measurements and conclusions was completed and dispatched to Paris and other scientific centres several months later than the others, which made among others Lalande suspicious about the reliability of his measurements.[17] Lalande considered Planman's measurements the most important ones made in Europe, a fact which naturally offended Hell. In 1772, when the quarrel was at its peak, Lalande even asked Lexell to mediate in the conflict (as Lexell mentions in a letter to Wargentin on 5 October 1772). Initially, Lexell had refused, wishing to stand aloof from petty quarrels, but soon he felt compelled to intervene.

What had started as a disagreement between Planman, Lalande and Hell had thus escalated into an outright dispute between the astronomers of Saint Petersburg and Vienna. In letters and books Lexell and Hell criticised one another in harsh words, but nevertheless, in formally correct Latin [3, 4]. Hell was suspicious about the methods of Euler and Lexell, while maintaining that his measurements were impeccable. Lexell, for his part, showed in his treatise [Lexell 23] that Hell's own arguments were groundless and false and that the logic of his deductions of the parallax was defective (Fig. 5.3). In the printed letter [Lexell 27] Lexell exposed more inconsistencies in Hell's report as to the moments of observation and the analysis of the results, but they were immediately refuted. Fearing that not only his own reputation was endangered, but also that of his superior Euler, Lexell wrote numerous letters to his friends Wargentin, Planman and Johann III Bernoulli, in

[17] In fact, a minor scandal arose a century later when the Viennese astronomer Karl Ludwig von Littrow (1811–1877) attempted to prove that Father Hell had falsified his results. However, later on Simon Newcomb (1835–1909) analysed Hell's notes anew and found no evidence of forgery [4].

which he tried to convince them of the solidity of his arguments and warned them against believing in those of Hell. In a letter dated 3 April 1773 (KVAC, Stockholm), he admits to Wargentin:

> The only thing I have wanted to prove is that Father Hell has erred when he has criticised the calculations of the 14th volume of the Commentaries and that his own calculations are so severely erroneous, that nothing can be concluded from them. I am delighted to have been able to show Father Hell some reason; as to Planman I am more in despair, although he admitted to me when I visited Åbo that there is no reason to believe Father Hell's observation to be invented.[18]

Through his support from Lalande, Planman had insisted on the correctness of his observations and deductions, as did Father Hell for his own part. Provoked by Planman's rather self-satisfied attitude Lexell wrote in a letter on 10 February 1774:

> Allow me to admit that I had hoped for a little more consideration from you in this matter, which is not the trickiest one and in truth requires more of a sound logic and critique than any sophisticated mathematics. Least of all had I anticipated that you, Herr Professor, would have resisted my reasons with a certain authority, which I, for all my appreciation of your personal character and qualities, cannot bring myself to approve. You may rely on it that not even Euler, the great Euler, is capable of convincing me on his mere authority, much less anybody else.[19]

Lexell always respected Wargentin's wise and diplomatic response, but was rather disappointed at the position Bernoulli had taken. Even when the heat of battle had somewhat cooled down, Lexell wrote in a bitterly ironic tone to Bernoulli, reproaching him for an unconcerned and ignorant attitude (HUB Ms. Coll. 171, letter dated 24 December 1775):

> Finally, I have received Father Hell's Supplement on the parallax[20]; I found it to be just as I imagined and even worse. You, Monsieur, must have read it very hastily when you wrote to me a year ago that I have reason to be satisfied with Father Hell. I would readily agree on this if I could imagine that it is out of kindness towards me that he persists in the objections he made against the calculations of the parallax in the 14th volume of the Commentaries; that he defends all the errors he himself has committed; that he has published one of my

---

[18]Det enda som jag welat bewisa, är at Pat: Hell haft orätt då han criticerat de räkningar som förekomma uti XIV Tomen af Comment: samt at Hans egna räkningar äro så swårt felaktiga, at af dem ingen ting kan slutas. Det fägnar mig, at jag kunnat bringa Pat: Hell til så mycket billighet, om Planman miströstar jag mera, likwäl måste han medge mig då jag war i Åbo, at ingen anledning är, at misstänka Pat: Hell observation för at wara updiktad.

[19]Emedlertid må Herr Professorn tillåta mig at upriktigt tilstå, det jag hade förmodat lite mera öfwerläggning af Herr Professorn hwad detta ämne angår, som wäl ei är af de aldra benigaste och i sanning mera fordrar en sund logica och riktig critique, än diupsinnig Mathematique. Aldra minst hade jag wäntat, at Herr Professorn emot mina skäl, allena tyckes wilja sätta en wiss auctoritet, som med al aktning för Herr Professorns person och egenskaper, jag icke kan finna mig uti at erkänna. Herr Professorn kan wara öfwertygad at Euler, den stora Euler, ingenting förmår öfwer mig blott genom auctoritet, mycket mindre någon annan.

[20]Maximilian Hell: *Supplementum ad Ephemerides Astronomicas Anni 1774 ad Meridianum Vindobonensem*. Viennae, 1773.

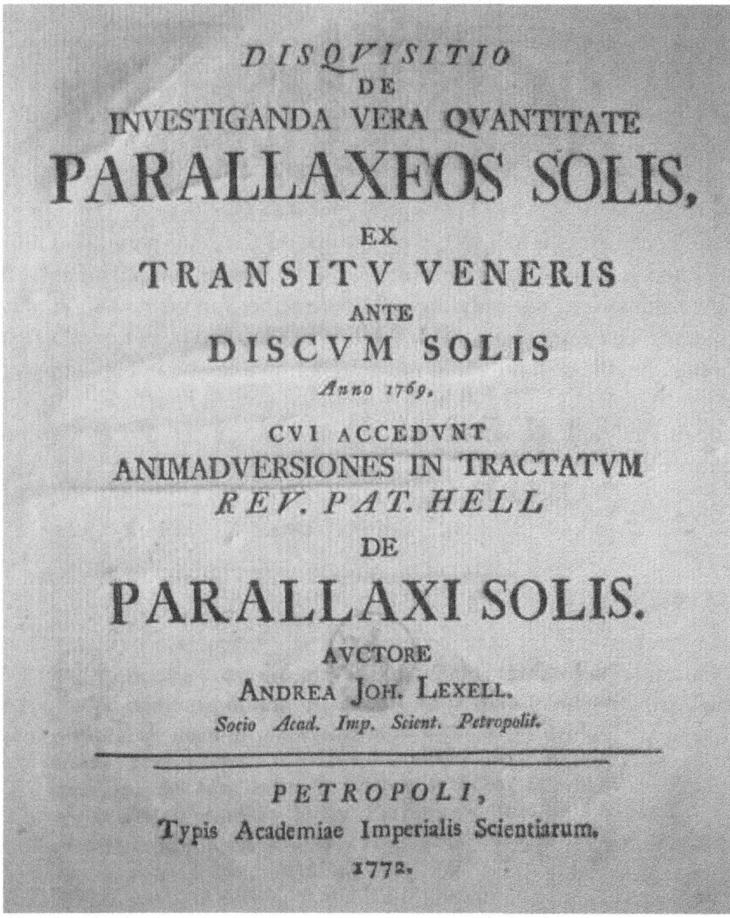

DISQVISITIO
DE
INVESTIGANDA VERA QVANTITATE
PARALLAXEOS SOLIS,
EX
TRANSITV VENERIS
ANTE
DISCVM SOLIS
Anno 1769,
CVI ACCEDVNT
ANIMADVERSIONES IN TRACTATVM
REV. PAT. HELL
DE
PARALLAXI SOLIS.
AVCTORE
ANDREA JOH. LEXELL.
Socio Acad. Imp. Scient. Petropolit.

PETROPOLI,
Typis Academiae Imperialis Scientiarum.
1772.

Fig. 5.3 Title page of Lexell's concluding work [Lexell 23] on the solar parallax. The title translates as: "Inquiry into the search for the true value of the solar parallax based on the transit of Venus in front of the disc of the Sun in the year 1769, with comments on the Honourable Father Hell's treatise *de Parallaxi Solis*" (Private collection. Photo: the author)

letters[21] without asking my permission for it; that he adds notes in it, partly trivial and for the most part absurd; that he reserves to himself the right to correct or rather pervert my calculations without understanding them; that he presents several odious insinuations and accusations against me. I say, if I were stupid enough to agree that this is all to my advantage, I would have much to praise him for. Be my judge, Monsieur, if you please. But let me also note the remarkable contrast between your conduct towards me and Father Hell. You approve the conduct of Father Hell without examination, and if I ask you for your view

[21]Referring to [Lexell 27]. We have included this letter in the list of Lexell's publications even if Lexell himself was opposed to publishing it. It is of course supplemented with Hell's own footnotes and refutations of Lexell's arguments.

> on a controversy between him and me, you excuse yourself by your lack of time. I did not ask you to say anything disadvantageous of the personal character of Father Hell, I only wanted to know if in your opinion he was wrong in an issue or not.[22]

In normal circumstances, a temperamental tirade of this magnitude would lead to a break-up in relations, but Bernoulli seems to have been cool-headed. The matter would no longer be discussed, and the correspondence continues in a respectful and polite tone.

The year 1775 marks a halt in Lexell's work on the solar parallax. Publications [Lexell 42] and [Lexell 43] written in 1775 contain the summit of his efforts in parallax determinations, not only the parallax of the Sun or the Moon, but of any distant object.[23] The computation of the solar parallax had been Lexell's first test as an astronomer, but the ensuing affair had taught him a lesson in human psychology: astronomers, like all scientists—himself included—were human beings, subject to a multitude of vices and weaknesses, ambitions and caprices, things which in many cases are more powerful than the noble pursuit of truth.

5.2 Perturbations of the Moon's Motion

Owing to the relatively rapid movement of the Moon in the sky—one revolution around the Earth in approximately 27.3 days—a popular method used to determine longitude at sea[24] was the measurement of the lunar distance[25] and the subsequent

---

[22]Enfin j'ai reçu le supplément de l'Abbé Hell sur la parallaxe; je l'ai trouvé tel que je me l'avois imaginé et même pire encore. Il faut bien, que Vous Monsieur, l'aviez parcouru bien à la hâte, lorsque Vous m'écrivîtes il y a un an, que j'aurai raison d'être bien content de l'Abbé Hell. J'en conviens volontiers, si je pourrois m'imaginer que ce soit par complaisance pour moi, qu'il persiste encore sur les objections, qu'il a faites contre les calculs sur la parallaxe dans le XIVe Tome des Commentaires; qu'il defend toutes les fautes qu'il avoit commis lui-même; qu'il fait imprimer une de mes lettres sans m'en demander la permission; qu'il y ajoute quantité des notes en partie triviales et pour la pluspart absurdes; qu'il s'approprie le droit de corriger ou plustôt pervertir mes calculs sans les entendre; qu'il propose plusieurs insinuations et imputations odieuses contre moi. Je dis, que si je serois assez bête pour me persuader, que tout ceci soit à mon avantage, j'aurois beaucoup à me louer de l'Abbé Hell. Soyez Vous-même Monsieur, mon juge s'il Vous plaît. Mais permettez aussi que je remarque le contraste singulier, qu'il y a entre Votre conduite envers l'Abbé Hell et moi. Vous approuvez la conduite de l'Abbé Hell, sans l'avoir examiné et quand je Vous demande Votre sentiment sur des choses controversées entre lui et moi, Vous vous excusez par Votre peu de temps. Je ne Vous ai demandé, que Vous disiez quelque chose au désavantage du caractère personnel de l'Abbé Hell, j'ai seulement voulu sçavoir si selon Votre sentiment il avoit tort sur une telle question, ou non?

[23]In fact, an expression for the apparent breadth of the Moon given in [Lexell 42] became known as "Lexell's formula" (*Lexellsche Formel, la formule de Lexell*) [160].

[24]The method was used at least to the end of the nineteenth century, when exact chronometers made it obsolete.

[25]The angular distance between the Moon and a celestial object (a fixed star).

comparison with lunar tables (for a detailed history of the problem, see [39, Vol. 6-2-1, pp. 621ff.; 155, 169]). However, for several centuries, the irregularity of the Moon's motion had preoccupied scientists. The problem is a particularly difficult one because of the complicated interplay between the gravitational forces of the Earth and the Sun respectively, and because the Moon and the Earth cannot be taken as punctiform bodies. A simple perturbational approach was not applicable, because the correction terms applied to the regular solution would not be as small as required [16, 155]. Like many a mathematician before him—notably Newton—and many of his contemporaries—such as Alexis Clairaut, Tobias Mayer and Jean d'Alembert—Leonhard Euler worked during all his life on the perfection of the theory for the motion of the Moon, by means of which its movement could be accurately predicted for years to come. The Moon's motion was so complicated that Clairaut at some point even doubted the correctness of the inverse square law of attraction.

At the end of the 1760s Euler was increasingly dissatisfied with his previous lunar theory from 1753, and so he engaged his new team of young collaborators Lexell, Wolfgang Ludwig Krafft as well as his son Johann Albrecht Euler to work out a new theory for the Moon [Lexell 19], which involved an immense computational effort in times when all calculations were executed only with the aid of mathematical tables [16]. With its nearly 800 pages, the book was a great undertaking even for the industrious Euler, who declared in the title that the work had been accomplished thanks to the "incredible zeal as well as the indefatigable labour of three academicians—"[...] *incredibili studio atque indefesso labore trium academicorum*". The minutes of the conferences mention:

> —18 March 1771 [141, Vol. III, p. 10] The Adjuncts Mr Krafft and Lexell presented: *Determinatio inaequalitatum motus lunaris*, a work which they have completed with much care and great efforts under the supervision of Mr Euler the Elder. It was then unanimously decided that it would be sent immediately for printing: the number of copies was set at 400, and the work would be printed *in quarto* like the Commentaries.

The venture involved the use of a system of rectangular coordinates rotating with the mean motion—instead of the fixed cylindrical coordinates which Euler had employed previously—and the solution of the differential equation of motion by means of the method of indeterminate coefficients and variation of constants. Nevertheless, Euler's new theory turned out to be less accurate than his first one from 1753, which had been converted into a semi-empirical tabular form by the astronomer Tobias Mayer[26] and was generally appreciated for its applicability and accuracy. In fact, as early as 1773, Lexell published his comparisons between Euler's, Mayer's and Clairaut's theories and found small discrepancies. Nevertheless, in his paper entitled in English "Comparison between the lunar theory of the illustrious Euler and recent tables of the celebrated Mayer" [Lexell 29], Lexell concluded:

[26]Mayer's book *Theoria Lunae* was published posthumously in London in 1767.

> When it comes to the tables of the illustrious Euler, they are such as those deduced by computation based on theory generally are, and therefore if they would be improved and corrected in like manner as the Mayerian tables, we do not doubt that with time they will yield the greatest exactitude that can be desired in this matter.[27]

However, this would not stop Euler from trying to improve his theory by purely theoretical means, using as little as possible empirical data to guide the theory. Yet, integration constants occurring in the derivations could only be determined empirically.

5.3 Occultations of Jupiter's Satellites

One of the most inventive ways of achieving terrestrial synchronisation in order to determine the difference in longitude between two observers on the Earth was provided by the occultations of the satellites (moons) of the planet Jupiter. The largest four satellites of Jupiter—the Galilean moons, so named in honour of their discoverer—can be easily observed using a relatively small telescope. The satellites orbit around Jupiter with a short period, and thus they can be seen to appear and disappear quite regularly in the shadow of the great planet. The disappearances and the reappearances are called immersions and emersions of the satellites, respectively.[28] These phenomena appear practically at the same instant when observed from different parts of the Earth, and thus, by comparing the exact moments in which they occur in different places, the difference in longitude can easily be determined. However, due to irregularities of the motion of the Jovian satellites,[29] including the motion of Jupiter itself, suitable moments for observing these occultations were difficult to predict.

To account for the irregularities of the motion of the Jovian moons, Pehr Wargentin had developed a statistical theory, which was widely recognised and used by contemporary astronomers.[30] Wargentin's tables had been found more accurate than those issued previously by the renowned Jacques Cassini in France. In his work,

---

[27]Page 567: "Tabulas vero Illustris Euleri quod attinet, eae fere tales sunt, quales per computum Theoriae superstructum deductae, quare si simili modo ac Mayerianae per observationes corrigantur et emendentur, non dubitamus quin aliquando praestiturae sint exactitudinem quae hoc in negotio desideratur, maximam."

[28]This method was invented and developed by the Danish astronomer Ole Rømer (1644–1710), who by observing seasonal variations of the frequency of these occultations was able to show that light arriving from Jupiter to the Earth takes time to propagate. By estimating the diameter of the Earth's orbit Rømer managed to determine the speed of light in 1676, for the first time in history.

[29]The irregularities are mainly due to the flattening of Jupiter's body, a fact that was not known at the time.

[30]Wargentin's studies of the moons of Jupiter began in his thesis entitled *Specimen astronomicum, de satellitibus Jovis* ... (12 December 1741), presided in Uppsala by Anders Celsius.

Lexell supported Wargentin's efforts and in numerous letters, often in a postscript in the margin, submitted some of his recent observations of the occultations of the Jovian satellites. Some of Lexell's observations of the Jovian moons and subsequent longitude determinations were also published in the Berlin *Ephemerides* edited by Johann III Bernoulli [Lexell 41].

5.4 The Great Comet of 1769

Comets are comparatively small celestial objects gravitating around the Sun [82]. Their orbits are typically elongated, the "outer" end reaching far beyond the orbit of Jupiter and the "inner" end coming close to the Sun. Their nucleus, consisting typically of ice, dust and rocky particles, range from some hundreds of metres to several kilometres in diameter. When the comets approach the Sun, their motion accelerates and while the intensity of the solar radiation increases, their volatile material starts to evaporate, thus provoking the characteristic *coma*[31] and tail. The eighteenth-century scientists had of course no means of knowing the constitution of the comets, as spectroscopy did not yet exist.[32] Neither could the astronomers determine their size or mass very accurately. However, seeing that the comets did not noticeably affect the motion of the nearby planets, it was understood that their mass must be comparatively small. Euler, for instance, had assumed that their mass would be comparable to or smaller than the mass of the Earth [16]. Unlike comets, asteroids (known since 1801) have a constitution more resembling small, irregularly shaped but solid planets. Thus, they do not lose material, in spite of moving in other respects like comets.

Ever since Newton's theory of gravity, comets have been understood to obey the same laws of celestial mechanics as the planets and their satellites. However, due to their short period of visibility, it is very difficult to compute their period, i.e. the time of one revolution around the Sun, from only one short observation series. The famous Halley's comet had been found to appear at regular intervals of 76 years; thus, it was tacitly assumed that all comets might follow such a regular pattern of behaviour.

An occasion to attempt the computation of the period of a comet was offered by the spectacular comet of the year 1769, whose modern code designation is "C/1769 P1". It was discovered on 8 August 1769 by Charles Messier at the Naval Observatory in Paris [82]. Having only recently accomplished the observations of the transit of Venus, the attention of astronomers everywhere was focused on

[31]Latin for hair; thus, a comet is like a hairy ball.

[32]Nevertheless, from the affinity of the tails of comets and the phenomenon known as the northern lights, Euler had speculated that the phenomena could be related, cf. "Recherches physiques sur la cause de la queüe des comètes, de la lumière boréale, et de la lumière zodiacale", *Mémoires de l'Académie des Sciences de Berlin*, 1746, pp. 117–140 (OO Ser. II, Vol. 31, pp. 221–238. E103).

this conspicuous object, whose tail was quickly growing. On 9 September its tail measured 60°, and 2 days later 90°. In the beginning of October, it had arrived so close to the Sun that it could no longer be distinguished (Fig. 5.4).

Christian Mayer also observed the comet in August during his sojourn in Saint Petersburg. On 16 October 1769, Lexell writes to Wargentin (KVAC; letter in Swedish)

> The comet has been observed here by Professor Mayer since 17 August old style until the 31st, but not all of these days. Last Sunday we encountered it anew in the [constellation of] *Monte Menalo*.[33] Since then it has been cloudy, but as the sky is now starting to clear up, we will probably see it again. Herewith I send to you data of the observations which have been done here so far, and humbly ask for yours. I have to admit that a part of Professor Mayer's observations of the comet appear to me doubtful, especially the one of 6 September new style, which does not comply with those preceding and following. From the observations on 3, 6 and 7 September we have calculated its elements, but they do not match the observations of Messier from the beginning of August and those made here last Sunday. The observation on 9 September seems to me quite good, since then the comet really went past the Equant and stood still along two fixed stars on the back of the Unicorn.

Computation of the orbit of the comet started almost immediately. In the first part of their work [Lexell 6], Euler and Lexell attempted to match the most reliable observations with the hypothesis of a parabolic orbit. Not being satisfied with this, however, in the second part of their study the authors extended their investigation to include other kind of conic sections as well. Having first determined the possible elliptical orbits which within a certain error of observation satisfy the actual positions of the comet on 21 August and 4 September 1769, they computed the true orbit as the one which satisfies the comet's position on 24 October. Hence they found the eccentricity of the orbit to be 0.998, the length of the semi-major axis 61.45 and the perihelion distance of the comet 0.1227 times the Earth-Sun distance. Based on three observations they determined the period of one revolution around the Sun to be 481.7 years; more specifically within the limits of 449 and 519 years, when an error of 1 min of an arc in the observations was allowed. Later on, the calculations became more precise, and in 1806, Friedrich Wilhelm Bessel, still a young astronomer in Lilienthal near Bremen,[34] determined the period to be much longer, about 2,090 years. The large discrepancy between the results reflects the fact that the plane of the orbit was seen at a very oblique angle with respect to the ecliptic, making it particularly difficult to arrive at the right conclusions from only a few observations. Bessel also surmised that Lexell had assumed the observed position of the comet on 8 August to be very precise, which apparently it was not.

[33] *Mount Maenalus* is a name, no longer used, of a stellar constellation situated between *Virgo* and *Boötes*.

[34] Friedrich Wilhelm Bessel (1784–1846), astronomer and mathematician, was appointed director of the Königsberg Observatory in 1810. His essay entitled "Untersuchung der wahren elliptischen Bewegung des Kometen von 1769" (published in the *Astronomisches Jahrbuch* (Berlin) for the year 1810, pp. 88–124) was awarded the astronomical prize of the Berlin Academy.

RECHERCHES ET CALCULS

Sur la vraie orbite elliptique de la Comete de l'An. 1769. et ſon tems periodique,

executées ſous la direction

de

Mr. LEONHARD EULER,

par les ſoins

de Mr. LEXELL,

Adjoint de l'Academie Imperiale des Sciences de Saint-Petersbourg.

A St. PETERSBOURG,

De l'Imprimerie de l'Academie Impériale des Sciences

1770.

6 RECHERCHES SUR LA COMETE.

Probleme.

Pour un tems propoſée quelconque $=T$, ayant le lieu de la Comete $=L$, (qu'on peut prendre tant pour la Longitude, que la Latitude) ſi pour un tems précedent $T-p$ on connoit le lieu de la Comete $=L-m$ et pour un tems ſuivant $T+q$, le lieu de la Comete $=L+n$, trouver le lieu de la Comete pour un tems indeterminée $=T+z$.

Solution.

Poſons ce lieu qu'on cherche

$$L+Mz(z+p)+Nz(z-q)$$

de ſorte que prenant $z=0$, ce lieu dévienne $=L$, mais puisque prenant $z=-p$, le lieu doit etre $=L-m$, nous aurons cette équation

$$L-m=L+Np(p+q)$$

d'ou l'on tire $N=\frac{-m}{p(p+q)}$. De la même maniere puisque poſant $z=q$, le lieu doit etre $L+n$, on aura cette équation

$$L+n=L+Mq(p+q),$$

d'ou l'on tire $M=\frac{n}{q(p+q)}$, par conſequent pour le tems indeterminée $T+z$ le lieu de la Comete ſera

$$=L+\frac{nz(z+p)}{q(p+q)}-\frac{mz(z-q)}{p(p+q)}$$

ou bien

$$=L+\frac{(npp+mqq)z}{pq(p+q)}+\frac{(np-mq)zz}{pq(p+q)}.$$

56 RECHERCHES SUR LA COMETE.

Pour cet effet poſons

$$u=\sqrt[3]{p}+\sqrt[3]{q},$$

pour avoir

$$u^3=p+q+3u\sqrt[3]{pq},$$

cette valeur étant ſubſtitué nous donne

$$p+q+3u\sqrt[3]{pq}=-3u+\frac{2N}{C}$$

d'ou il eſt clair que

$$1^{\circ}.\ \sqrt[3]{pq}=-1 \text{ et } 2^{do}.\ p+q=\frac{2N}{C},$$

de là on voit, que $pq=-1$, donc poſant $p=-\text{tang.}\,\alpha$ on aura $q=-\text{cot.}\,\alpha$ et partant l'autre équation devient $\text{tang.}\,\alpha-\text{cot.}\,\alpha=\frac{2N}{C}$ ou bien

$$\frac{\text{ſin.}\,\alpha^2-\text{coſ.}\,\alpha^2}{\text{ſin.}\,\alpha\,\text{coſ.}\,\alpha}=-\frac{2.\,\text{coſ.}\,2\alpha}{\text{ſin.}\,2\alpha}=-2\,\text{cot.}\,2\alpha=\frac{2N}{C}=2\,\text{cot.}\,(180^{\circ}-2\alpha)$$

de ſorte que $\text{cot.}(180^{\circ}-2\alpha)=\frac{2N}{2C}$, poſant donc $180^{\circ}-2\alpha=2\mu$ pour avoir $\text{cot.}\,2\mu=\frac{2N}{2C}$, d'ou l'on trouvera aiſément l'angle μ et enſuite on aura $p=\text{cot.}\,\mu$ et $q=-\text{tang.}\,\mu$, qu'on cherche une autre angle ν de ſorte que

$$\text{tang.}\,\nu=\sqrt[3]{\text{tang.}\,\mu} \text{ pour avoir}$$

$$\sqrt[3]{p}=\text{cot.}\,\nu \text{ et } \sqrt[3]{q}=-\sqrt[3]{\text{tang.}\,\mu}=-\text{tang.}\,\nu$$

et partant notre quantité cherché

$$u=\text{cot.}\,\nu-\text{tang.}\,\nu=\frac{\text{coſ.}\,\nu^2-\text{ſin.}\,\nu^2}{\text{ſin.}\,\nu\,\text{coſ.}\,\nu}=\frac{2\,\text{coſ.}\,2\nu}{\text{ſin.}\,2\nu}=2\,\text{cot.}\,2\nu.$$

Voila

136 RECHERCHES SUR LA COMETE.

I. Longitude du ☋ $=11^{s}.\,25^{\circ}.\,4'.\,41''$

II. Inclinaiſon de l'orbite $=40^{\circ}.\,49'.\,33''$

III. Le demiparametre de l'orbite $=0{,}2451294$

IV. L'excentricité $=0{,}9980036$

V. Diſtance du Perihelie au Soleil $=0{,}1226851$

VI. Elongation du noeud deſcend. au Perihelie $=149^{\circ}.\,10'.\,51''$

VII. Tems du Perihelie Oct. $7^{j}.\,15^{h}.\,37'.\,37''=7^{j}.65112$.

CX. On voit donc que l'orbite de cette Comete eſt en effet une ellipſe; il vaudra donc bien la peine de determiner tant l'axe de cette ellipſe, que le tems periodique de la Comete. Pour cet effet, ayant trouvé la diſtance du Perihelie $\odot\Pi=\frac{b}{1+e}$, on n'a qu'à la diviſer par $1-e$, pour avoir le demiaxe de l'orbite $\frac{b}{1-ee}=a$ et alors $a\sqrt{a}$ donnera le tems periodique exprimée en Années. D'ou on ſera le calcul ſuivant:

$$\text{Log.}\,\odot\Pi=9.0887920$$

$$L(1-e)=7.3002476 \qquad a=61{,}4531843$$

$$\text{Log.}\,a=1.7885444 \qquad a\sqrt{a}=481{,}7$$

$$L\sqrt{a}=0.8942722$$

$$La\sqrt{a}=2.6828166$$

dont le demiaxe de l'orbite eſt $61{,}45318$ et partant la diſtance de l'aphelie au Soleil $122{,}7836835$, enfin le tems periodique doit être de $481{,}7$ Années.

CXI.

Fig. 5.4 The title page and some excerpts from Euler's and Lexell's work on the comet of 1769 [Lexell 6]=[E389]. First, the problem is set up (*top right*), a cubic equation emerging in the analysis is solved (*bottom left*), and finally the results of the analysis are summarised. The work is printed in small 4° (19.5 × 24 cm) and contains 160 pages and two figure plates (Helsinki University Library)

Almost 3 years after its discovery, Lexell was still occupied with the comet of 1769, when he wrote to Wargentin in a letter dated 2 March 1772:

> I have made many calculations of the comet of 69 in addition to those given in Euler's book *Recherches sur la comète*. I attach here 12 different elements. Many of these result from your observations. Although the *tempus periodicum* [time of one revolution] is uncertain, I think it is quite an achievement to have determined something as precisely as this, the more so when I can show clearly that if an error of 2″ is made in the determination of the inclination, it will consequently give an error in the period of 1 year. Having determined the *distantia perihelii* [the closest distance of the object to the Sun] quite accurately, I was able to determine the eccentricity using only two observations, and if I was allowed to choose the best of observations, I believe that I would be able to determine the *tempus periodicum* with an accuracy of 10 years. But I would not dare to embark on such a lengthy work.

However, soon afterwards the first disappointing review of the work appeared in the *Allgemeine deutsche Bibliothek* (1772, Band XVI, 2. Stück, pp. 657–659). The anonymous author of the review, signed "Ik", begins:

> It is seen from this work that Mr Euler, who due to problems of eyesight cannot write himself, has asked Mr Lexell to take the pen and to write what he should calculate in numbers, mostly with the help of logarithmic tables. Except for the calculations and the spelling, everything else is Eulerian viz. the style, the arrangement of the numbers, the treatment of the equations, and the following improvements of what was not sufficiently considered at the start. For Mr Euler can let unsuccessful attempts in the calculations appear in print just as the successful ones, and perhaps may have his special reasons to do so.[35]

Obviously, Lexell was upset and complained to Wargentin in a latter dated 5 October 1772 (KVAC, Stockholm)

> When Professor Euler's treatise on the comet of 1769 was reviewed in the *Allgemeine deutsche Bibliothek*, the completely unknown reviewer wanted to create an impression of my person, which I do not deserve, only because my name is mentioned in front of this work. His uttering goes as follows: this work is dictated by Mr Euler to Mr Lexell, who has taken the pains to calculate by using logarithmic tables. Although I do not have the right to judge the thoughts and meanings of the hearts of other people, I believe I am not wrong in thinking that the reviewer has attempted to say that I have not done anything else, not only in this work, which I admit to be correct and am ready to confess to the world, but also that I am incapable of doing anything else. Whether the latter conclusion holds true or not is not my intention to look into, the question is only how the reviewer has been able to present such an unfavourable thought of someone whose capacity he surely does not know personally, and whether it has not been without the slightest reason? If the reviewer had had the foreknowledge that I was affecting the honour of having my name printed on the title page of a work entirely due to Professor Euler, I would grant him the right to degrade me; But I can guarantee that the title of this work has been written entirely in my absence and I

[35]Man sieht aus dieser Schrift, dass Hr. Euler, der wegen Mangel des Gesichtes nicht selbst schreiben konnte, dieselbe dem Herrn Lexell in die Feder angegeben, und ihn vorgesagt, was er in Zahlen und meistens mit Hülfe der logarithmischen Tabellen zu rechnen habe. Ausser dieser Rechnung und der Orthographie ist alles übrige Eulerisch, nemlich der Styl, die Anordnung des Wertes, die Wendungen in den Rechnungen, und die nachgeholten Verbesserungen dessen was anfangs nicht genug überdacht war. Denn Hr. Euler läßt fehlgeschlagene Versuche im Rechnen eben so wie die gelungenen in Druck erscheinen, und mag vielleicht seine besondere Gründe dazu haben.

> solemnly swear that I did not affect this honour, but on the part of the reviewer, whoever he might be, it seems to point to jealousy and reproachfulness that he has uttered such words about my insights. I readily admit that Professor Euler in this matter, as well as in other calculations I have made under his supervision, has wanted to grant me more honour than I deserve; in my view my merit in such occasions is to have done this great man a favour and somehow prove my gratitude, in return for the manifold kindnesses towards me.

Surmising the author to be the sharp-tongued mathematician Abraham Gotthelf Kästner[36] from Göttingen, Lexell writes to Johann III Bernoulli on 2 May 1773:

> Dans le Vol. XVI de la *Bibliothèque Universelle Germanique* on a donné l'analyse du traité de Mr Euler sur la comète de l'An 1769, ou je suis maltraité d'une manière si indigne, que je n'aurai dû m'attendre des pis de mon plus grand ennemi. L'auteur de cet exposé me fait connoître comme un homme dont toutes les connoissances se bornent à faire des calculs numériques avec le secours des tables logarithmiques. Mais quelqu'il soit cet auteur, il ne me connoît certainement si particulièrement, qu'il a pû risquer un jugement si désavantageux, sans se tromper grossièrement. Je suis fort persuadé qu'aucun des Académiciens de Berlin ai fait cet exposé, puisqu'il contient des choses qui font connoître que l'auteur soit fort peu versé dans la science des calculs sur les comètes. J'avoue que mes soupçons tombent uniquement sur Mr Kaestner, parce que je ne connois pas aucun autre, qui aye une si forte disposition de critiquer et de chicaner sur ce que les plus célèbres mathématiciens ont fait, quoiqu'il ne soit pas en état d'apprécier les mérites et les avantages de leur travaux.[37]

In one way or another Kästner must have been informed of Lexell's suspicions against him, because in the April 1773 issue of the *Journal des Sçavans*, he declared himself with regard to this review:

> Dans le journal allemand [...] on a parlé de ces travaux d'une manière injuste, en rendant compte de l'ouvrage de Mr Euler sur les comètes, dont les calculs ont été faits par Mr Lexel. L'auteur du journal le représente comme un élève, dont le mérite se borne à écrire des calculs numériques sous la dictée de Mr Euler, à additioner des logarithmes, et qui n'avoit presque d'autre part à l'ouvrage, que de l'avoir écrit par de sa main. Cependant Mr Euler a fait mettre le nom de Mr Lexel dans le titre même de l'ouvrage comme un nom qui ne deparoît pas le sien.

In other words, for Kästner it was clear that Euler had included Lexell's name on the front page not because of some caprice or a wish to promote Lexell without any merit, but simply because Euler wanted to recognise Lexell's contribution as essential for the work. Kästner's declaration does not prove his innocence as to

[36] A. G. Kästner (1719–1800), born in Leipzig, was Professor of Mathematics at the University of Göttingen. Kästner was feared for his sarcastic attitude, and although not much renowned for original research, he educated many first-rate mathematicians and physicists. He also published histories of mathematics [32] and translated the proceedings of the Royal Swedish Academy of Sciences into German. When Lexell met with Kästner personally in 1780 (see the letters to Wargentin in Sect. 9.4), he no longer seemed to hold a personal grudge against Kästner.

[37] Lexell says that he is quite certain that none of the Berlin academicians have written this review since it reveals that the author is very little versed in the science of calculations relating to comets. He confesses that his suspicions fall solely on Mr Kaestner because he knows no one else with such a strong disposition to criticise and quibble about what the most famous mathematicians have written, without himself being capable of assessing the merits and benefits of their work.

the review, and Lexell's reaction to it remains unknown. The true author of the original anonymous review was apparently never divulged, but the memory of this unpleasant episode haunted Lexell for years to come.

5.5 The Mysterious Comet of 1770

Just like the comet of 1769, the comet of 1770 was discovered by Charles Messier [82]. On the evening of 14 June 1770, when Messier as usual was scanning the sky with his telescope, he caught sight of a new object in the constellation of *Sagittarius*, which he realised was a comet. Its luminosity increased rapidly during the month of June till it became visible to the naked eye. At that point, astronomers all over the world had observed the comet. However, its appearance and motion were uncanny from the start: it had a weak but exceptionally large coma—over 2° in breadth—practically no tail (possibly because it was hidden behind the coma), and its orbit seemed unlike that of most comets. The diameter of the coma was estimated to about 96,000 km in modern units, which is not exceptional (Fig. 5.5).

This comet "D/1770 L1" (where the 'D' means that it has disappeared), also known as Lexell's comet, occupies a special place in the history of astronomy because of its peculiar orbit, which posed a serious problem of celestial mechanics at that time. The laws of mechanics and Newton's theory of gravitation had been applied mainly to the orbits of the planets and their moons, but not actually to predict and explain the imminent disappearance of a comet. Like for instance the comets Halley and Encke, Lexell's comet was named after the astronomer who computed its orbit, while many other comets are named after the person(s) who first discovered them, cf. [95, 96]. In fact, most comets are known only by a code designation.

On 1 July 1770 the comet passed the Earth at a distance of only 2.3 million kilometres, equivalent to 0.0146 AUs, which is the closest encounter of a comet in recorded history (asteroids may have been closer). At that point the diameter of its coma was estimated to be 2.4°, i.e. four times the apparent diameter of the Sun or the Moon. In the middle of July it was already too close to the Sun to be observed, but in August it reappeared while receding from the Sun. At that point it exhibited a faint tail, 1° long, meaning that it was definitely a comet with an active nucleus, and not an asteroid. During the month of September it finally disappeared (Fig. 5.6).

Without delay, astronomers around the world—Lexell of course among them—started to compile observations and compute its orbit. It was soon understood that a normal parabolic orbit could not fit the observations, and that its orbit was necessarily elliptical and of a short period. Long and tedious calculations by many astronomers confirmed that its period had to be approximately 5.58 years. This was a strange and unexpected result. No previously known comet had such a short period. The comet should have returned to its perihelion in 1776 but could not be observed because of sunlight. At its following supposed appearance in 1781, the astronomers could not detect it. What had happened?

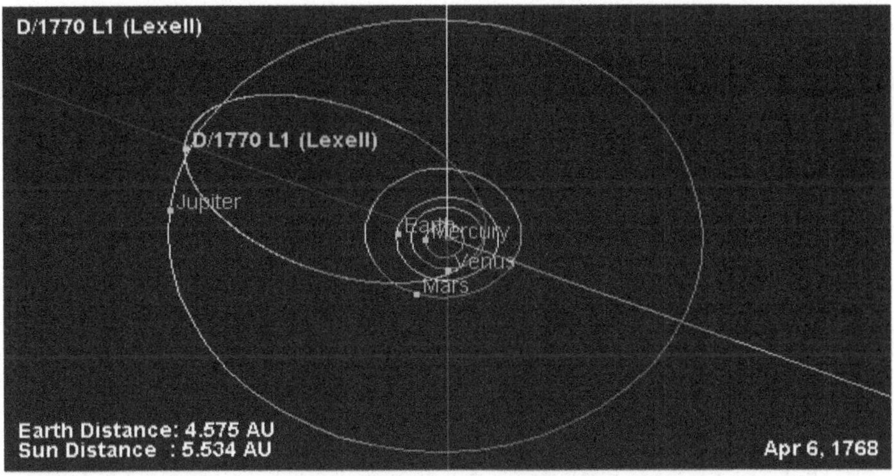

Fig. 5.5 Orbit and position of Lexell's comet at the moment when it was captured by the action of Jupiter's gravity and sent on a short-period orbit. Screen image of the OrbitViewer-program available at the JPL-NASA website (with consent of Osamu Ajiki, Astroarts Inc., and Ron Baalke, JPL. OrbitWiewer software by Osamu Ajiki, Astroarts Inc, and Ron Baalke, Jet Propulsion Lab., available at: http://neo.jpl.nasa.gov/orbits/)

By means of painstaking calculations which lasted several years ([Lexell 49], [Lexell 55] and [Lexell 56]), Lexell came to realise that through its close encounter with Jupiter in 1767, the comet had been thrown into a temporary orbit of 5.58 years. Lexell also showed that, after completing two revolutions, in its 1779 re-encounter with Jupiter it would probably be ejected into a different orbit. Observations proved that Lexell was right: the comet has probably not been sighted since. It is now orbiting the Sun somewhere in outer space, where it escapes our observation.[38] Whether it will appear again is uncertain; neither does it seem possible to prove that the comet "*d* 1889" was actually identical with Lexell's comet, as suspected by the American astronomer Seth Chandler [27].

In a public assembly of the Petersburg Academy of Sciences held on 24 October 1778, Lexell presented the results of his computations relevant to the motion of the comet of 1770 entitled *Réflexions sur le temps périodique des comètes en général, et principalement sur celui de la comète observée en 1770* [Lexell 64]. From his researches he concluded that in 1767, before arriving near Jupiter, the comet had another (longer) period, but that the gravity of Jupiter had thrown it into a different orbit with a short period (5.58 years). The period of Jupiter is about 12 years, which is nearly twice that of the comet. When the comet in 1779 would again approach Jupiter and enter its gravitational field, whose strength at that distance from the Sun

[38]Lexell's comet belongs to the Jupiter family comets, such as comet Churyumov-Gerasimenko, which is the target of the European Space Agency Rosetta mission. (Personal communication: Professor Anny-Chantal Levasseur-Regourd).

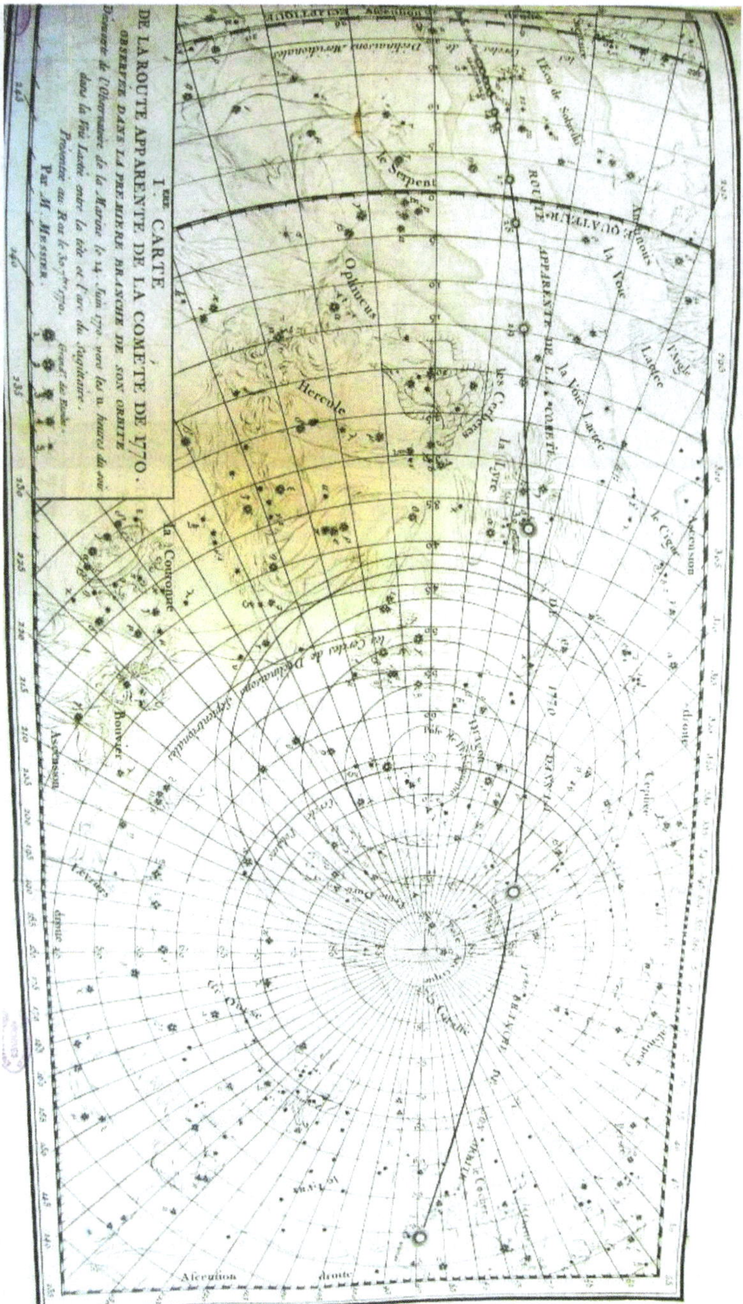

Fig. 5.6 The apparent route of Lexell's comet as recorded by its discoverer Charles Messier in [Lexell 49] (Photo: the author)

surpasses many hundred times that of the Sun, it would be thrown out into another unknown trajectory.[39] Thus, the orbit of the comet changed at least two times due to its close interaction with Jupiter. The mathematical treatment of this problem posed a lot of difficulties, as it involved two attracting bodies at the same time [166], namely the Sun and Jupiter; and especially considering that the mass of Jupiter in relation to the Sun's was not accurately known. The historian of astronomy J. H. Mädler [114, pp. 473, 493] calls this Lexell's finest achievement. The general solution of this complicated problem, where the Sun and Jupiter alternately, and in fact simultaneously, acted as central bodies, has become reasonably manageable only in the age of computers (Fig. 5.7).

In the *Journal des Sçavans* of 1780 (p. 689), the renowned Joseph Jérôme de Lalande praised Lexell's work as the most detailed of all the reports that had been published on the strange comet, saying: "[T]here are shown the comparisons of calculations with observations and the testing of different assumptions or calculations, by means of which Mr Lexell has ensured himself that no other orbit could represent the observations as well [as the one with the period of 5.58 years]".[40] Lalande cited Lexell's letter, but did not mention Lexell's conclusions about the likely disappearance of the comet. This prompted Lexell to reproach Lalande in a letter, which has not been found, but which Lexell refers to in one of his reports to Wargentin from Paris in 1781.

It was only around 1800 that Lalande suggested to one of his young students, Johann Karl Burckhardt (1773–1825), that he should perform the calculations anew. In 1806, using the methods of Laplace, Burckhardt came to the same conclusion as Lexell. Later on, Urbain Le Verrier (1811–1877), the astronomer who discovered Neptune [58], in his researches of 1844, 1848 and 1857, confirmed Lexell's results, adding that it was impossible for the comet to have been captured by Jupiter and become its satellite at the second close encounter of the two celestial bodies. The researches made by both Laplace and Le Verrier indeed suggest that the comet had been passing not far away from Jupiter both in 1767 and in 1779, although it has never been orbiting around it (as was the case, by 1990–1994, for comet

[39]Lexell writes: "À cause de l'action de Jupiter, il pourra même devenir douteux, si, à l'avenir, on a la satisfaction d'observer la comète dans la même orbite qu'elle parcouroit en 1770; car si les élémens que nous venons d'établir étoient tout à fait exacts, la prochaine conjonction de Jupiter avec la comète se feroit l'an 1779 le 23 d'Août à 12 heures à peu près, la longitude de ces astres étant alors 6s.3° 34′. Or le calcul prouve, que pour cette longitude, la distance de la comète à Jupiter est à peu près la 491eme partie de sa distance au Soleil, d'où il s'en suit que l'action de Jupiter surpassera celle du Soleil 224 fois, ce qui ne manqueroit pas de produire un changement total dans le mouvement de la comète. Quoiqu'on ne puisse pas compter sur la plus scrupuleuse exactitude de cette conclusion, vû que des petites variations dans les élémens peuvent donner des résultats très différens; néanmoins toutes les circonstances bien considérées, on peut soutenir, qu'au moins dans l'une ou l'autre des conjonctions de Jupiter avec la comète du 1767 ou 1779, l'orbite de la comète a dû souffrir des changemens sensibles, par l'action de Jupiter."

[40][O]n y voit les comparaisons des calculs avec des observations & les différentes suppositions ou les essais de calculs par lesquels M. Lexell s'est assuré que toute autre orbite ne représenteroit pas aussi bien les observations.

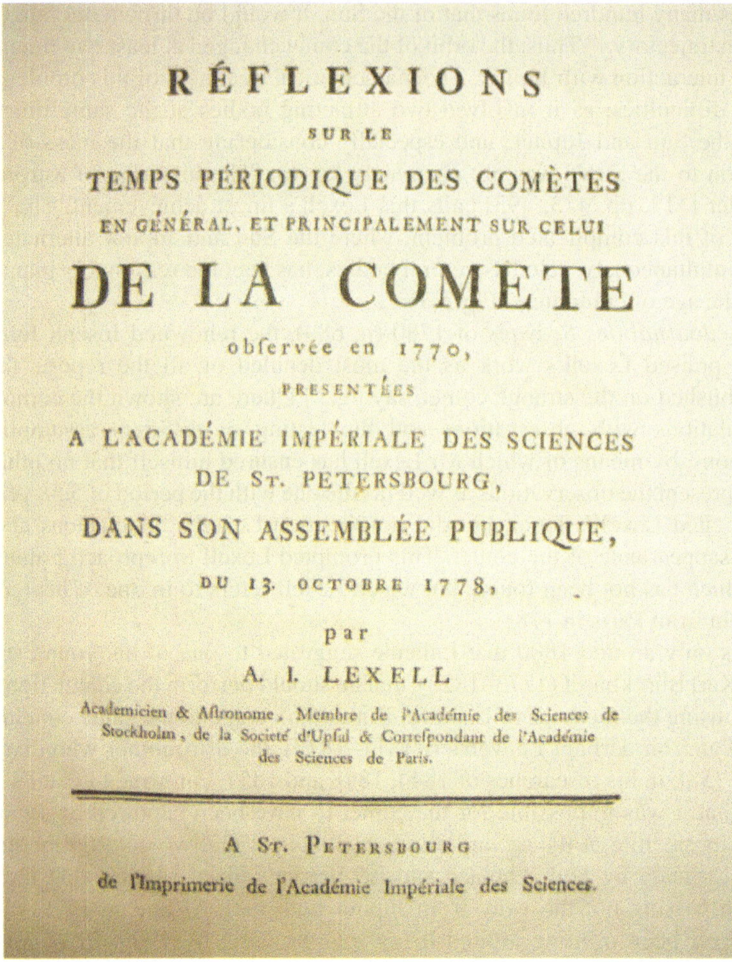
RÉFLEXIONS

SUR LE

TEMPS PÉRIODIQUE DES COMÈTES

EN GÉNÉRAL, ET PRINCIPALEMENT SUR CELUI

DE LA COMETE

obſervée en 1770,

PRESENTÉES

A L'ACADÉMIE IMPÉRIALE DES SCIENCES

DE St. PETERSBOURG,

DANS SON ASSEMBLÉE PUBLIQUE,

DU 13. OCTOBRE 1778,

par

A. I. LEXELL

Academicien & Aſtronome, Membre de l'Académie des Sciences de Stockholm, de la Societé d'Upſal & Correſpondant de l'Académie des Sciences de Paris.

A St. Petersbourg

de l'Imprimerie de l'Académie Impériale des Sciences.

Fig. 5.7 Title page of Lexell's *opus maius* concerning the comet of 1770 (Helsinki University Library. Photo: the author)

Shoemaker-Levy 9). Gravitational perturbations may have significantly modified its orbit, with the comet being moved into a hyperbolic orbit or having too large a perihelion distance to be again observable. Positional observations in the 1770s were not accurate enough to allow a precise determination of its orbit and thus of its further development [82, pp. 447–451]. One of the problems Lexell had to tackle was the interpretation of the observed positions in a statistically correct way, when appropriate statistical methods did not yet exist.[41] In practice, he had to make

[41] As in the processing of the observations of the Venus transit, the method of least squares was not available.

numerous calculations with different initial parameters to arrive at the most likely solution. For an overview of Lexell's celestial mechanics we refer the reader to Chap. 8.

With Euler before him, Lexell was among the first to come to conclusions on the size and mass of the comet.[42] From the fact that the satellites of Jupiter did not seem to be in the least affected by the close interaction with the comet he concluded that its mass (and probably also its size) must be comparably small. In the light of present-day knowledge it is remarkable that Lexell's theory of the interaction of Jupiter anticipates the method which is used in our times to accelerate space vessels and satellites, namely to use the gravitational field of large planets as a "swing-by", i.e. as a kind of gravitational slingshot.

5.6 The New Planet: King George's Star

As will be seen in Chap. 9, while staying in London in 1781, Lexell was one of the first to know of Herschel's discovery of a new celestial object, which was initially believed to be a distant comet.

Friedrich Wilhelm Herschel (1738–1822), known also by the name William Herschel, was a musician and astronomer from Hanover (then connected with Britain through the Royal House of Hanover), who had moved to England in 1757. There, besides composing symphonies, he developed a particular talent in observational astronomy. He became famous for his discovery of the planet Uranus, which he named *Georgium sidus*, King George's star. He also discovered two of its major moons (Titania and Oberon) as well as two moons of Saturn. Later on, he was the first person to discover the existence of infra-red radiation. His younger sister Caroline Lucretia Herschel contributed significantly to his astronomical researches.

On the night of 13 March 1781, Herschel noticed a curious object while surveying the part of the night sky around the constellation of *Gemini* with his seven foot reflector [58]. The object had a "magnifiable disc", contrary to the fixed stars, and a movement of its own with respect to the fixed stars. Accordingly, Herschel communicated the discovery of a new comet to the Royal Society in a letter on 4 April 1781. In his observation journal Herschel had noted on "Tuesday, 13 March 1781" the following words: "In the quartile near ζ Tauri, the lowest of two is a curious either nebulous star or perhaps a comet".

The discovery of the "comet" was announced on 26 April 1781, and among the first to hear about it was Lexell, who at the time was visiting London. Owing to his skills in theoretical astronomy, Lexell was particularly well prepared for the challenge of determining the orbit of the new object. In fact, based on Herschel's

---

[42] A discussion of the mass of the comet Halley (which had re-appeared in 1759) is found in Johann Albrecht Euler's prize-winning essay entitled *Meditationes de perturbatione motus cometarum ab attractione planetarum orta*, St. Petersburg, 1762.

and Maskelyne's[43] observations, he quickly concluded that the orbit conformed approximately to a circle, in other words, it moved like a distant planet (cf. Sect. 9.6). At the time of his return to Saint Petersburg in the autumn of 1781, there was more data available on the new slowly moving object obtained from Wargentin, Johann Elert Bode (1747–1826) and Christian Mayer, but not enough to prove convincingly that the object moved in a circular orbit and not in a parabolic one. Instead of waiting for a couple of years to attain conclusive proof, Lexell began to look for possible earlier sightings of the star. Indeed, the Berlin astronomer Johann Elert Bode had noticed that an object in the constellation of the *Pisces*, which had been observed in Göttingen by Tobias Mayer, was no longer there, and it was not to be found in the catalogue of John Flamsteed either.[44] Lexell interpreted this as an earlier sighting of the object, and using this data he calculated the exact orbit, which also proved to be an ellipse. Here is Lexell's original account [Lexell 82], [Lexell 101]:

> To complete our proof, it only remains for us to prove that all elliptical orbits whose eccentricity is somewhat larger cannot fit in with the motion of the new planet; but as for the parabolic orbits which approximate the observations best one only finds errors of three minutes or less. Similarly one may imagine that the errors for elliptical orbits would be even smaller and for this reason we need observations over several years to determine the true value of the eccentricity. All we can do in the meantime is to exclude successively several kinds of ellipses and continuing this work we will not fail in the end to find the right one which fits the observations. However, there is a way to shorten this process considerably by using the very important observation made by Mr Bode, astronomer in Berlin, who by taking pains to examine several fixed stars in the Zodiac listed in the catalogues, noticed that one of these, which the illustrious Mr Mayer had observed in the year 1756 in Göttingen in the constellation of Pisces, was no longer in the same place where Mayer had seen it. By taking into account the elements of the new planet, it indeed seems probable that it was found on the 25th of September 1756 at that place, where Mayer had observed a star, which is not to be found in the catalogues of Flamsteed or any other known at the time, at least the difference which lies between the calculated and the observed position can be explained in part by the uncertainty of the average distance and also partly by the equation of the centre of the yet unknown planet. Thus, this observation facilitates the research of astronomers on the planet, and one may be confident that the eccentricity will eventually be determined fairly accurately, since the location of the planet for this observation has advanced an angle of more than 100 degrees from that of the observation of 17 March 1782.

Further, noticing that Mars was in the apparent vicinity of the new planet at that time, Lexell was able to estimate the size of the object [Lexell 82], [Lexell 101]:

> ...as the planet in the month of April and May last year happened to be in the vicinity of Mars, I examined which one of the two [stars] was the largest, and I discovered that the diameter of Mars exceeds considerably that of the planet. However, Mars being at this time around its apogee [=the most distant point of the elliptical orbit of the planet from the Earth], its diameter did not exceed 5″ [seconds of arc], from which I conclude that the diameter of the new planet is certainly smaller than 5″, and I even think that it is no greater

[43] Nevil Maskelyne (1732–1811), Astronomer Royal at Greenwich.

[44] The British Astronomer Royal John Flamsteed (1646–1719) had catalogued and designated over 3,000 fixed stars (*Historia coelestis Britannica*, 1725). Apparently Uranus figured among them, erroneously designated as "34 Tauri".

> than 3″. So, conjecturing that this diameter is actually 3″, and because the planet is seen from a distance almost 19 times greater than that of the Sun to the Earth, if the planet was as far the Earth as the Sun, it would extend to an angle of approximately 57″, from which it may be concluded that it exceeds the other planets except Jupiter and Saturn in size and that it is roughly 36 times larger in volume than the Earth.

In conclusion, Lexell computed the volume of the object to be about 36 times that of the Earth and the semi-major axis of its orbit as 19 times that of the Earth. The current figures are 63 times the volume of the Earth and 18.4–20 AUs, respectively. Consequently, the time of one revolution of the object around the Sun was estimated to be between 82 and 83 years.

The result agreed fairly well with the semi-empirical Titius-Bode law,[45] expressing the distances of the planets to the Sun. According to this law the mean radius of an orbit in AUs is

$$r(n) = 0.4 + 0.3 \cdot 2^n,$$

where Mercury corresponds to $n = -\infty$, Venus to $n = 0$, the Earth to $n = 1$, Mars to $n = 2$, the asteroid belt (Ceres) to $n = 3$, Jupiter to $n = 4$ and Saturn to $n = 5$. For each planet in turn, the distance is almost double of that of the preceding one. In fact, Uranus (corresponding to $n = 6$) was found at a mean distance of 19.2 AU, when the Titius-Bode law gives $r(6) = 19.6$ AUs. In spite of the close agreement, Lexell does not refer to the law in his report, neither to any hypotheses on the size of the solar system.

From the movement of the new planet Lexell further surmised that there might still exist other more distant planets and that the size of the solar system could extend to several hundred Earth-Sun distances, as suggested by the orbits of the comets. However, contrary to what is sometimes claimed, Lexell in no way anticipated, or even hinted at, the discovery of the planet Neptune (as did later Friedrich Wilhelm Bessel), although he recognised the possibility of finding other planets further away in space.

[45]Named after Johann Daniel Titius (1729–1796) and Johann Elert Bode. Titius formulated the rule in 1766, Bode in 1772. The law has not been proven theoretically.

As to the name of the new planet, he finally writes (op cit):

> ...Mr Herschel, who as a discoverer of the new object has incontestably the right to name it, has chosen the name of Georgium Sidus, and Mr Bode in Berlin has proposed to name it Uranos, which concurs fairly well with its position in the sky, being more distant than Saturn. The astronomer Mr Prosperin[46] in Uppsala has proposed the name Neptune. Although the astronomers are of course free to give it a name which they deem suitable, it must be admitted, that the name of Georgium Sidus is not quite convenient, as the word Sidus refers rather to a fixed star than to a planet; For this reason the new planet could be named the Neptune of George III, or the Neptune of Great Britain, to eternalise the memory of the great efforts of the English during the last few years. A very competent mathematician and astronomer of Dresden Mr Koehler[47] has proposed to give it the sign of *Platina del Pinto* (Platinum), which could be expressed by one of the characters ... [variations of the symbol ♅] ... which seems quite convenient.

As it was, the name Uranus and the symbol of Platinum were quickly accepted by astronomers worldwide, except in England, where the *Nautical Almanac Office* insisted on the name "Georgian planet" until 1850. Lexell's role in the discovery of the new planet has been abundantly discussed in scientific literature. We may mention, in particular, that the philosopher of science Thomas Kuhn in his somewhat controversial book *Structure of scientific revolutions* (1962) elucidated scientific paradigms (Kuhn's term) in the process of scientific discovery by mentioning Lexell as the first one to recognise the possibility that the newly discovered celestial object was in fact a planet, as it clearly moved in a nearly circular orbit around the Sun. So, with good reason one may ask whether Lexell in point of fact discovered the planet, or whether it was Herschel or one of the numerous astronomers who had spotted it before but mistaken it for a comet or a fixed star. It is a matter of debate, not for astronomers, but for philosophers.

[46]Erik Prosperin (1739–1803), *Observator Regius*, Melanderhjelm's successor in the Chair of Astronomy at Uppsala and a specialist in orbit calculations [127].

[47]Johann Gottfried Koehler (1745–1801), German astronomer, inspector of the "Mathematisch-Physicalischer Salon" (the Royal Instrument Cabinet) in Dresden.

Chapter 6
Professional Relations and Correspondence

Scientific contacts between the Imperial Academy of Sciences of Saint Petersburg and other learned institutions in Europe were primarily built on personal relations. Lexell's role in this context was noteworthy owing to his network of academic contacts in Sweden. His most notable correspondents were Pehr Wargentin, to whom he wrote 112 long and intimate letters in 1768–1783, Carl Linnaeus in Uppsala and Anders Planman in Åbo. Outside Sweden, Johann III Bernoulli in Berlin seems to have been his principal contact. All these men played an important role in Lexell's life and career, but for rather different reasons.

The letters written by Lexell reveal a highly sensitive personality, but at the same time an outspoken, remarkably fearless and self-confident academician. His opinions of his contemporary scientists and colleagues, although biased and sometimes even inconsiderate, reflect underlying prejudices and tensions between the scholars. His correspondence is precious material for a biographer and valuable also in connection with the sociology of science.

6.1 Pehr Wilhelm Wargentin

Pehr Wilhelm Wargentin had a special place in Lexell's professional relations with his home country. Wargentin was born on 11 September (o.s.) 1717 in Sunne, in the province of Jämtland, Sweden [125]. His grandfather Wilhelm Wargentin had married Magdalena Wittfooth, a sister of the grandfather of Anders Johan Lexell's mother; their son—Pehr Wilhelm Wargentin's father—Wilhelm Wargentin the younger was born and brought up in Åbo, where he was ordained as a priest. Subsequently he served 3 years as a vicar in Jomala parish on the Åland Islands. During the ravages of the "Great Northern War", Wilhelm Wargentin's home had been destroyed and his first wife had died. He then fled to Sweden, where he remarried and had a son, Pehr Wilhelm. The young Pehr Wilhelm, showing interest in astronomy at an early age, was sent to Uppsala, where he was instructed by

J.C.-E. Stén, *A Comet of the Enlightenment*, Vita Mathematica 17,
DOI 10.1007/978-3-319-00618-5_6,

Anders Celsius, Samuel Klingenstierna and Mårten Strömer. As a subject for Wargentin's master's thesis, Celsius had suggested the complicated motion of the four satellites of Jupiter, which, as has been indicated, was an important topic at the time because of its connection with the longitude problem. Wargentin's semi-empirical tables of the eclipses of the Jovian satellites made him internationally famous; and he continued to perfect them throughout his life.

In 1749, the already renowned Wargentin was appointed Permanent Secretary of the Royal Academy of Sciences in Stockholm, stationed from 1753 onward in the Stockholm Observatory. Because of his official position as Secretary and also owing to his pleasant and suave character, Wargentin became a central figure in Swedish astronomy and science in general. He had wide interests, ranging from astronomy to meteorological observations as well as population statistics, in which he was a pioneer [25]. Though as a theoretician he was second-rate, owing to his lack of a proper mathematical foundation, he diligently collected data and willingly assisted his friends, thereby creating an exceptionally large scientific network [125]. From 1760 onwards he was a foreign member of the Petersburg Academy of Sciences, receiving formal remuneration for his active cooperation. Being also a corresponding member of the Paris Academy of Sciences, Wargentin was engaged in an extensive cooperation with his French colleagues [127]. On a French initiative in 1751, he launched the Swedish campaign to observe Mars,[1] Venus and the Moon, and in the 1760s he was charged with organising the Venus transit observations. It is remarkable that Wargentin, despite his international reputation, never travelled outside Sweden. The Swedish historian of science Sten Lindroth summarised Wargentin's influence in these terms (translated and somewhat abbreviated from the original [100, pp. 49–51]): Wargentin's workload grew endlessly, but he never seemed pressed for time. As a Secretary of the Academy, he organised the meetings and took the minutes, dealt with the growing amount of correspondence, edited the Academy's proceedings (the *Handlingar*, for which he wrote 71 scientific articles) and the almanacs, negotiated with the authorities and the royal court, supervised the academic library and the observatory, performed astronomical observations and so on. He was the very soul of the Academy; surrounded by illustrious scientists, he was the incarnation of Swedish natural science at its peak. Wise and authoritative, well-liked by everyone, he even found the time to help and encourage independent researchers and amateurs in the country. Wargentin was Linnaeus's intimate friend, and many were the lecturers and priests he advised and whose lamentations he listened to like a father confessor. He was the administrator and father figure who rooted the Academy into Swedish society (Fig. 6.1).

It is clear that Lexell had reason to feel gratitude towards Wargentin, without whom he possibly would not have been employed in Saint Petersburg. The 112 letters Lexell wrote to Wargentin are a testament to his profound respect for him not

---

[1]Based on his own and Abbé Nicolas Louis de La Caille's (1713–1762) simultaneous observations of Mars on the Cape of Good Hope, Wargentin determined the solar parallax of approximately 10 arc seconds, cf. KVAH, 1756, pp. 60–72.

Fig. 6.1 Carl Fredrich Brander's portrait of Pehr Wilhelm Wargentin (KVAC, Royal Swedish Academy of Sciences, Stockholm. Published with permission)

only professionally, but also personally and intimately. In a letter to Anders Planman in Åbo dated 4 December 1771, having praised La Caille as the most dependable and truthful of the French observing astronomers,[2] Lexell expressed his appreciation of Wargentin in the following words:

> In Sweden, we have one man who could be as useful for astronomy [as de la Caille] and as purehearted. Here, you will without doubt recognise Secretary Wargentin.[3]

Planman forwarded the letter to Wargentin and obtained the following response in a letter dated 15 May 1772 (HUB Ms. Coll. 171):

---

[2]Abbé de La Caille was also considered one of the most industrious astronomers of his times. On his journey to the Cape between 1750–1754, he catalogued thousands of stars of the southern hemisphere.

[3]Wi äga i Swerige en man, som med honom kan tänkas i bredd lika nyttig för astronomien och af lika hederligt hierta. Herr Professorn känner wäl straxt häruti igen Sec: Wargentin (HUB Ms. Coll. 171).

> I apologise for my delay in returning the letter from Lexell, which has amused me a lot. The praise he heaps upon me was unexpected and undeserved. I am one of those who shine somewhat at a distance, but cannot be touched at a closer distance. Unfortunately I neglected to learn the solid foundation when it was the time. Thank God, I have nevertheless managed to pull my weight and do some good. Good will and a steady health have helped me there.

Lexell's letters to Wargentin concerned not only scientific matters. In his difficult choices and problems, Lexell usually asked him for advice and moral support. For instance on 5 August 1771 he writes as follows:

> I presume that you, Herr Secretary, have received our academic work of the last year. I have to admit that, in spite of all caution in the calculations of the solar parallax, some faults have slipped in as for example one in the calculation of the observations in Orsk. It comforts me that since the readership knows that it is I who have executed the calculations, all such faults have to be blamed on me and not on old Euler; because by no means would I like to see such a great man to be held responsible for my negligence. For my own apology I can only plead that *Opere in longo fas est obrepere somnum*.[4]
> With the mail I got yesterday from Finland I have received the deplorable news that Professor Wallenius[5] has become the victim of serious mania, and that there was little hope of his recovery. The friend of mine, who told me this, also asked me if I, in the case that some substitute was to be appointed to manage the *Matheseos* Chair, would like to take over this position *pro tempore* for half the salary.
> Given my love of the Fatherland, as well as my present place of living here in Petersburg, I would of course not fail to use any possibility to get an established position at home, but on the other hand I have some serious doubts, which I now wish to present by asking you, Herr Secretary, to express your opinion about them and to give me some good advice as to how to respond to them. The first inconvenience, although perhaps the smallest, is that I would not be able to leave Petersburg without any reproaches of irresolution, when I only recently have been elected ordinary member [of the Academy of Sciences], since it might be considered ungrateful to leave a place so quickly, where one has been shown so much favour and courtesy. The other difficulty is that my engagement here obliges me to announce one year in advance if I want to leave my post at the Academy, whereas in Åbo it seems to be a matter of urgency to appoint somebody to the Chair of Mathematics. If Count Orlov would now be present this difficulty could probably be revoked, but right now it does not seem possible, and in any case I find it difficult to break off my obligation. But the major inconvenience arises with my future situation in Åbo. I° As I could not be appointed to anything more than an acting professor, in the event of Wallenius recovering, I would be deprived of both allowance and job. II° Should his condition not improve, he could still live for many years, and during this time I would have to serve as an acting professor if I did not wish to be transferred to some other Chair, which I perhaps do not see fit to do. III° If he should die, it would be the worst alternative, because then his allowance would go to an older professor, and as there are in Åbo many professors and many more to come in the future, I might be obliged to work 20 years without a salary.
> It really vexes me that this will perhaps be my only chance to return to my Fatherland, and that it will slip through my hands.

In the spring of 1773, by the time when Lexell was received as a foreign member of the Royal Swedish Academy of Sciences, he was already recognised in the learned circles of Sweden and the whole of Europe. Consequently, it became

[4]In a long work sleep will intrude. Horace

[5]Professor Martin Johan Wallenius, Lexell's teacher of mathematics in Åbo, died on 22 March 1773.

a matter of urgency to attract Lexell back from the cosmopolitan city of Saint Petersburg, before it was too late. In fact, it already seemed to be.[6]

Despite Lexell's love for his country, family and friends, as time went by, he started to feel more and more at home in his new environment of inspiring scientists. From this point of view, rural Åbo with its small university had little interest to offer a researcher at his level. He had also received messages from his friends and colleagues in Åbo, especially the Professors Porthan and Pehr Adrian Gadd,[7] indicating that not everybody in Åbo looked forward to his return. Lexell especially was astonished to learn about the unfriendly attitude of his former teacher, Professor Jacob Gadolin, who apparently saw his return to the university as undesirable. Perhaps jealousy from on the part of his former colleagues was not so unexpected, in view of the reputation he had established for himself by his work in Saint Petersburg.

Lexell's original intention was to stay in Saint Petersburg for no longer than it was necessary for him to obtain a comfortable position in his native country. Thus, led by his sense of duty he applied for the Chair of Mathematics in Åbo, which was announced as vacant after the death of Wallenius, and not surprisingly, from seven applicants he was selected and appointed to the chair in 1775. However, being at that time already deeply involved with numerous projects, and feeling a strong moral obligation to finish his commissions for the Imperial Academy of Sciences, he had included in his application a *sine qua non* condition of a temporary adjournment of 1 year, which was granted by the King and twice extended until 1780. During all this time of vacancy, Lexell offered half of his salary from Åbo to his self-elected deputy in Åbo, Johan Henrik Lindquist,[8] the other half to be used for the acquisition of modern astronomical instruments for the university. Moreover, he made considerable donations to the University Library in Åbo [165]. The salary for his chair in Åbo was less than half of what he received in Saint Petersburg, but on the other hand, he knew that living in the Russian metropolis was equally more expensive. This special arrangement was tailored for Lexell exclusively and shows how eagerly the government tried to entice him back to Sweden. Although the arrangement was proposed by Lexell himself, he felt more and more uncomfortable

[6]Wargentin's position in the issue can be judged only from Lexell's letters, as his replies are for the most part lost. Actually, it seems that Wargentin could well understand and support Lexell's point of view.

[7]Pehr Adrian Gadd (1727–1797), Professor of Chemistry at Åbo, where he became known for his research in agriculture and economy.

[8]Johan Henrik Lindquist, born in 1743 in Nystad (in Finnish: Uusikaupunki) [88,129] was a student of M. J. Wallenius. Before being appointed lecturer in mathematics in Åbo, he had lectured in astronomy in the fortress of Sveaborg in 1770. Among his early works may be mentioned *De integratione fluxionum formae*: (sin. Z)^m.(cos. Z)^n dZ (1769), and *De motu proiectilium in aëre* (1770). He was *pro tempore* professor in 1770–1781 and full Professor of Mathematics in 1781 as well as Rector of the Academy in 1790–1791. He presided over and supervised 24 theses in mathematics and physics and was elected a member of the KVA in 1785. Lindquist died in Åbo in 1798.

with it and finally, in 1780, he abandoned once and for all his plan to return to Åbo. The reasons for this will be discussed in Chap. 9.

6.2 The Swedish Mathematicians

Although Lexell had considered a position in the Uppsala Academy at least three times, and the only time he actually applied for such a position he was clearly defeated, as he saw it, by two less qualified applicants, he tried his best to stay on good terms with his Swedish colleagues, Daniel Melander (to become Melanderhjelm), Fredrik Mallet and Erik Prosperin. However, only two letters by Lexell to these men have been recovered, and both of them are addressed to Fredrik Mallet, Royal observer at the Uppsala astronomical Observatory in 1757–1773 and Professor of Mathematics at Uppsala university in 1773–1794 [126]. Mallet was born in Stockholm in 1728 as the son of a tobacco manufacturer of French Huguenot origin. Mallet's teachers at Uppsala, Professors Samuel Klingenstierna and Mårten Strömer, recognised his talent in mathematics. In 1752, he defended his thesis *pro gradu magisterii* entitled *De Mercurii theoria*, concerning the orbit of the planet Mercury. He spent the years 1754–1757 on a *grand tour* in England and France, and during his period in London he acted as an intermediary between the famous telescope manufacturer John Dollond and Klingenstierna, who were collaborating on the development of an achromatic telescope [30, 128]. On his return to Sweden, Mallet was appointed *Observator Regius* at the Uppsala Observatory. However, the position was underpaid and Mallet, who was a hypochondriac, as well as depressive and in poor health throughout his life, suffered deeply as a result of the situation. Apparently unaware of Mallet's agony, Lexell wrote a letter dated 25 January 1771 (UUB, A596fol; in Swedish):

> Honourable Herr Observator
> Since the correspondence which I had the honour of maintaining with you, Herr Observator, was interrupted a few years ago partly because of your travel to Lapland,[9] partly because of my move to Petersburg, I request your permission to re-commence it now.
> Secretary Wargentin has without doubt told you about my activities here in Petersburg; thus, I do not need to repeat them. Though I have to mention that since last September I have been working with Professor Euler on his new theory of the Moon, a subject he says he is considering for at least the twentieth time. His ambition is to bring the theory to the highest possible certainty and accuracy, and thereupon calculate new lunar tables. I have to admit that if my journey to Russia has not been useful otherwise, it has certainly made me at home in astronomical calculation, which I was not familiar with when I arrived. Adjunct Krafft, who has returned in November from his astronomical expedition, is now alternating with me on this mentioned work, which is the more pleasant for me as it gives me some freedom to do other things.

[9]Fredrik Mallet had travelled to Pello, Lapland, to observe the transit of Venus in 1769. His observations failed due to bad weather [126].

Secretary Wargentin has some time ago assured me that you, Herr Observator, would be willing to calculate the observations on the solar eclipse of 1769; I would be delighted to study your results. I have also calculated that solar eclipse using two different methods, the former is Professor Euler's and is given in the IInd part of the Commentaries for 1769. I have not yet arranged the calculations using the latter method, but I will do it at the earliest possible opportunity, and they will be printed in the Commentaries for 1770. You have without doubt also been occupied with calculations of the solar parallax. I cannot understand la Lande's procedure when, by comparing Abbé Chappe's[10] and Father Hell's observations, he managed to deduce the parallax as 8.9″, I for my part cannot make it higher than 8.7″. I long to hear your opinion.
In the Commentaries of the last year, a treatise of mine will be published *de criteriis integrabilitatis formularum differentialium*, for which I found the inspiration in the following remarkable theorem. If, when $\mathrm{d}y = p\,\mathrm{d}x$, $\mathrm{d}p = q\,\mathrm{d}x$, $\mathrm{d}q = r\mathrm{d}x$, etc., V is a function of x, y, p, q, etc., of such a kind, that for $\mathrm{d}V = M\,\mathrm{d}x + N\,\mathrm{d}y + P\,\mathrm{d}p + Q\,\mathrm{d}q$ etc. one has

$$N - \frac{\mathrm{d}P}{\mathrm{d}x} + \frac{\mathrm{d}^2Q}{\mathrm{d}x^2} - \frac{\mathrm{d}^3R}{\mathrm{d}x^3} + \text{etc.} = 0,$$

then the differential formula $V\,\mathrm{d}x$ will be integrable. Professor Euler discovered this theorem long ago and proved it by means of the elements of variational calculus, but as they seem less suitable here, the idea of searching for another proof which only relies on the elements of differential calculus came to my mind. Having found this proof it shed some more light on the same subject, so that it motivated me to discover all the characteristic criteria by means of which one may immediately determine, for any differential formula whatsoever, whether it is integrable.
Since you, Herr Observator, have been willing to become responsible for the globes, which have been ordered from the engraver Åkerman[11] for the Petersburg Academy, may I now kindly remind you of them so that they are all ready next spring and transported here by the first possible ship. I humbly ask you to pay my respects to Professor Melander and I am etc.

Although written in a friendly tone, the letter apparently did not elicit a reply; thus, in a letter to Wargentin, Lexell indicated that he suspected that Mallet had been taken ill. Be that as it may, in the second and apparently the last letter to Mallet, which is undated and incomplete, Lexell refers to an article by Mallet in the *Kongliga Vetenskaps Academiens Handlingar* entitled "Remark in response to d'Alembert on the diffraction of light rays".[12] In this article, Mallet refutes d'Alembert's objections regarding Klingenstierna's demonstration against the Newtonian law of refraction [128]. These objections, which d'Alembert had published in his *Opuscules mathématiques* (Vol. III, pp. 359ff.), concerned the question whether the chromatic dispersion of light in refraction could be

[10]Jean-Baptiste Chappe d'Auteroche (1728–1769), ecclesiastic and astronomer, known for his observations of the 1761 Venus transit in Tobolsk, Siberia. He observed the transit of 1769 at the tip of the California peninsula but was infected by fever and died on his observation site [136].

[11]Anders Åkerman (ca. 1721–1778), engraver and maker of globes in Uppsala [139]. His globes enjoyed much popularity in the Petersburg Academy of Sciences and the Russian administration, and Lexell was acting as the principal agent for these acquisitions through Wargentin.

[12]*Anmärkning, emot Herr d'Alembert, om ljus-strålars spridning*, KVAH, 1771, pp. 138–152.

eliminated or not. Klingenstierna and Euler had independently of each other discovered that, contrary to what Newton had maintained, it was indeed possible to circumvent chromatic aberration in refracting telescopes by specially designed compound lenses containing materials of different indices of refraction. In the letter, Lexell details some errors which Mallet had committed in his demonstrations, but in the main issue he took Mallet's and Klingenstierna's side in the debate against d'Alembert. Apparently, this letter to Mallet did not elicit a reply either.

When the vacancy of the Chair of Mathematics in Uppsala was announced in 1772, the applicants were, besides Lexell, the observer Mallet, the Adjunct of Mathematics and Physics Erik Prosperin and the Adjunct Daniel Hallencreutz. Of these four candidates for the position, Mallet was the favourite and besides Lexell the most competent one [126]. However, Mallet was feeling increasingly threatened by the 12-years-younger Lexell: so in a letter to Wargentin dated 22 December 1772 (KVAC, Berg. brevs. Vol. 19, p. 421) he openly solicited Wargentin's support in the election. While he admitted that Lexell deserved to be considered on his merits, it was nevertheless he himself who should be proposed on the grounds of his many years of service at the Uppsala Academy. Conscious that Lexell's international reputation was constantly growing, Mallet's colleague Daniel Melander, Professor of Astronomy in Uppsala, wrote a letter to Wargentin dated 23 February 1773 (KVAC, Berg. brevs. Vol. 18, p. 153), stating his concern that Lexell would be proposed before his favourite colleague Mallet. Moreover, Melanderhjelm asked Wargentin to influence Chancellor Scheffer[13] to Mallet's benefit. Most of all, Melanderhjelm was afraid that Linnaeus "... has been meddling again and has begun to support Lexell", as it is a known fact that Linnaeus was not on very friendly terms with the younger generation of Uppsala mathematicians [126]. In fact, of the letters addressed to Wargentin by Melanderhjelm and Mallet, at least ten were concerned with Lexell and his competence for the position in Uppsala [26].

Eventually, everything went as planned, that is, Mallet won the majority of votes and was appointed to the Chair of Mathematics. Pleased at the outcome, Melanderhjelm wrote to Wargentin on 6 July 1773 (KVAC, Berg. brevs. Vol. 16, p. 457) that he still hoped that Lexell would succeed Duraeus in the Chair of Physics, and that he was certain that mathematics would be in good hands in Uppsala if Lexell was appointed "as the third professor". However, as we now know, Lexell would never again apply for a position in Uppsala.

As shown by the following extract from a letter by Lexell to Wargentin, dated 23 November 1772, Melanderhjelm was indeed right concerning Linnaeus's "meddling":

---

[13]Count Ulrik Scheffer (1716–1799), statesman and military commander, Chancellor of the Royal Academy of Åbo and former Swedish Minister in Paris.

An unexpected piece of news, which Archiater[14] von Linné was pleased to tell me,[15] is that Captain Meldercreutz has applied for resignation [from the Chair of Mathematics], and that he (the Archiater) asks me if I am interested in that position. I may now announce to you, Herr Secretary, that I have sent my application right away, which by now should have been submitted and put on record. I have no idea how this will be received in Uppsala, but if Herr Secretary likes my undertaking, I am satisfied. I have to say, first and foremost, that I have not applied for this post in order to really have it; I would, at least at present, fall much short in this post. And if I, by means of the greatest merits in the world, could obtain the post, I should not covet it if I knew that someone more competent and merited was to apply for it. And in Uppsala the Observator Mallet is in this respect evidently the man for the post. The only purpose of my application is to be assured what my fellow countrymen may think of me, as there has been until now much doubt about it. Possibly candidates will also come forward who may cause me to be eliminated from the short list of proposed candidates, but in Uppsala there are only two men who can surpass me, namely the Observator Mallet and Adjunct Prosperin, the rest of the docents at this Academy are *si cetera paria essent* [although in other respects alike] much younger than me. If the favour was shown to me of being proposed for the post, I am quite anxious not to win.
I suppose that, as I have not recommended myself to anybody close to His Majesty the King, it might happen that someone will eventually come to think of me. Nevertheless, to be sure I have to ask you, Herr Secretary, to convince either His Excellency the Chancellor Scheffer or some of the other gentlemen nearest to His Majesty the King, that I have applied only in order to be assured of the good-will of my compatriots in the future, in the case that some position in my Fatherland would be vacant, which I could take without blaming myself of being an obstacle to somebody more competent.
I do not believe that the fact that I have not been employed at the Uppsala Academy [the University] can be used against me; that would make a strange contrast to the benefits I have enjoyed abroad. However, using precisely this argument, many more competent men of greater merit have been excluded in Uppsala and many other Swedish places of learning. I have not suffered from any injustice in Uppsala as I have not applied for anything, but I have the clearest evidence that if I had applied for any position whatsoever, I would have lost this position to someone of inferior merit. In 1763 I was planning to apply for the lectureship in astronomy, but when I noticed that this was being made increasingly difficult, I refrained from applying. In 1767, when the position of Adjunct was vacant in Uppsala, I hesitated so long with my application that it finally arrived too late. But by taking account of the thoughts of some gentlemen there, I was convinced that even if my application had arrived in time, I would have been excluded from the list of proposed candidates, and that

---

[14]Honorary title of the Swedish court physician.

[15]Lexell writes to Linnaeus on 7 May 1773: "För den ynnest Wälborne Herr Archiatern behagat hafwa för mig wid voteringen til förslaget af Mathesos Professionen, aflägger jag min ödmiukaste tacksägelse. Huru wida Adjuncten Prosperin för mig war berättigad til andra rummet på förslaget är en sak som icke anstår mig at yttra mig öfwer; uti egen sak måste min tanke altid anses för partisk. Det fägnar mig at syslan nu är bortgifwen och Herr Professor Mallet nämnd, jag skulle wara högst obillig om icke jag erkände honom för den skickeligaste och mäst förtienta dertil. Om han tänker något så när billigt om mig wet jag icke: för min afresa til Petersburg war jag ofta i correspondence med honom, sedan mit wistande här, har jag skrifwit honom twänne gånger til, utan at han wisat mig den höfligheten, at åtminstone låta weta det mina bref framkommit".—In brief, Lexell thanks Linnaeus for voting for his application to the position in Uppsala, saying that Mallet is fit for the job. However, he deplores the fact that Mallet has not deigned to reply to his letters, not even to acknowledge their receipt.

by Wetterquist[16] and Hallencreutz,[17] the merits of which gentlemen I do not wish to ponder, except to say that they were less merited than me. I should only thank God that this was so to be, because otherwise I would perhaps not have thought about Russia, where I have been shown more favour than I perhaps deserve.

6.3 Linnaeus and the Swedish Naturalists

If for anything, eighteenth-century Swedish science is known for its school of natural history founded by Carl Linnaeus (after ennoblement von Linné), considered the founder of binomial nomenclature, taxonomy, and in a sense even ecology (Fig. 6.2). Equally famous were his so-called "apostles", a group of disciples and students who travelled all over the globe to apply Linnaean taxonomy to the *flora* and *fauna* and minerals of the world. Equipped with Linnaeus's *Systema naturae* (1735) and *Philosophia botanica* (1751), the apostles kept in touch with their Master in Uppsala, dispatching to him scientific reports, plants, insects and seeds. The expeditions were risky, and many of the apostles perished of tropical diseases. One of the most celebrated apostles and Linnaeus's favourite was Pehr Kalm (1716–1779), who became Professor of Economy at the Royal Academy of Åbo. Kalm's popular narrative of his expeditions in North America, *En resa till Norra America* (Stockholm, 1753–1761) was translated into German, French, Dutch and English. Peter Forsskål (1732–1763), another talented Finnish-Swedish natural scientist and apostle of Linnaeus, was less fortunate: he died of malaria on an expedition in Yemen.

The five major research expeditions of the Petersburg Academy of Sciences to southern Russia and Siberia, which took place between 1768 and 1775, were commissioned to collect botanical, mineralogical, ethnographical and geographical data, and not least to estimate the potential natural resources of Russia. The main organiser of the expeditions was the German-born natural historian Peter Simon Pallas [171], who himself led an expedition to the environs of Volga, the Southern Urals, Western Siberia and to the Baikal region. The second expedition was led by the Russian botanist and geographer Ivan Lepyokhin (1740–1802), a student of Linnaeus, who travelled to the lower Volga, and then went to Archangel and the White Sea, and finally to Belarus. One of Linnaeus's Swedish disciples, Johan Peter Falck, travelled to Orenburg and Astrakhan. Falck had moved to Saint Petersburg in 1763, where he became Professor of Medicine and Botany at the Medical School (*Collegium Medicum*) and Curator of its medical gardens.[18] Falck's expedition to

[16]Olof Wetterquist (1733–1809), fortification officer, student of Melanderhjelm. He had defended his thesis in astronomy in 1764 [127].

[17]Daniel Hallencreutz (1743–1816), a student of Melanderhjelm, had defended his thesis in astronomy in 1763 [127].

[18]Falck suffered from bad health and depression all his life but was nevertheless a diligent botanist, in constant correspondence with Linnaeus [157].

Fig. 6.2 Alexander Roslin's (1718–1793) portrait of Carl Linnaeus (von Linné), painted ca. 1775, the time when Linnaeus was corresponding with Lexell (KVAC, Royal Swedish Academy of Sciences, Stockholm. Published with permission)

Orenburg was accompanied by the Pomeranian chemist Johann Gottlieb Georgi (1729–1802), who had studied under Linnaeus with Falck. The fourth expedition, to Astrakhan, the Caspian Sea and the Caucasus, was led by Samuel Gottlieb Gmelin, who, like his uncle Johann Georg Gmelin, corresponded with Linnaeus. The second expedition to the same region was led by Johann Anton Güldenstaedt. These expeditions were undertaken in very difficult circumstances: the risk of being captured by violent, rivalling tribes was ever-present. In fact, two of the expeditions met with a tragic end, and in both cases it fell to Lexell to break the sad news to Linnaeus. The main personality here was Johan Peter Falck, whom Lexell had never met, but whom he supported against the attempts of a group of academicians wishing to call him back on the grounds that he had failed his mission. Even after Falck's death by his own hand, Lexell was involved in the quarrel between Laxman and Georgi concerning the responsibility of editing and publishing Falck's manuscripts.

Early in the 1770s, Linnaeus seems to have been more and more concerned about the lack of information from his colleagues in Russia, especially Falck, who was usually a regular correspondent. At the end of 1771, he approached Lexell in a letter, whose exact date is unknown, because the original has disappeared, but which apparently Lexell himself translated into French, so it could be understood by the whole Academy. This first letter (or a fragment of it translated into French) by Linnaeus to Lexell, which is preserved in Saint Petersburg, Russia (АРАН Фонд

1, опись 3, дело № 58, листы 102–103) has been reprinted and commented in [130].[19] Lexell's reply, dated 13 March 1772, is as follows:

> Honourable Herr Archiater and Chevalier
> It is a great honour for me to have been honoured with a letter from you, Herr Archiater, who for a long time have been the foremost among the scholars of our country. I regret that my knowledge of Natural History is so limited, that I am perhaps not able to provide information in this science which is as reliable as it should be. However, I will not fail to report to you from time to time anything which may seem remarkable in the journeys of our physicists [=natural scientists].
> At the time when I had the honour of receiving your letter, Professor Laxman was still in Petersburg.[20] I informed him therefore briefly about your thoughts about the Chinese bush, which has bloomed in the gardens of Saint Petersburg. For some time now he has described the bush and given it a new *genus*, namely that of *Koelreuteria*, thus I do not know whether this opinion of his could have pleased you. On that occasion he also promised to have the honour to write to you, Herr Archiater, before his departure, which he must have forgotten because he is now since Tuesday last, or 28 February old style, departed. As he has not said farewell to his friends, I have not had the opportunity to further remind him of it. Similarly, he told me that he had a small collection of plants, which he felt obliged to introduce to you, and promised to send it to me so that I might be able to prepare for its transportation next spring, but this he has also neglected.[21] Finally he promised to send one specimen of the *Cynoglossus* which Pallas called *Rindera*, but as he did not perform the task, I do not know any other collector in Petersburg who could do it. Otherwise, he was perfectly convinced of the correctness of your thoughts concerning this plant.
> I have communicated to our Imperial Academy the sentiment of respect you, Herr Archiater, have expressed for the remarkable favour, which our illustrious Empress has shown the Sciences and especially natural history. Even if this grace is so precious, that its value cannot be increased through praise, I am convinced that it would considerably please Her Majesty to know that the finest scholars appreciate Her special generosity in a way which it truly deserves. Likewise, it is a great honour for the members of the Academy that their attempts to fulfil Her Majesty's praiseworthy intentions have received your approbation and acclaim. That Gmelin in the Proceedings of Saint Petersburg [i.e. the *Novi Commentarii*] has presented well-known, indeed quite general plants as new ones is something we have come to expect from him. Besides the inconsistency of his mental condition, he seldom writes except when slightly intoxicated. No reports have been received from him in half a year, the last one came from Rest, a small town by the Caspian Sea, from where he wanted to proceed to Tauris or the old Ecbatana,[22] but he intends to return to Astrakhan before the winter. As this has not taken place, we are in complete ignorance whether he has been

[19]A total of seven letters by Lexell to Linnaeus have been preserved. At least three letters were written by Linnaeus to Lexell, of which only the copy of the first one has survived in Saint Petersburg. The letters are part of the collection bought by the British physician and botanist James Edward Smith (1759–1828) from Linnaeus's widow Sara Elisabeth Linnaea (1716–1806). They are now preserved at the Linnaean Society of London and accessible online (linnaeus.c18.net/Letters/).

[20]Erik Laxman was commissioned to travel to Moldavia, Wallachia and the Crimea to establish a Mint [62, 87, 130].

[21]Laxman had in fact delegated this task to his assistant and compatriot Mikael Holmberg. The plants were dispatched in 1772. Suffering from a guilty conscience for having misinformed Linnaeus about the reason for Laxman's delay, Lexell asked Wargentin, in a letter dated 22 June 1772, to correct his error vis-à-vis Linnaeus.

[22]Ancient capital of Parthian kings, situated in what is now Iran.

unlucky in some way, which without doubt would be an irreparable loss for natural history. Gueldenstaedt is still in Grusia [Georgia], whence he will return to Astrakhan no earlier than next winter. Lepechin has stayed in Archangel since autumn, from where he has sent some of his colleagues to the islands of the White Sea, where they will collect whatever may seem remarkable. Falck has been called back and he may return to Petersburg in the autumn. It may be true, that Professor Falck has not quite fulfilled the desire of the Academy by not sending reports from his expedition as often as he has been instructed to do, and which the other travellers have done, but then on the other hand it cannot be denied that the Academy, by calling him back, is guilty of the crime of condemning him without hearing him first, especially as the reasons for this are by no means valid. I° It is believed that he has done nothing. The objection has been raised against this accusation, that if it was really true, it could have been either because he was not able to, or because he did not want to do something. That both these alternatives are groundless can be fully confirmed for many reasons, in particular your opinion of him in his letter. II° [The Academy] believes to have done him a favour by calling him back, when nevertheless he himself a year ago, by way of a less decent submission begged the Academy not to commit this fault. Councillor Müller[23] in Moscow is the one who he has to thank for this token of friendship. Whether he himself [i.e. Müller] is the author of it or whether he has acted on Pallas's recommendation (which Falck should believe), I do not know. The pharmacist Georgi, who was ordered to accompany Falck, has probably also played a part in the game.[24] Falck was reduced to a state of shock on receiving this message and was then taken severely ill, but later he seems to have come round. He will keep his position at the *Collegium Medicum* with the same benefits as before his departure. The journey Pallas has planned to China cannot be undertaken, therefore he intends this year to travel across Siberia partly by himself, partly by sending out his emissaries. He will probably return before the autumn next year.

At present there is no vacant position for König[25] in the Academy, but in a couple of years there will certainly be, since Gmelin and Pallas will soon resign after their return.[26] I do not doubt that the Academy on that occasion will much reflect on your opinion in König's favour.

Quite recently Her Majesty the Empress has commissioned the Academy to deliver an opinion about the offer, which Mr Adanson[27] has made in a letter to Her Majesty partly of his natural collection, partly of his precious description, which he intends to make of it, with 30000 plates engraved in copper appended. He says that he has been encouraged to it by the grace which Her Majesty has shown his compatriot Mr Diderot by buying his library for 30000 Roubles, with a dispensation to use it for the rest of his life. Mr Adanson alleges to have worked with these 30000 drawings no less than 32 years, every day 16 to 18 hours. The collection contains 15000 plants, 5000 insects, and the rest in smaller proportion.

---

[23]Professor Gerhard Friedrich Müller (1705–1783), historian, explorer, and a founding member of the Petersburg Academy of Sciences.

[24]Lexell implicates Pallas, Georgi and Müller as being in a plot to remove Falck from his commission. As early as 1769, the Academy was anxious about the silence surrounding Falck's expedition [141, Vol. II, pp. 708, 721]. However, Falck had replied that he was too ill to return.

[25]Johan Gerhard König (1728–1785), Baltic-German pharmacist, whom Linnaeus recommended for employment at the Petersburg Academy of Sciences.

[26]Actually, Pallas and Gmelin did not resign from their office in Saint Petersburg.

[27]Michel Adanson (1727–1806), French naturalist and explorer. The "description" in question, which Adanson unsuccessfully tried to sell to the Academy, was an illustrated catalogue of his natural collection in the spirit of the *Encyclopédie*. In telling this derogatory anecdote about Adanson, Lexell was undoubtedly aware of the opposition of the French naturalist to the Linnaean taxonomy and Latin nomenclature.

13000 plates have already been drawn, thus he wants to start the publication of this work immediately. The intention is that Her Majesty would be so gracious as to make an advance payment of the expenses, where perhaps 500000 Roubles would not be enough. You may easily surmise the statement by the Academy. Without being a natural historian, I dare say that Mr Adanson must be the greatest charlatan ever, but also a shameless fool, who ventures to make so incomparably strange propositions to such a great and enlightened monarch, that no one with even a modicum of *bon sens* would like to accept them. To illustrate Mr Adanson's character I have to add that some years ago it was planned to invite him here on the basis of his reputation, but that he had presented some unreasonable demands. He was offered a pension of 1200 Roubles a year, but he demanded 4000 Roubles as a small compensation for I do not know what pensions and gratifications he expected from Paris. Moreover, he wanted the Academy to buy his Cabinet for 60000 Livres, so that he could be raised to the nobility and not be obliged to stay more than a year, if the climate in Petersburg did not happen to please him.

Next spring you will receive the XVth volume of our Commentaries, which is ready, as well as a translation into German of Lepechin's journey.[28] The Commentaries of the physical section contain the following articles: I. *Rariorum avium expositio* Gmelin, some of which will be known previously. II. *Descriptiones avium* Lepechin. III. *Descriptio cyprini rutili* Koelreuter.[29] IV. *Descriptio piscis e coregonorum genere* Koelreuter. V. *De leone observat: anatomicae* Wolff. VI. *Novae plantarum species* Laxman. These are *Veronica pinnata, Spiraea altajensis, Dracocephalum altajense, Robinia spinosissima* and *Trifolium Dauricum*. The second part of Pallas's *Journey* is in press, but will not be ready before the autumn.

With deep respect and perfect esteem I remain,
Honourable Herr Archiater and Chevalier,

Your

most humble servant
And: Joh: Lexell

St. Petersburg, $\frac{2}{13}$ March 1772

The letter shows that Lexell, who initially was only an observer in the issue concerning Falck's expedition, was rather keen on giving Pallas, Georgi, and ultimately the "grey eminence" Müller in Moscow, the blame for calling Falck back against his will. In fact, in another letter to Linnaeus, Lexell suggests that Müller's actions are expressions of his jealousy of the Swedes.

In the letter to Linnaeus dated 18 September 1772 Lexell continues his account of the journeys of the Russian explorers:

Professor Pallas continues his journey towards the Chinese border and he has probably already passed Selengin[t]sk, or rather he should now be on his way back. In Siberia, if I remember correctly in Krasnoyarsk, he has found considerable amounts of solid iron, which he will transport in the next suitable winter conditions. More northward in Irkutsk, on a river bank he has discovered a head as well as the front and the rear part of a rhinoceros. Despite the long time it has been preserved in the permafrost, there is still skin and hair on its head. This will be a rather *curieux pièce* in our Natural Cabinet. Professor Gmelin

[28] Ivan Lepyokhin's journal of his expeditions to Volga, Caspian Sea and the Ural Mountains was translated from the Russian and published in two volumes as *Tagebuch der Reise durch verschiedene Provinzen des Russischen Reiches in den Jahren 1768 und 1769* (Altenburg, 1775).

[29] Joseph Gottlieb Kölreuter (1733–1806), botanist, born in Sulz am Neckar, student of J. G. Gmelin in Tübingen and Adjunct of the Petersburg Academy of Sciences in 1756–1761.

has returned happily to Astrakhan having undergone many hardships, and although he has lost a large part of his natural history specimens with the death of most of his company, he has nevertheless retained the descriptions rather complete. He believes he will be able to supplement the missing pieces later on. After his return to Astrakhan he married, and seems inclined to settle down there permanently, which is why he has raised for consideration the possibility of establishing a botanical garden there. Next year he envisages visiting the eastern part of the Caspian Sea, but everybody here deems that this journey will be much more adventurous than the Persian one, and that only sheer luck will permit him to return from there alive and well. Güldenstaedt has all the time sojourned in Georgia with Czar Heraclius, but by now he should have returned to Kizlyar. Next year, his intention is to travel via Kuban to Crimea to investigate the natural objects of this peninsula, whence he will return to Petersburg via the Ukraine. Lepechin is in Archangel, from where he has visited Russian Lapland and the islands of the White Sea. He has dispatched a great deal of rare sea animals. We are expecting from him a skeleton of a whale, which like many of its comrades, was deposited on the shore not far from Kola. All these men except Gmelin are expected back in Petersburg by the end of next year, or at the latest in 1774, the earliest moment when natural history will truly benefit from their investigations. About Professor Laxman we hear only rumours, because he has not written anything himself, not even to his wife but a single time on his journey from Tula. About Falck we do not know whether he is dead or alive. Nevertheless, he should be in Petersburg by now. Although it is true that he has suffered an injustice by being called back, he should on the other hand also have hastened his return, because if he is still absent for one more winter, the consequences might be less agreeable for him. In fact, he has nothing to gain by staying away so long, but worsens only his state of mind more and more.
The XVIth volume of our Commentaries is now ready except for the plates, thus it is not sure whether it can be delivered this year. However, I wish to communicate a list of the dissertations in natural history which it contains. I° *De corde Leonis Auctor* Wolff. II° *Anatome accipenseris Rutheni Auctor* Koelreuter. III° *De hermaphrodito ad sexum virilem pertinente Auctor* Lepechin. There are three of these brothers near Archangel, two of them are married, one is unmarried. The oldest one has had three children. Their parents have had more sons beside them, but nothing of this sort has been noticed. IV° *Salmo leucichthys et cyprinus chalcoides Auctor* Gmelin. V° *Kraschennikovia nov. Plant. genus Auctor* Gueldenstaedt. VI° *Koelreuteria. Auctor* Laxman.

The account continues on 7 May 1773:

Our natural historians are now approaching Petersburg. Professor Lepechin was already here, but now he has travelled for a fee to the newly conquered provinces [in Belarus], to investigate their natural history. Professor Pallas has returned from Siberia and is now in Casan,[30] where he has found Professor Falck, who has been stranded there the whole winter. Only by the greatest efforts did Pallas make him pay his respects to the local governor. It would be best for Falck to hasten his return, but he has proposed going to Georgia to benefit from the warm baths there; which journey he rather should abandon, considering that he could return easily along the Volga to Astrakhan and Kizlyar, but to go to the baths from there is something of an adventure, because such a journey needs an escort, which Falck at the moment has no hope of obtaining. This summer Pallas is going to stay in the Volga area, to return to Petersburg next winter. Gmelin is now said to have started his journey on the east coast of the Caspian Sea. Gueldenstaedt was not in as great a danger as we feared, he is now said to be on his way to Crimea.

[30] Kazan in the present-day Republic of Tatarstan, Russia.

A year later, on 19 May 1774, Lexell announces to Linnaeus the death of one of his most diligent disciples:

> As I have not had the honour of paying my respects to you, Herr Archiater, in a letter for a long time, I regret to say that a rather regrettable event now brings me to write.
> Our poor Falck is dead, and the most dreadful thing is that he has taken his own life.
> I take the liberty of reporting the way this happened, notwithstanding how awful this subject may be. Last spring he had taken the decision to go to the warm baths in Grusia [=Georgia], which plan he realised last summer with such a good result for his health that he returned from there quite happy and sane. But on his way back he collapsed again in Casan around the time of New Year and from then, until the beginning of April, he ate nothing. All this time he suffered from haemorrhoidal pains, and one week before his death, he was affected by a fever. When the pharmacist Georgi arrived in Casan in the end of March he [i.e. Falck] was found totally dehydrated, only skin and bone was left, and his eyes were glowing wildly, although his domestics confirmed that it had been going on for a long while. At that point Falck had already started to avoid any company, or to put it more precisely, nobody came to him, as it was clear that it only made him anxious. Nevertheless, he was able to tolerate Georgi. On 31 March Georgi came to him for the third time, when he found Falck locked in, but when Georgi had identified himself, [Falck] had engaged himself in a conversation with Georgi and little by little become happier, and started talking with much reason, clarity and calmness on a number of issues, so that Georgi was not able to notice anything suspicious. After Georgi had left, [Falck] started walking back and forth on the floor, urging that his hunter should go to rest, which he also did; on his return at 4 o'clock in the morning, he [the hunter] found [Falck] still alive. But half an hour later, when the maid brought in the tea water, she found him fallen down lifeless in his bed. Underneath him was found a razor knife by means of which he first had cut an opening in the throat, and a pistol by means of which he had shot himself dead, the shot had entered the throat and exited through the neck; next to him there was a pound of gunpowder, some of which he probably had used for loading the gun. From the pillows a place could be seen in the bed where he had hidden the pistol, thus he had probably been contemplating this dreadful deed for a long time. That was miserable Falck's tragic end.
> Knowing that you, Herr Archiater, have tender feelings for him, I thought that I could keep secret neither this event, nor the surrounding circumstances, although I think they might considerably affect you, Honourable Herr Archiater, when they have even affected me a great deal, who did not have any other connection with Falck than to be his compatriot. Although no valid reasons for such a deed have been given, I nevertheless assume that only a few may persist in a state as pitiful as that of Falck himself, before taking such a dreadful decision. God, who is his judge, examines his deed after the intent, and all men must remain silent.
> I suppose that you, Honourable Herr Archiater, will let his brother[31] know of his death, but as to the way to do it, you will surely find the most prudent way, so that he [i.e. Falck's brother] will not be too much depressed by this news. As soon as [Falck's] property arrives in Petersburg and everything has been thoroughly investigated, the Academy of Sciences will probably hand it over to the Swedish Minister at this place, so that it may come into the possession of the relatives.
> This year has been unfortunate for our expeditions in natural history, because even Professor Gmelin has had the misfortune of being captured by a small prince on the border between Persia and Georgia a few miles from Derbent, who calls himself Usisnoi-Chan.[32] As a ransom for Gmelin's person he demands from the Russian Crown both people and money.

---

[31] Anders Falck (1740–1796), schoolmaster and astronomer in Skara, Sweden.

[32] Amir-Hamza, ruler of the Kaitag tribe.

Fig. 6.3 Among a dozen plants conserved by Lexell and sent to his colleagues in Sweden was this *Polygonum viviparum* (Alpine Bistort), which has survived in the herbaria of Peter Jonas Bergius, Bergianska trädgården, Stockholm (www.bergianska.se. Published with permission)

In a letter dated 17 September 1774, Lexell mentions that it was by then established as a fact that Gmelin had died in imprisonment on 27 July 1774.

In his last letter to Linnaeus, dated 19 April 1776 in Saint Petersburg, Lexell congratulates Linnaeus on the occasion when he was one of the eight foreign members of the Imperial Academy of Sciences to receive a medal as part of the celebrations of the peace treaty signed in 1774 between Russia and Turkey (by which Russia got access to the Black Sea). Also he announces that Professor Güldenstaedt in the latest Commentaries has published a description of the animal called the *schacal* [jackal], which had been brought to Saint Petersburg from Astrakhan, and which proved to belong to the genus of *canis* [dog-family].

This letter ends Lexell's partly dramatic correspondence in natural history with Linnaeus, who died on 10 January 1778. Lexell also corresponded on natural history with the Swedish botanist P. J. Bergius, whose herbaria contain several plants which he had received from Lexell (Fig. 6.3).

6.4 Pallas, Georgi and Laxman

In spite of Lexell's misgivings about the honesty of his colleagues Pallas and Georgi vis-à-vis Falck, he was able to continue his cooperation with them, in particular by arranging contacts with his Swedish colleagues. During his journey in Siberia, Pallas found the iron meteorite fallen in 1749 near the city of Krasnoyarsk and had it transported to Saint Petersburg—a considerable task, as the meteorite originally weighed about 700 kg. When Pallas was distributing specimens of the meteorite to his colleagues in the learned world, Lexell was commissioned to send a sample to the renowned Swedish physicist and chemist in Uppsala Torbern Bergman,[33] to whom Lexell wrote on 21 April 1777 (UUB G21. In Swedish):

> Professor Pallas has promised me a sample of the pure iron. I also have hopes of obtaining the lead spar and the magnets, either through Pallas or Georgi.

Another commission concerned the Journal of the Swedish botanist Carl Peter Thunberg[34] from Japan, which Georgi had proposed translating into German. In his letter to Thunberg (UUB G300q) dated 15 January 1782, Lexell discreetly warns Thunberg about Georgi's initiative. In a subsequent letter to Wargentin, Lexell states more openly that he feared that Georgi wanted to undertake the translation only because of the money.

Lexell's attitude towards the natural scientist and explorer Erik Laxman (1737–1796) was perhaps even less approving than his judgement of Pallas and Georgi. This may seem strange since, after all, Laxman was his only compatriot colleague at the Imperial Academy of Sciences (cf. Laxman's biography [87]).

Erik Laxman was born and raised in Savonlinna (in Swedish: Nyslott), at that time a small village around a late medieval castle[35] in Savo, a province of Eastern Finland. Laxman's childhood was tragic: he had witnessed the ravages of the war and the ensuing Russian occupation in 1741–1743. In spite of this, he was sent to school in Rantasalmi on the Swedish side of the border, and further to the *gymnasium* in Borgå (Porvoo). In 1757, he entered the Royal Academy of Åbo, but had to abandon his studies after a couple of weeks because of extreme poverty following the death of his father by drowning. Despite his interest in natural history, especially botany, he decided to secure his livelihood by becoming a Lutheran priest in the part of Finland attached to Russia after the war of 1741–1743, which suffered from a lack of clergymen. In 1762, Laxman moved to Saint Petersburg, where he soon became known to the Petersburg Academy of Sciences as an able botanist. Thus, when Laxman had been appointed priest to the German congregation in

---

[33]Torbern Bergman (1735–1784), Professor of Chemistry in Uppsala.

[34]Carl Peter Thunberg (1743–1828), Swedish botanist and apostle of Linnaeus, known for his expeditions to South Africa and Japan.

[35]The *Olofsborg* (St. Olaf's castle), an old Swedish stronghold against Russia, fell into Russian hands by the peace treaty of Åbo in 1743. Today, it is the venue of the annual Savonlinna opera festival.

Barnaul, Siberia, he was also elected correspondent for the Academy. After many years of fruitful journeys across Siberia he became, in 1770, an ordinary member of the Academy and Professor of Economics and Chemistry. However, rather than writing scientific reports and books his mind was set on travelling and exploring, and in 1780 he resigned from the Academy to be commissioned as a Councillor of Mining at Nerchinsk near the Chinese border. In 1784, he moved to Irkutsk, continuing his explorations in Siberia and also founding a glass factory near Lake Baikal. He died on a journey to Tobolsk in 1796.

Judging by the correspondence, Lexell's attitude to Laxman was a thoroughly disapproving one, such that their encounters must have been rather strained and unpleasant for both [87]. In fact, Lexell seldom mentions his compatriot colleague in Saint Petersburg without referring to him in a less than flattering manner. Either he is forgetful in his correspondence (especially with Linnaeus) and negligent in his commissions (the prolonged editing and publication of Falck's manuscripts[36]), or he is unreasonable towards his colleagues (when he dismissed his young assistant, the compatriot Mikael Holmberg[37]), or gossiping (cf. Lexell's account of the visit of King Gustav III to Saint Petersburg). Whether or not there were grounds for the discontent with Laxman, it is not easy to comprehend Lexell's hard and disapproving attitude towards his colleague, and it is rather to Laxman's credit that he, as far as we know, did not pay Lexell back with the same coin. As Laxman's letter to Archbishop Mennander suggests (letter in Swedish, dated St. Petersburg, 30 November 1778), Laxman did not bear Lexell a grudge, neither did he blame Lexell for his own difficulties [132]:

> [Professor Lexell] would be more useful in a university closer to the [Russian] border than Åbo. Through him, foreign students could be attracted to the Swedish universities, especially if there were more of his kind in the other sciences. Fortunate Lexell can leave Saint Petersburg as soon as he is badly treated. He is entirely worthy of this happiness. I, for my part, who have been struggling since my childhood in poverty and destitution, and without any support from benefactors have made my own way with much toil, will have to carry my own yoke even in the future, and forever remain a stepson. May the Lord teach me to revere His Providence and to feel my nothingness!

6.5 Recognition from Foreign Academies

For Lexell, his reception on 19 April 1771 as an ordinary member and professor into the Petersburg Academy of Sciences meant a conclusive approbation of his talents. The news of Lexell's success and his rapid promotion was certainly noticed

[36]Laxman was in possession of Falck's papers until 1780, when he was commissioned to Nerchinsk. The editing work was continued and completed by J. G. Georgi in three massive volumes of *Beyträge zur Topographischen Kenntniss des Russischen Reichs*. St. Petersburg, 1785–1786 [157].

[37]Mikael Holmberg (1745–1813), laboratory assistant and pharmacist, returned to Finland in 1774 [81].

with much astonishment in his home country, which he had left only 3 years earlier for want of career prospects. Thanks to Wargentin's influence, Lexell was elected in 1773 a foreign member of the Royal Swedish Academy of Sciences (*Kungliga Vetenskapsakademien*), and owing to the influence of Melander and Linnaeus, a year later, in 1774, he became a foreign member of the Royal Society of Sciences in Uppsala (*Kungliga Vetenskaps-Societeten*). The memberships were of course in the first place honorary recognitions of a great talent, as these societies, unlike for instance the Petersburg Academy of Sciences, did not pay any compensation for the services rendered to them. Nevertheless, the Royal Swedish Academy of Sciences encouraged their members to participate actively in its pursuits, and those members who did not live up to expectations could actually lose their membership.

Lexell's assiduousness and the fact that his contributions appeared in the Saint Petersburg Commentaries independently of Euler's did not pass unnoticed in the scientific societies of Europe. However, to become a member of the illustrious societies of Paris, London and Berlin, it was first necessary to establish personal contact with their most influential members. Knowing Lexell's ambitions and talents, he did not fall short in these endeavours. We know that Lexell was engaged in frequent correspondence with Johann III Bernoulli in Berlin, one of the prodigious sons of Johann II Bernoulli, who earned his PhD at the age of 14 and was appointed director of the Royal Berlin Observatory at 19. Rather than expertise in practical astronomy, the young Bernoulli's talents were better suited for correspondence, networking, publishing and scientific journalism. He also took care of a part of the letters and manuscripts of his famous relatives as well as those of Johann Heinrich Lambert. From Lexell's side, the correspondence begins by a letter dated 27 April (o.s.) 1773 (UBB Mscr. L I a 703, f° 108r–158v), apparently encouraged by a letter from Bernoulli (which has not been found), where Bernoulli asks for permission to publish excerpts from his letters in the *Ephemeriden* (AJEB) of Berlin. Lexell answers the request by saying:

> Si Vous voulez avoir la bonté de rendre une partie de remarques publique, il dépend de Vous, Monsieur, de choisir la voye, qui Vous paroît la plus convenable. En cas qu'entre ces remarques, il y en auroient de telles, qui contiennent de jugements sur les recherches des autres astronomes, comme par exemples sur les calculs de Mr Pingré,[38] je Vous supplie, Monsieur, d'en adoucir un peu les expressions, puisque en disant trop ouvertement son sentiment, quelque vraie qu'il puisse être on se fait souvent des ennemis.

In other words, while allowing publication of his results, Lexell nevertheless asks Bernoulli to moderate his outspoken and sometimes temperamental remarks on other astronomers, which are characteristic of Lexell's correspondence with his closest associates, Wargentin and Planman. Johann Bernoulli obviously recognised Lexell as a reliable source of information and published during the 1770s several of Lexell's astronomical reports and articles translated into German (their correspondence was written consistently in French). Bearing this in mind, it seems a bit surprising that the Prussian Academy of Sciences did not invite Lexell to be a

[38] Alexandre-Guy Pingré (1711–1796), theologian and astronomer in Paris.

foreign member, but on the other hand, as far as we know Lexell did not express a desire to become one either. The last of the 16 letters Lexell wrote to Johann Bernoulli, dated 25 April 1782, does not hint at any imminent crisis or disruption in the relationship. Lexell's relations with his younger associate Bernoulli seem to have been close but strictly professional. Lexell certainly could show his defiant and arrogant side, too, especially when he was discontented with Bernoulli, whose friendship he nevertheless seems to have valued highly.

Lexell likewise made several friendly advances towards the Royal Society of London through Doctor Charles Morton (1716–1799), physician and librarian and a Secretary of the Royal Society of London, as well as Astronomer Royal Nevil Maskelyne (1732–1811) and the scientific "networker" and developer of instruments Jean-Hyacinthe de Magellan.[39] Some of Lexell's letters to the two former were published in the *Proceedings of the Royal Society of London*, but the coveted recognition in the form of a fellowship in the Society never arrived. The Royal Society of Edinburgh elected Lexell as its member in 1784, but he would barely survive its acknowledgement. A similar recognition was given to him the same year by the *Accademia delle Scienze di Torino*, which the King of Sardinia Victor Amadeus III, after many doubts and delays had elevated to the status of a Royal Academy in 1783 [116, p. 130]. Among its most eminent members in the eighteenth century was Joseph Louis Lagrange, born in Turin in 1736, and most likely the one who in his capacity as its Honorary President proposed Lexell as a member to succeed the newly deceased Leonhard Euler. In the year of Lexell's death 1784, the "Commission for the Discovery of the Longitude at Sea" (in short, the "Board of Longitude") of the British government invited him as its member, with a free subscription to all of its publications, in particular the *Nautical Almanac*.

The Paris *Académie Royale des Sciences* recognised Lexell's merits more rapidly. The preparation for his election may be traced in the correspondence between Leonhard Euler and the Marquis de Condorcet,[40] which took place in a short period of time, in 1775–1776 [54]. At the time the young nobleman, having recently completed his 25 articles on mathematical subjects for the *Supplément à l'Encyclopèdie*, was working intensively on the problem of integrability of differential equations (see Chap. 8) and at the same time shouldering more and more responsibilities at the *Académie Royale des Sciences*: he was its Adjunct Secretary since 1773, and from August 1776 onwards, its Permanent Secretary. He also took

[39]Magellan (1722–1790), known as João Jacinto de Magalhães in his native Portugal, developed and manufactured scientific instruments in London [69,163]. Although not a scientist of distinction, his sociability and language skills made him acquainted with the scientific élite of the day. His correspondence was voluminous [55,69].

[40]Marie-Jean-Antoine-Nicolas de Caritat, Marquis de Condorcet (1743–1794), wrote his first mathematical publication *Essai sur le calcul intégral* in 1765, dealing with many theorems originating from Euler's work. As a *protégé* of d'Alembert's [7], he became increasingly involved with politics. After the French revolution in 1789 he took an active part in social reforms; his *Esquisse d'un tableau historique des progrès de l'esprit humain* was published posthumously in 1795.

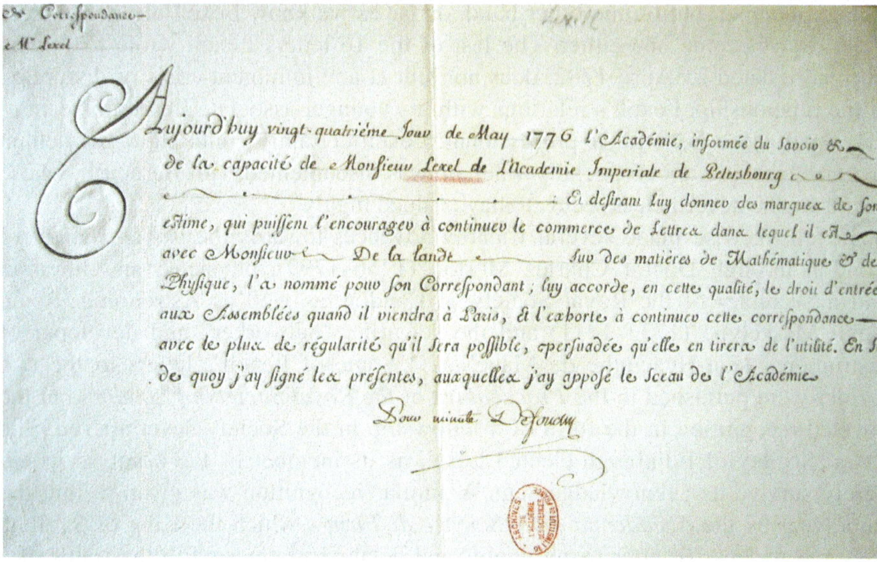

De Correspondance
de M. Lexel

Aujourd'huy vingt-quatrième Jour de May 1776 l'Académie, informée du Savoir &
de la capacité de Monsieur Lexel de l'Academie Imperiale de Petersbourg
Et desirant luy donner des marques de son
estime, qui puissent l'encourager à continuer le commerce de Lettres dans lequel il est
avec Monsieur De la Lande sur des matières de Mathématique & de
Physique, l'a nommé pour son Correspondant; luy accorde, en cette qualité, le droit d'entrée
aux Assemblées quand il viendra à Paris, & l'exhorte à continuer cette correspondance
avec le plus de régularité qu'il sera possible, persuadée qu'elle en tirera de l'utilité. En foy
de quoy j'ay signé les présentes, auxquelles j'ay apposé le Sceau de l'Académie

Pour minute Defouchy

Fig. 6.4 Diploma announcing Lexell's nomination as a corresponding member of the *Académie Royale des Sciences*, signed by Jean-Paul Grandjean de Fouchy (1707–1788), astronomer and Permanent Secretary of the Academy (This copy of the diploma is preserved at the Archive of the Academy in Paris. Photo: the author)

part in politics in the reformation work of his friend Turgot,[41] who in 1774 was Minister of the Navy in the government of Louis XVI. Condorcet wanted to take the opportunity to present to his friend two military *œuvres* of Euler, which he particularly admired, namely the *Théorie complette de la construction et de la manoeuvre des vaisseaux*[42] and *Neue Grundsätze der Artillerie*.[43] Turgot proposed to the King that a new edition of these works be published for the use of the French military schools. The proposition was well received. To the new Paris edition of the former work on the construction of ships, Lexell had proposed some improvements in a letter to Condorcet, which with Euler's approval were attached as a supplement [Lexell 48]. In the unpublished Post Scriptum to the letter, dated 13 December 1775, Lexell writes:

> Il y a déjà deux ans que j'avois prié M. de la Lande de me procurer l'honneur d'être associé à l'Académie des Sciences de Paris, en qualité de son correspondant, mais comme je n'ai

---

[41]Anne Robert Jacques Turgot, Baron de l'Aulne (1727–1781), economist, politician and *encyclopédiste*. While visiting Paris in 1780–1781, Lexell went to his house with the Marquis de Condorcet (see Chap. 9).

[42]"Complete theory of the construction and steering of ships", originally published in St. Petersburg, 1773 (OO Ser. II, Vol. 21, pp. 80–222. E426).

[43]Euler's edition (translated and augmented) of Benjamin Robins's *New principles of gunnery* (1742). Berlin, 1745 (OO Ser. II, Vol. 14, pp. 1–409. E77).

> reçu aucune réponse de lui sur cet article, j'ose m'adresser à vous Monsieur, pour vous demander en faveur en cas que vous ne la trouvez pas trop au dessus de mes mérites.

Apparently, Lexell had already asked Lalande to support his membership in the Paris Academy of Sciences in 1773, but did not receive an answer. Condorcet's affirming response to the request came the following summer, when, in a letter to Euler dated 10 July 1776, he announced [54]:

> J'ai été assez heureux pour contribuer à faire donner à M. Lexell le titre de correspondant de notre académie. Il honore le titre par ses talens, et par son attachement pour vous qui doit le rendre respectable à tous ceux qui aiment le génie et la vertu.

In fact, Lexell had been appointed *membre correspondant* for Joseph Jérôme de Lalande on 24 May 1776 (Fig. 6.4).

Chapter 7
Academic Events in Saint Petersburg

The time which Lexell spent in Saint Petersburg coincided with the most productive and glorious period in the history of the Imperial Academy of Sciences. The annual publications of the Academy, the *Novi Commentarii* and subsequently the *Acta* and *Nova Acta*, were filled with important articles in Latin and French on various scientific topics. A significant part of the mathematical papers were written by Leonhard Euler. The solemn, official part of the publications was the "History" section, containing the highlights of the academic year as well as lengthy abstracts of the ensuing scientific articles. However, in many ways the more interesting part of the events took place behind closed doors. Indeed, Lexell's personal correspondence with for instance Pehr Wargentin reveals many intriguing, sometimes even embarrassing details in the working dynamics of the organisation, on which subject the official minutes of the academic conferences are silent. On the other hand, the records of the academic conferences sometimes divulge controversial events, which Lexell would rather not have mentioned even to his friends. In the following, we will describe a selection of events based on the accounts found in Lexell's letters and other sources.

7.1 Diderot's Visit

The history of mathematics, like every branch of history, has its share of tall stories. One of them is the well-known and widespread rumour about the alleged encounter in 1774 between Denis Diderot (1713–1784) and Leonhard Euler at the imperial court in Saint Petersburg. Owing to the project of the *Encyclopédie*, its editor-in-chief Diderot had become one of the best known philosophers of the French Enlightenment. He was engaged in correspondence with Catherine II of Russia and nominally became her librarian in exchange for a generous stipend [117]. At the beginning of the 1770s, when Diderot had finished the *Encyclopédie*, he finally agreed to travel to Russia to meet the illustrious *Protectrice des Philosophes*, but

J.C.-E. Stén, *A Comet of the Enlightenment*, Vita Mathematica 17,
DOI 10.1007/978-3-319-00618-5_7,

he also had personal interests to visit Saint Petersburg: on his recommendation, the Empress had commissioned his friend, the renowned French sculptor Étienne Maurice Falconet (1716–1791) to execute a monument of Peter the Great, a colossal equestrian statue called the "Bronze horseman", situated on the Senate Square.

However, not everybody in Saint Petersburg anticipated Diderot's arrival as enthusiastically as the Empress. The members of the Petersburg Academy of Sciences were in fact rather unfavourably disposed to him, despite maintaining seemingly cordial relations in order to please the Empress. This resistance is understandable since Diderot was very closely identified with a materialistic and antireligious philosophy, which in particular Euler despised. The resistance clearly manifested itself when Director Orlov at the conference session on 25 October (o.s.) 1773 raised the question of receiving Diderot and his friend Baron Grimm[1] as its foreign members [141, Vol. III, pp. 104–105]. None of the academicians showed any enthusiasm, but realising that it was the will of Catherine, they bowed dutifully to the decision. Only Leonhard Euler as the leading academician uttered his opinion, and without mentioning Diderot he called to mind the forcible election of La Mettrie[2] to the Prussian Academy of Sciences, in which he had unwillingly taken part. In spite of Euler's misgivings, the majority of the academicians were in favour of the association of Diderot and Grimm as foreign members of the Academy [36]. The two philosophers were consequently invited to the conference, which was held on Friday 12 (1) November 1773, in the presence of the Director and the Eulers. On the occasion Diderot communicated and read to the Academy a number of questions on the natural history of Siberia, on its metallic ores, mountains and salt lakes, snakes and vodka made from fermented mare's milk [117]. The task of responding to these questions was given to the natural historian and mineralogist Erik Laxman.

The story about the confrontation between Diderot and "a senior member of the Petersburg Academy" first appeared in Dieudonné Thiébault's[3] *Souvenirs de vingt ans de séjour à Berlin* (1804), whereupon the British mathematician Augustus de Morgan romanticised the anecdote in his *Budget of paradoxes* (1872), quoting it as follows:

> He [Diderot] conversed quite freely, and gave the younger members of the Court circle a good deal of lively atheism. The Empress was much amused, but some of her councillors[4] suggested that it might be desirable to check these expositions of doctrine. The Empress did

---

[1]Friedrich Melchior, Baron von Grimm (1723–1807), German-born French philosophical author and diplomat.

[2]Julien Offray de la Mettrie (1709–1751), French physician and proponent of a materialistic philosophy, became a member of the Prussian Academy of Sciences in 1748.

[3]Dieudonné Thiébault (1733–1807) was a French author, Professor of French Grammar in Berlin and a *lecteur* (reader) at Frederick's court.

[4]Possibly the Councillor of the Court Franz Ulrich Theodosius Aepinus (1724–1802), a member of the Imperial Academy of Sciences, known for his pioneering theory of electricity and magnetism [64, 69] and for being the tutor of the Grand Duke Paul (later Paul I of Russia) [55]. In fact, after his appointment as tutor in 1765, Aepinus did not even attend the conferences of the Academy.

> not like to put a direct muzzle on her guest's tongue, so the following plot was contrived. Diderot was informed that a learned mathematician was in possession of an algebraic demonstration of the existence of God, and would give it him before all the Court, if he desired to hear it. Diderot gladly consented, though the name of the mathematician was not given, it was Euler. He advanced towards Diderot, and said gravely, and in a tone of perfect conviction: *Monsieur,* $(a + b^n)/n = x$, *donc Dieu existe; répondez.* Diderot, to whom algebra was Hebrew, was embarrassed and disconcerted; while peals of laughter rose on all sides. He asked permission to return to France at once, which was granted.

Although de Morgan's account follows closely that of Thiébault, who—may it be pointed out—did not witness the incident himself, but learned of it at Frederick's court, one should not place too much confidence in the description.[5] Moreover, there is no reason to underestimate Diderot's knowledge of mathematics, although he never showed any enthusiasm for it [117, p. 154].

Nevertheless, some kind of incident or confrontation between Diderot and an eminent member of the Academy had certainly taken place. Six years later, when Lexell visited Paris and met his colleague-astronomer Joseph Jérôme de Lalande, with whom he was dragged into an argument on religion, he alludes to the incident in question by writing to Johann Albrecht Euler from Paris on 10 November 1780 (АРАН Фонд 1, опись 3, дело № 65, листы 107–114):

> ...Then I told [de Lalande] that Mr Diderot had been clearly defeated by Aepinus in the well-known dispute which they had at Count Orlov's house.[6]

Perhaps the famous debate had not taken place in the court at all, as may be surmised from Lexell's account, but in Count Orlov's house.[7] And instead of Euler, Lexell mentions Aepinus.

Some details from Johann Albrecht Euler's correspondence complete the picture of the sequel. Unlike Johann Albrecht Euler, who despite his position of Secretary of the Academy, generally kept his distance from the court (undoubtedly on the recommendation of his father), Lexell seems to have moved with ease and been well-regarded in the higher political spheres. Known for appreciating courteous and enlightened company, Lexell was a frequent guest of Prince Aleksandr Mikhailovich Golitsyn[8] (1718–1783), who as acting Governor of Saint Petersburg at the time was in charge of passports. At the same time, Golitsyn was on good terms with Diderot, and so J. A. Euler reports to Formey [36]

---

[5]The King did not hesitate to use every opportunity to tell malicious anecdotes about his former employee Euler [43]. Technically, it is possible that Frederick himself injected Euler into the story.

[6]Ensuite je lui aussi faisoit entendre, que Mr Diderot avoit été mis bien bas par Mr Aepinus dans le dispute si connue, qu'ils ont eue chez le Comte Wladimir Orlow.

[7]This has been noted by the French historian Georges Dulac [35], who has studied Diderot's sojourn in Saint Petersburg in the winter 1773–1774, especially in the light of the correspondence between J. A. Euler and Samuel Formey.

[8]The fact that Lexell owned a portrait (engraving) of the Prince [47] suggests that their relations were close. Prince Golitsyn's father had been field marshal and Governor of Finland during the "Greater Wrath", Prince Mikhail Golitsyn.

> [...] Professor Lexell told me at the Academy that he had learned that Mr Diderot will depart from here under the name of Mr Denys; and as nobody can leave the country without a permission signed by the Prince Vice-Chancellor, whom Lexell visits quite often, Mr Diderot had begged the Prince to write his passport for the name Denys instead of Diderot, because, he interjected, if I would travel with my own name, all the Princes of Germany would certainly detain me for some days when passing through their domains, especially the King of ... [Poland], who certainly would arrange a party to see me. Here, it seems to me that M^r^ L[exell] has somewhat embroidered the story.

Be that as it may, Euler seems to have interpreted the story as an unmistakable mark of the great vanity of the French philosopher.

7.2 Kepler's Manuscripts and Czar Peter's Horoscope

At the conference of the Petersburg Academy held on Friday 7 May (26 April o.s.) 1773 [141, Vol. III, p. 90], the Secretary J. A. Euler read a letter from Christoph Gottlieb von Murr, a patrician from Nuremberg (Nürnberg) dated April 8, 1773. Von Murr had appended a list of manuscripts that had belonged to the astronomer Johannes Kepler (1571–1630), which were for sale in Frankfurt am Main in the house of a widow named Trümmer. In the letter he offered the manuscripts to Catherine II for 2,000 Roubles.

This collection of documents, the *Manuscripta Keppleriana*, had been bound in 20 volumes by their former owner Michael Gottlieb Hansch, a scientist and a friend of Leibniz. Before that, the manuscripts had survived many dangers (including a fire) while passing through many hands and owners until 1765, when they were discovered in the trunk of the Warden of the Mint of Nuremberg. Von Murr was now commissioned to offer them to every academic society in Europe to find a worthy owner for the works of one of the greatest astronomers of all times. The response was negative, however, since the price was considered too high, and it was doubtful if the manuscripts could be of any use. Scientists were then, just as they are often today, more interested in advancing their own discipline than in the thoughts of their ancient predecessors.

The Secretary continued:

> As His Excellency Mr Chamberlain Rzhevsky,[9] by sending this letter to the Conference, had let it be announced that Her Majesty ordered the Academy and in particular Professor Euler the Elder to inform Her in writing an opinion of this collection of manuscripts, the Secretary also read a letter which the same Mr Murr had addressed especially to him containing certain supplementary details about this purchase. It was decided that Mr Euler the Elder would communicate in the first assembly his opinion whereupon the other Academicians form theirs.

[9]Alexey Rzhevsky (Алексей Андреевич Ржевский; 1737–1804) had been, since 10 June (30 May o.s.) 1771, Vice Director of the Academy acting during Vladimir Orlov's absence.

Indeed, in the next meeting of the Academy on 29 April 1773 [141, Vol. III, p. 91], Leonhard Euler pronounced his opinion about these manuscripts, to which the other academicians gave their consent. There was a deal.

In the summer of 1774, after the manuscripts had arrived at the imperial court in Saint Petersburg [141, Vol. III, p. 144–145, August 22, 1774], an urgent message from the court was read in the academic conference: "The will of Her Majesty being that the Academy makes use of these manuscripts by publishing those which could interest the learned, the conference charged Mr Krafft & Lexell to peruse them and to select those articles which merit being rendered public". A year later, on 12 October 1775 [141, Vol. III, p. 210],

> The Secretary of the conference read a letter from the Director, which he had obtained from Moscow, dated 5 October. Mr de Domashnev[10] informs that Her Majesty the Empress desires to learn of the contents of the manuscripts of the famous astronomer Kepler, which She had purchased some time ago for the Library of the Academy and that She especially desires to know if the Academy could make any use of it [...] whereupon Professors Krafft and Lexell, who had already skimmed through a part of these manuscripts, were charged with presenting in the next meeting their report (on 16 October), which will then be sent to the Director.

A hand-written sketch of Lexell's report with a preliminary classification of Kepler's manuscripts is preserved at the Archive of the Russian Academy of Sciences in Saint Petersburg (АРАН; see the list "Source Material in Saint Petersburg" in the index part of this volume). Despite the century-long good intentions to publish the manuscripts, the project has not advanced since.

Another rather special commission had been given to Lexell only a month earlier [141, p. 202–203]. A letter from State Secretary Müller from Moscow was read out at the conference of 11 September 1775. In the letter, Müller, who was compiling the history of Peter the Great from his birth to the wars against Sweden, asked for a copy of a letter which the Dutch man of letters Johann Georg Graevius (1632–1703) wrote to his colleague, the scholar and poet Nicolaas Heinsius (1620–1681) at the time of the birth of the Czar (in 1672). The letter concerned a peculiar astronomical phenomenon at this time, and therefore Müller requested the astronomers to compute the positions of the planets and the celestial bodies at this time and to explain the phenomenon in such a way that

> [...] by the precepts of astrology one could draw some conclusions which would have some relevance to the marvellous acts of Peter the Great. Mr Lexell was charged to perform the calculations and undertake the researches.[11]

[10]Sergey Domashnev (1743–1795) was a military officer and poet, educated in Moscow. He served as Director of the Petersburg Academy of Sciences until 1782.

[11][...] si par les préceptes de l'Astrologie on pourroit tirer quelques conclusions qui ayent du rapport aux actions merveilleuses de Pierre le Grand. M. le Prof. Lexell se chargea de ces calculs et recherches.

A week later, on 18 September 1775:

> Professor Lexell delivered to the Secretary the geocentric positions of all the planets which he had calculated for the days of the conception and the birth of Peter the Great, as requested in the meeting on 11 September; He also promised to furnish the entailing astrological predictions and to draw up a sort of horoscope. For the rest, the Academician reported that he has not been able to find among the letters of the famous Graevius a reply about the memorable phenomenon on the sky, which had apparently been seen in Moscow at the time of birth of Peter the Great.

That astronomers were commissioned to make up horoscopes was not an unheard-of phenomenon as such. After all, scientific icons like Kepler and Newton did the same, and not only for their own amusement, but quite earnestly. However, nothing suggests that Lexell attached the slightest importance to this subject, which he performed as a matter of course. Nowhere in his correspondence does he mention being bothered or embarrassed by this commission.

7.3 Incidents and Scandals

From time to time Lexell filled his letters with details from current events to gossip from the Academy and the imperial court. The life behind the scenes of the Academy was often less placid and harmonious than one might have imagined [37]. The directors of the Academy were ambitious noblemen, often more interested in promoting their own career than in scientific work and practice, and always susceptible to flattery and praise. The academicians, on the other hand, could also turn out to be ambitious and competitive, and sometimes indomitable. Thus, there prevailed an almost continuous tension between the academicians and their chief, who seldom participated in the weekly conferences, unless his attendance was absolutely necessary. Lexell was extremely careful in expressing his opinion of his superior in his personal correspondence with Wargentin, but judging by the more outspoken letters written by Johann Albrecht Euler to the Secretary of the Prussian Academy of Sciences in Berlin, Samuel Formey, the Director Count Orlov was regarded as a "very singular man with his caprices", and nicknamed *le Froid* or *l'Ours* [36] (Fig. 7.1). In general, Lexell seems to have been more obedient and respectful to the will of the aristocracy and the authorities than the Eulers were.

The following incident, which took place late in 1773, and is colourfully described by Lexell (letter to Wargentin dated 11 June 1775, see also [37]), throws light not only on the dynamics between the academicians and the authoritarian director of the Academy, but also on Lexell's own attitude to the set-up. The background was an article that appeared in the *Allmänna Tidningar*, a gazette edited by C. C. Gjörwell in Stockholm, and which concerned the calendar published by State Secretary von Stählin. Lexell, who evidently kept up with news from Sweden and Finland, was certainly attentive to the reports from Russia, where he obviously had access to first-hand information. As Lexell considered the article in question to be particularly erroneous, he decided to deliver a complete account of the events to

Fig. 7.1 Count Vladimir Grigoryevich Orlov, Director of the Imperial Academy of Sciences of Saint Petersburg 1766–1774. Engraving after contemporary painting (Stählin [154])

Wargentin, "... which Herr Secretary may communicate to Librarian Gjörwell, on condition that it will be kept a secret".

> At the beginning of last year Minister Stählin decided that an engraving of the Grand Duchess[12] would be included in the next Calendar of the Court. A skilled local engraver had also caught the image of Her Imperial Highness successfully, but Stählin who knew better wanted to refine on the work of others, thus spoiling the work altogether, so that it did not bear any resemblance at all. However, Stählin believed he had succeeded very well, and with this conviction on New Year's Day he removed, more or less by force, twelve copies of the Calendar from the Academic Bookshop, which he bound the same day or the following (I do not know which) and presented them to Their Imperial Highnesses, Count Panin,[13] the foreign ministers and many of the distinguished persons. Later on Count Orlov, who as usual was going to present the Calendar for the New Year to Her Majesty the Empress and the Grand Duke, received a dry compliment from the latter, saying that he already had the Court Calendar. This was made with a mien which for Orlov suggested something disagreeable. Thus, he started to investigate immediately who had dared to present the Calendar to the Grand Duke. Now the crime of Stählin became known. I call it a crime since according to the regulation of the Academy, no books or printed papers may be published before they have been presented to the Empress by the Director or his representative. Now Stählin had dared to anticipate the Chief of the Academy. He was then, as they say, summoned by the Count

[12]Princess Wilhelmina Louisa von Hessen-Darmstadt had shortly before been married to Grand Duke Paul. She died in 1776 after childbirth.

[13]Count Nikita Ivanovich Panin (1718–1783), Russian statesman.

> where he was confronted with the Inspector of the Academic Bookshop, and where the Count was said to have addressed him very brusquely, even threatening him with arrest etc. Meanwhile, Her Majesty had let it be known that the difference between the appearance of Her Imperial Highness and the engraved portrait was quite remarkable, thus all the published copies of the Court Calendar were recalled. This circumstance did not exactly improve Stählin's case, but it is not clear if it worsened it either. The Count could well have been satisfied with the lesson he had given Stählin, but he also decided to demonstrate his authority by punishing Stählin for his crime by making him pay with one month's salary, meaning about 125 Roubles. This resolution, which he had made at home, was sent to the Academic Commission[14] to be filed and carried out. The members of the Commission next formulated a proposition to the Count requesting him to withdraw this severe resolution, but without success. The second question to the Count was whether this resolution should be implemented, even if it was not signed by the members of the Commission, and the reply was that even if they thought otherwise, they should sign the will and decision of the Chief. Then both the Eulers declared not only that in their opinion the Count had no authority to punish unilaterally and without due investigation a member of the Academy, but also that they were not obliged to sign the Count's decision against their conviction. In the meantime, Stählin uttered nothing and took it calmly like a good boy, although he normally is prone to manifest his status as the foremost member of the Commission. Rumovski and Kotelnikov explained that, although they realised the unjustness of the Count's decision, it was nevertheless their duty to sign. Then Professor Euler the younger let slip a remark: *Vous signerez donc par bassesse* ["You sign by baseness, then"]. Rumovski took this as an offence so grave that he immediately went to the Count, who, although not having visited the Academy for a year, arrived immediately and let Professor Euler the younger be called in from the hall of the Academic Conference to the room of the Commission, where only the Count [Orlov], Rumovski and Protasov[15] were present. The Count then told Euler that he [Euler] had offended Rumovski, and that [Rumovski] had the right to demand satisfaction. In reply to this Euler admitted having uttered something in the heat of the moment, which he would not recount in *sang-froid*, but that he thought that the Count had abused his power as a Chief, when he wanted to force the members of the Commission to sign a decision, which they by their conviction did not approve. To this the Count answered that the members of the Commission were there only to execute his decisions etc. The next day old Professor Euler sent to the Count a demand for removal from the Academic Commission of both him and his son, but as they both had been appointed to this duty by Her Majesty, the matter was submitted to the Empress, who accepted the proposal. In his application Euler had claimed the reasons to be his old age and illness, but it may easily be surmised that Her Majesty had been briefed of the underlying events. In my opinion *l'Esprit de la constitution* [the spirit of the Academy's statutes] requires that Euler should not have acted contrary to the Count; that everybody should be loyal to his Superior, who is in charge. The outcome of the story was that both Eulers resigned from the Commission and Stählin lost his salary for one month.

Lexell further regretted to Wargentin that of all the members of the Academy, von Stählin was

> ...the best known to the distinguished people in Saint Petersburg; but that is not to his advantage, because most of them see him as a fool, and this opinion is so common, that it extends from Her Majesty the Empress to our soldiers at the Academy. He is a man of a

[14] Board of Direction of the Academy presided over by the Director.

[15] Alexey Protasov (1724–1796), anatomist, extraordinary academician, Secretary of the Academic Commission.

rather cheerful temperament, but of an unbearable vanity, which makes him seem ludicrous and despicable.[16]

As Johann Albrecht Euler mentioned to Formey [36], after this episode, neither he nor his father and von Stählin would ever again step into the academic commission [20]. As a result, Leonhard Euler withdrew his participation in the conferences of the Academy altogether, as did the Director Count Orlov. More than a year would pass without a proper director of the Academy until Catherine II appointed her young favourite, Chamberlain of the Court Sergey Domashnev, as the new director. Domashnev proved to be intelligent and ambitious, but in the same time overbearing and disrespectful, and thus turned out to be an even worse choice for the Academy, as we shall see later [20].

On 30 January 1777, Lexell describes the celebrations of the 50th anniversary of the Imperial Academy of Sciences of Saint Petersburg:

> Three of our fellow-countrymen were appointed foreign members of this Academy, namely the Professors Wallerius, Melander and Bergius.[17] As Professor Pallas in the last mail has already notified Professor Bergius about his appointment, I ask you, Herr Secretary, only to verbally congratulate him on my behalf for a well-merited distinction. Both the attached letters to Wallerius and Melander contain notifications of this. The other foreign members who were proclaimed on this occasion were: His Majesty the King of Prussia, who has declared to the Academy the grace of letting his name be the first among its foreign members, Baron Haller,[18] Count Buffon,[19] Director Margraf[20] and la Grange,[21] Chevalier Pringle,[22] Maskelyne, Marquis de Condorcet, Messier, Gleditsch,[23] Johann Bernoulli, Burman,[24] von Born,[25] Daubenton,[26] Sigaud la Fond,[27] Toaldo,[28] Lorgna,[29] and Mallet in Geneva. All of these except Maskelyne and Burman were on the list proposed by

---

16... [I]ngen af Academiens ledamöter är så allmänt känd hos de förnäma som Stählin; men det är ei til hans fördel, ty de fleste hålla honom för en narr, och denna öfwertygelse är så allmän, at den begynner sig ifrån Hennes Majs^t Keiserinnan och slutar sig med wåra soldater wid Academien. Han är en man af tämmeligen godt sinnelag, men en odrägelig vanité som gör honom hos alla människor löjelig och föraktelig.

17 Johan Gottschalk Wallerius, Daniel Melander (Melanderhjelm) and Peter Jonas Bergius.

18 Albrecht von Haller (1708–1777), Swiss anatomist, physician and botanist.

19 Georges-Louis Leclerc, Comte de Buffon (1707–1788), French naturalist known for his monumental work *Histoire naturelle* (1749–1788).

20 Andreas Sigismund Marggraf (1709–1782), chemist and director of the physics section of the Prussian Academy of Sciences in Berlin.

21 Joseph Louis Lagrange (1736–1813).

22 John Pringle (1707–1782), army surgeon, royal physician.

23 Johann Gottlieb Gleditsch (1714–1786), Professor of Botany in Berlin.

24 Johannes Burman (1707–1779), Dutch botanist.

25 Ignaz Edler von Born (1742–1791), Hungarian mineralogist.

26 Louis-Jean-Marie Daubenton (1716–1800), French naturalist and physician, who collaborated with Count Buffon.

27 Joseph-Aignan Sigaud de Lafond (1730–1810), French experimental physicist.

28 Giuseppe Toaldo (1719–1797), Italian priest and astronomer.

29 Antonio Maria Lorgna (1735–1796), Italian physicist and engineer.

the Director, of which I had suggested the former and Professor Pallas the latter. The proposals were well received by the majority as well as the Director. However, some of those appointed did not receive enough votes, namely Daubenton, Sigaud la Fond, Toaldo, Lorgna and Messier, of which neither of the three middle ones had received three votes, but the Academicians could not turn down the proposal of the Director, when also Messier and Daubenton were suggested; the former was among the mathematicians and the latter among the physicists. So, in an accord between the Director and the Academicians, these five were appointed. This anecdote you, Herr Secretary, will be pleased to keep to yourself.
The ceremony was opened by Professor Euler as Secretary with a complimentary speech, which was followed by a rather lengthy speech by our Director in Russian, thereupon Professor Gueldenstädt gave a lecture about the improvements that can be made for the products of Russia, in French. After that, once the program concerning the Academy's prize competition had been announced, the foreign members were proclaimed. Then Professor Rumovski read an outline of a work on the geography of Russia, and finally the names of the recently appointed *membra honoraria* were announced, among which His Imperial Highness[30] was the foremost. He was present with his consort, but due to indisposition the Empress herself could not arrive. Professor Euler distinguished himself on this occasion to his own credit and to that of the Academy. The speech of our Director was undoubtedly quite elegant, but it was too long and annoyed the Grand Duke [...]

7.4 Visit of the "Count of Gothland"

In 1771, Crown Prince Gustav succeeded his father King Adolf Fredrik as Gustav III of Sweden. Gustav's mother, the ambitious and intelligent Lovisa Ulrika, a younger sister of Frederick II of Prussia, was keenly interested in science, philosophy, and aesthetic trends, a fact that was also reflected in the young Gustav's taste [10]. Less than two years after taking power, Gustav achieved his mother's goal through the *coup d'état* of 1772, which ended the short era of parliamentarianism in Sweden. At the same time, Russia lost its influence in Swedish politics, which greatly troubled Gustav's cousin,[31] Catherine II of Russia. There were reasons to fear that Sweden could reclaim what she regarded as her rightful territories in eastern Finland, and sure enough, a war broke out between Sweden and Russia in 1788. Ever since the beginning of his reign, Gustav had been waiting for a chance to show his good intentions, but the Russian Empress had at first been reluctant to meet the King. However, when Gustav announced that he would travel incognito under the fictitious name of the Count of Gothland, the visit was agreed [74], and in the summer of 1777 Gustav sailed from Stockholm to Saint Petersburg to meet Catherine and negotiate with her.

Early in the morning on 16 June 1777 the ship carrying the royal visitor arrived at the fortress of Kronstadt on the island of Kotlin (in Finnish: Retusaari) on the sea approach to Saint Petersburg. Having dined later that day in Saint Petersburg

[30]Grand Duke Paul, son of Catherine II.

[31]Catherine's mother, Johanna Elisabeth of Schleswig-Holstein-Gottorp (1712–1760), was the sister of Gustav's father, King Adolf Fredrik of Sweden.

with the Foreign Minister Nikita Panin and the Swedish Minister Johan Fredrik von Nolcken,[32] the King with his royal entourage joined the Empress at the imperial summer residence of *Tsarskoye Selo* some 25 km south of the capital. The first meeting was a cordial family gathering; the two cousins saw a theatrical performance—the King was a keen lover of the theatre—played a game of chess and enjoyed supper together. There would be many more negotiations during the 4 weeks, from 16 June to 16 July, which the King stayed in Saint Petersburg. On the second day of his visit, he assisted the Empress in laying the foundation stone of the Chesme Church near Saint Petersburg to commemorate the Russian victory over the Turks in 1770 at Çesme Bay in the Aegean Sea. Throughout his public appearances, Gustav was very particular about preserving his pretence of anonymity and insisted on not wearing his usual regalia and decorations.

Besides visiting the imperial palaces, the King attended a meeting of the Academy of Sciences. Every academician was present at the solemn conference presided by Director Sergey Domashnev on Friday, 4 July (23 June o.s.) 1777 [141, p. 308]. After two lectures on natural history, the royal visitor was guided through the halls of the Academy, the curiosities of the *Kunstkammer*—where among other things a wax effigy of Czar Peter was on display—and even to the top of the building to admire the view of the Neva. The King was thereupon guided to a special pavilion in the yard exhibiting the famous *Globe of Gottorf*, a seventeenth century celestial globe on the inside and a map of the world on the outside—in fact, an early planetarium, about 3 m in diameter— constructed on the initiative of the Duke of Holstein-Gottorp. During the Great Northern War, Duke Christian August of Holstein-Gottorp[33] had offered the globe as a present to Peter the Great, who subsequently ordered it to be transported to Saint Petersburg.[34]

Then there followed a visit to the medal collection and to the mineral cabinet, the library and the printing office, where the King halted at a working printing press: he picked up a sheet of paper which, as if by accident, slipped out of the machine, showing his own portrait with some complimentary verses (see Fig. 7.2). Finally the King, content with what he had seen and heard, presented Director Domashnev with a golden box, and also presented Pallas, Leonhard Euler and Lexell each with a gold medal struck to commemorate the establishment, in 1775, of the court of appeal in the town of Vasa in Ostrobothnia, Finland [87, p. 116].

Lexell's account of the visit is found in a letter to Wargentin dated 25 July 1777:

> When the first news of the forthcoming visit of His Majesty the King of Sweden arrived, somebody proposed to the Director that when the King honours the Academy with a visit,

[32] Johan Fredrik von Nolcken (1737–1809), Swedish envoy in Saint Petersburg in 1773–1788 [129].

[33] Christian August (1673–1726) was a grandfather of Catherine II.

[34] The globe, which had already been damaged during the removal, was placed in the tower of the *Kunstkammer* building, but was almost completely destroyed in a conflagration in 1747. Reconstruction of the wooden parts had commenced almost immediately and the globe was restored and gradually upgraded [74, 141].

Fig. 7.2 Portrait of the "Count of Gothland" alias Gustav III of Sweden printed on the occasion of his visit to Saint Petersburg (Kungliga biblioteket, Stockholm. Published with permission)

I should give a lecture on the remarkable comet [of 1770], and this suggestion was at first approved by the Director, but then he realised that this subject was not interesting enough. It was finally Professor Pallas who had the honour of delivering a lecture, and the subject was in fact quite interesting; but he talked so quietly, that no one could hear anything. Our Director delivered a very elegant, ceremonial speech, which I shall send [to you] with the next ship, in the meantime it can be obtained from Archiater Dahlberg.[35] In the same speech, he honoured me unexpectedly with much praise, by which he presumably, as the courtier he is, tries to please the person he talks about; because I can rather well realise that Director Domashnev's friendship for me can be no deeper than my respect for him. Archiater Dahlberg promised that he would propose him as a member of the Swedish Academy of Sciences [KVA] immediately upon his return to Stockholm, but I forgot to tell the Archiater that I hoped the Academy could admit him as a member without any delay, as was the case for Prince Kurakin,[36] without obeying the waiting period provided by law; at least it will please Director Domashnev all the more, the sooner he will be the object of this courtesy. He has this weakness of vanity and ambition without the slightest moderation. Not

[35]Nils Dahlberg (1736–1820), royal physician.

[36]Prince Alexander Kurakin (1752–1818), Russian statesman, was elected a foreign member of the KVA in 1776.

satisfied with obtaining the Order of Vasa, he also wanted the Order of the Sword, which was denied to him, however. On the occasion of the visit of His Majesty the King to the Academy, he [the King] showed grace to three of its members by bestowing upon them a golden medal, namely old Professor Euler, Professor Pallas and me. However agreeable it is for me to have received this token by the grace of my King, I have to admit that I would rather have been deprived of it, as I can clearly perceive that many of my colleagues appear to believe that I cannot lay claim to such a favour any more than they. And to admit the truth, I must concede that there can be no other reason for which I have shared this favour than the fact that I have the advantage of being a subject of His Majesty. A slight misunderstanding from Baron Nolcken and an indiscretion from my side have given occasion to this jealousy, as well. The same day, when the King visited the Academy, Baron Nolcken said that he had brought four medals to be bestowed upon those who are members of the Swedish Academy of Sciences. I then explained to him that we were five members of the Swedish Academy of Sciences, and that it is not Professor Euler the Elder, but the son who is a member.[37] He then withdrew the medals, and that for a good reason, since two of them had been struck to commemorate the Revolution[38] and can therefore be given only to Swedish subjects. Also Baron Nolcken had believed that Laxman was Swedish by birth. As I believed this scheme was to be followed, I was imprudent enough to entrust something about it to Mr Laxman, asking him not to tell anybody, which he has done nevertheless. Then the King changed his intentions, partly in consideration of Mr Euler the Elder, to whom he was particularly inclined to show his favour, and partly because there were no more medals but those which had been struck to commemorate the Revolution. Ever since the first project was disclosed, I have noticed some discontent among those who were preparing to receive this distinction, and one of them has questioned me directly about it, to which I replied that it was simply untrue that every member of the Swedish Academy would receive the medal. It is my fault to have informed Laxman about it, otherwise no one would have known, and now they might even believe that the plan was changed on my initiative, which is not true.
[...] I presume that His Majesty with his suite has already arrived in Stockholm, since the Cornet Borgenstierna came yesterday as a Courier with the news that the King was in Helsingfors [Helsinki] on Monday, and he brought with him letters to Her Majesty the Empress and their Imperial Highnesses.

Evidently, Lexell was very careful not to reveal any signs of approval or disapproval of the King's personality in his description to Wargentin. The King was young and eager to show his good intentions, and Lexell had reason to believe in the best of his capacities. However, while Gustav's personal tastes were in art, theatre and literature, the sciences were largely neglected, a fact that Lexell could perceive for himself when he visited Sweden on his return from the European Continent in 1781.

[37]Lexell seems to have been unaware that Leonhard Euler had in fact been elected a foreign member (seat 33a) of KVA in 1755, while his son Johann Albrecht was elected a foreign member in 1771 (seat 79). The reason for the misunderstanding seems to be that Euler's name did not appear regularly in the list of members occasionally published in the proceedings of the Royal Swedish Academy of Sciences.

[38]Gustav III called his *coup d'état* in 1772 the "Revolution".

7.5 On Euler and His Family

As a close friend of Euler and his family and as a frequent guest in his house, it is not surprising that Lexell, in his correspondence, speaks of his Master only with admiration and respect (Fig. 7.3). As regards Euler's eyesight, Lexell writes to Wargentin on 21 October 1771:

> A piece of information which without doubt will delight all those who value learning and excellent service, but which will be most welcome for all mathematicians as well, is that our respected Euler has been successfully operated on for the cataract in his left eye, and that he now sees rather well. As yet, he has not gone so far as to read or write himself, which in any case will not be possible without using reading glasses, but otherwise he seems capable of distinguishing all objects. By chance I happened to be present at the operation. It is hard to imagine what kind of effect it had on his family. Before [the operation] had been completed, one could perceive in them a suppressed anxiety and fear on account of the possibility of it not going well. But when the operation was over they were all crying for joy. He has in fact a rather honourable family and I can for my part easily concur with what the other Academicians say of his house, that it is 'blessed by Heaven': *c'est une maison bénie du Ciel.*

On 20 August 1773, Lexell continues on the subject, when Euler's vision had deteriorated anew:

> Professor Euler's vision is rather similar as it was before he had the cataract of his eye operated. He can perceive some daylight, but nothing quite distinctly. It is fortunate, however, that he can have someone to stay with him, helping him in his work. Now he has got a young Swiss, who lives with him and keeps him company every day, as his Amanuensis. Thereby I and Professor Krafft are freed from working for him. Last time I saw him I noticed that his hearing had begun to deteriorate. There could be no harder fate for him than to become deaf.

The recently arrived young Swiss in question was Nikolaus Fuss (1755–1826), whom Daniel Bernoulli had recommended as an assistant to Euler.[39]

On 10 December (29 November o.s.) 1773, Lexell writes to Wargentin regarding the death of Katharina Euler (*née* Gsell) [43, p. 125], the spouse of Leonhard Euler:

> Old Professor Euler has recently suffered the loss of his honourable wife, with whom he spent the last 40 years in perfect harmony and love. She was quite a venerable woman of an incomparably pious and well-meaning disposition, which makes this loss the more considerable for her family and friends. The loss has evidently touched him deeply, but besides the consolation he gets from his sincere worship of God, he also has a powerful means of distraction in his unremitting work.

[39] Nikolaus Fuss helped to prepare several hundred of Euler's papers and books. His first own works in mathematics, optics and astronomy were initiated and partly guided by Euler. Later he made more independent contributions to spherical trigonometry in Euler's and Lexell's footsteps and wrote many textbooks for schools [140]. He became Secretary of the Petersburg Academy of Sciences after J. A. Euler (1800) and was married to a grand-daughter of Euler's. A short biography of Fuss has been published in Russian [112].

Fig. 7.3 Leonhard Euler painted by Joseph Darbès (1747–1810) in 1778 [16]. Oil on canvas (Musée d'art et d'histoire, Ville de Genève, inv. N° 1829-8. Photo: Bettina Jacot-Descombes. Published with permission)

Three years later, to the great surprise of everybody and to the considerable bewilderment of Euler's family, Euler decided to marry again, as Lexell details in his letter to Wargentin dated 7 August 1776:

> A most strange piece of news from Petersburg is that old Professor Euler on the next Thursday, that is, tomorrow, shall marry the sister of his late wife, an old *Mamsell* 53 years of age: This event is for everybody quite unexpected and odd.

Euler's new spouse, Salome Abigail Gsell, was a half-sister of his late wife Katharina. When Euler had announced his plans of marrying again, his children had reacted with a mixture of anger and disappointment, no doubt because of the fear of a shrinking inheritance, cf. Gleb K. Mikhailov's account in [43].

In a letter to Wargentin dated 5 January 1776, Lexell mentions the recent consolidation of the Petersburg Academy of Sciences by the addition of chemist Johann Gottlieb Georgi and the mathematicians Nikolaus Fuss and Mikhail Golovin.[40] When Lexell was commissioned to examine the quality of Fuss's own work, he was struck by the feeling that the true author of the work was not Fuss but Euler himself. As this letter shows, Lexell clearly found it difficult to hide his reservations:

> Our new Director, *Kammerjuncker* [Chamberlain of the Court] Domashnev, has now entered his office more solemnly than usual; because he delivered a speech on that occasion, to which Professor [Johann Albrecht] Euler as Secretary of the Academy delivered a reply. He

[40]Mikhail Golovin (1756–1790) was a nephew of Mikhail Lomonosov and a disciple of Euler. He prepared many textbooks in mathematics and a Russian translation of Euler's *Théorie complète de la construction et manoeuvre des vaisseaux* (1778). He also edited Lomonosov's works.

> [Domashnev] is still a young man and seems to have much ambition to be well informed of the state of the Academy, and shows much zeal to make every possible improvement that would lead to progress in the sciences. Professor Rumovski, who during Count Orlov's times, as well as during his leave of absence, has been the most powerful member, may still maintain his influence, but it looks doubtful whether the two Eulers will continue their membership in the Academic Commission, with which they have flattered themselves hitherto.
> It is said that the Academy will receive reinforcements after New Year [old style] in the shape of three new Adjuncts, namely the pharmacist Georgi, and Mr Fuss and Golovin, both disciples of Euler the Elder. By the order of the new director they should first produce some proof of their skills, and Mr Georgi and Fuss have already given theirs. As to Mr Georgi's written specimen, I cannot judge anything, but so much I can say that the Latin has surely been revised by Professor Pallas, because in that subject Georgi is like pharmacists usually are. The specimen delivered by Mr Fuss is beautiful and reveals an exceptional genius in the mathematical sciences. I have to wonder how he, as the young man he is, in such a short time has absorbed not only the way of thinking of his great Master, but also the way of writing so perfectly, that unless the name of Herr Fuss appeared on the front page of the thesis, I would swear that it was Euler's. If he is the true author of this thesis himself, then the Academy could undoubtedly expect something good and remarkable from him, but if it has been composed with the help of old Euler, then I readily admit that the behaviour of this great man seems to me quite peculiar, since he has quite expressly given assurances of not knowing what his students were planning to write. Such learned loans are by no means unusual in Euler's family. For all the written pieces which have been submitted for the prize competitions in the name of the sons have been written entirely by the father.[41] Even so, I do not want to judge Mr Fuss to his disadvantage. It is only my opinion, that if he himself is the author of his thesis, he fully deserves, although still young, the encouragement of being promoted to become an Adjunct, the more so if he probably will be the one who can fill the place of the great Euler; but if he is not, it would have been better for him to wait for a while.

Despite Lexell's respect for Euler as a scientist, he obviously did not uncritically identify him as a paragon of virtue. The measures Euler sometimes took to secure the livelihood of his closest relatives—although perfectly understandable—nevertheless fulfil the characteristics of nepotism. Even when Euler did not participate in the weekly conferences of the Academy from 1773 onwards, he continued to influence the Academy's decisions through his son Johann Albrecht, who was appointed its secretary. As a matter of fact, Euler was until his death the most eminent member of the Academy and also headed the opposition against its more or less despotic directors. In such power struggles, Lexell turned out to be more conciliatory than the Eulers, who persistently resisted the authoritarian rule of the Academy, especially when it did not please them.

In the light of the surviving correspondence, Lexell appears remarkably silent about the characters of his closest associates in the circle around Euler, especially about his son Johann Albrecht as well as Nikolaus Fuss and Wolfgang Ludwig Krafft. Surely he must have entertained colourful opinions of his closest colleagues. Especially interesting would have been to hear his view on J. A. Euler, who as

[41] Referring to the winning essays of Johann Albrecht and Karl Johann Euler (1740–1790) in the prize competitions of Paris Academy of Sciences.

Fig. 7.4 Portrait of the young Johann Albrecht Euler (1734–1800) by Emanuel Handmann (1718–1781), ca. 1756 (Private collection, Switzerland. Reproduced by consent of the owners)

Permanent Secretary of the Academy was a key person in the events and intrigues of the Academy (Fig. 7.4). Why this silence? No doubt it was because of loyalty.

We have seen various instances of the Petersburg academicians being divided most of the time on the best strategy to adopt to cope with the authoritarian leadership and turn the situation to their benefit. Traditionally, it was the Russian versus the Western academicians who were opposed to each other. In this constellation, Fuss and Krafft were presumably more loyal to Leonhard Euler than to the directors, whereas Lexell generally did not accept coteries and was more pliable in surrendering to the rule of the authorities. On the other hand, divisions and tensions also arose from an undue competition of an individual academician for the director's favour.

Whether Fuss and Krafft at any point appeared as competitors or as a threat to Lexell vis-à-vis Euler is not easily judged, but regarding the depth and permanence of Lexell's mathematical and astronomical work, it is unlikely that such a situation ever occurred. On the contrary, everything this far suggests that Lexell was well-liked and enjoyed a wide respect within the Academy not only for his knowledge but for his pleasant character. Nonetheless, future archive work may cast some additional light on that subject, as there still remain archival sources in Russia which the author has not been able to peruse.

Chapter 8
Lexell's Work in Mathematics

As a scientist Lexell is undoubtedly best remembered for his work in celestial mechanics. For, according to a letter to Anders Planman dated 4 December 1771, in his own judgement

> ...old Euler is the only one [in Saint Petersburg] who appreciates the sublime speculations. The rest here thinks *Quae supra nos nihil ad nos.*[1] Had I not been needed as an astronomer, I would never have become a professor.

In these rather discouraging terms Lexell confesses to his colleague that Euler was without equal in the Academy. Nevertheless, Lexell's appreciation of himself seems rather harsh. In fact, Lexell was quite competent as a mathematician and one of the few at the time who was actually capable of understanding Euler's way of thinking. Even if the main part of his mathematical contributions may indeed have been suggested by Euler, it does not mean that his work was entirely superficial and self-evident, let alone that he was merely an epigone of Euler. In particular, Lexell's independent research in spherical geometry and polygonometry testifies that he was himself quite capable of grasping new and original fields of research.

The history of mathematics *qua* history of science dates back to Jean-Étienne Montucla's (1725–1799) pioneering work *Histoire des mathématiques*, the first edition of which appeared in Paris in 1758. However, Montucla died before completing the third volume, comprising the Age of Enlightenment, and the work was continued by Joseph Jérôme de Lalande and Charles Bossut. Some of Lexell's results were included in the didactic and historically well organised treatises on calculus and series [83, 84] by Sylvestre-François Lacroix (1765–1843). The first comprehensive and profound account of the development of mathematics is Moritz Cantor's (1829–1920) *Vorlesungen über Geschichte der Mathematik*, published in four volumes and completed in 1908 [23]. Although the latter work was criticised by the Swedish historian of mathematics Gustaf Eneström (1852–1923) for its occasional errors and omissions [32], it has the advantage of giving a fairly accurate

[1] What is above us is not for us (attributed to Socrates).

J.C.-E. Stén, *A Comet of the Enlightenment*, Vita Mathematica 17,
DOI 10.1007/978-3-319-00618-5_8,

outline of the different periods of mathematical research. Lexell features in several chapters of the fourth volume covering the years 1759–1799. Some of Lexell's findings in geometry are identified in [39], as well as in the surveys by Anton von Braunmühl [19] and Max Simon [150]. Additionally, the Russian historian V. I. Lysenko has elucidated Lexell's mathematical research in a series of comprehensive articles in Russian [109–111, 113]. In this chapter we give an overview of the principal areas of research which engaged Lexell as a mathematician and fill in some lacunae in the existing literature.

8.1 Early Work in Differential Geometry

Lexell's first independent work in mathematics was the thesis entitled *De methodo inveniendi lineas curvas, ex datis radiorum osculi proprietatibus* [Lexell 3], which he presented in Uppsala in 1763 (for the circumstances of this dissertation, see Sect. 2.3). Although the work does not contain any references, it draws essentially on Jakob Bernoulli's *Theorema aureum* for the radius of curvature.[2] A likely source is therefore Jakob Bernoulli's *Opera* (Geneva, 1744), which at some point was available at the University Library in Åbo [165]. According to the list of Lexell's books [47], he owned a private copy of it, but it seems unlikely that he could have possessed such a valuable book during his time in Åbo. Another source, or route for the transfer of knowledge, was Lexell's teacher Martin Johan Wallenius, who during a stay in Sweden in 1752–1754 had been influenced by Klingenstierna, who in turn was a student of Johann I Bernoulli.

Expressed in present day terminology, Lexell's thesis in Uppsala deals with a method of determining a curve from the knowledge of its radius of curvature. Lexell introduces the thesis by establishing 14 theorems, among which we may mention the following: If on a plane curve passing through the point $C = (x, y)$, the radius of curvature (*radius osculi curvae*[3]) is $R(C)$, $\mathrm{d}s$ is an element of the curve at C, and $p = \sin(\angle \frac{y}{R})$, then (*Theorema I:mum*)

$$\mathrm{d}x = -R\mathrm{d}p, \quad \mathrm{d}y = -\frac{Rp\mathrm{d}p}{\sqrt{1-p^2}}, \quad \text{and} \quad \mathrm{d}s = -\frac{R\mathrm{d}p}{\sqrt{1-p^2}}.$$

If $P = CE$ denotes the line drawn from C perpendicularly to E on the horizontal line drawn through the centre of osculation L, then on the basis of Theorem I (*Theorema 2:dum*)

[2] Jakob Bernoulli's elegant formula for the radius of curvature $\rho = \mathrm{d}^3s/(\mathrm{d}x\mathrm{d}^2y)$, where s is the arc length, appeared first in *Acta Eruditorum* for 1694.

[3] The osculating circle (*circulus osculans* is Latin for "kissing circle") of a curve at a given point C on the curve is defined, according to Leibniz (cf. *Acta Eruditorum* for 1686), as the circle passing through C and two additional points on the curve infinitesimally close to C.

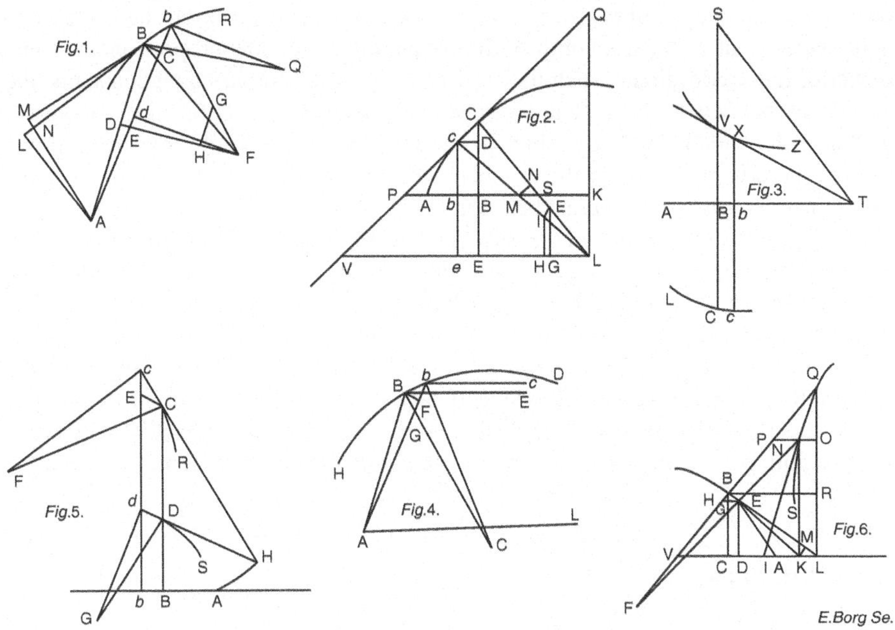

Fig. 8.1 Illustrations pertaining to Lexell's Uppsala dissertation in differential geometry (Lexell 3)

$$\mathrm{d}x = -\frac{P\mathrm{d}p}{\sqrt{1-p^2}}, \quad \mathrm{d}y = -\frac{Pp\mathrm{d}p}{1-p^2}, \quad \text{and} \quad \mathrm{d}s = -\frac{P\mathrm{d}p}{1-p^2}.$$

Denoting $LE = V$, one further obtains (*Theorema 3:ium*)

$$\mathrm{d}x = -\frac{V\mathrm{d}p}{p}, \qquad \mathrm{d}y = -\frac{V\mathrm{d}p}{\sqrt{1-p^2}}.$$

Hence, if the part of the tangent CV at C, which intercepts the horizontal line LEV at V, be denoted T (*Theorema 4:tum*)

$$\mathrm{d}x = -\frac{Tp\mathrm{d}p}{\sqrt{1-p^2}}, \qquad \mathrm{d}y = -\frac{Tp^2\mathrm{d}p}{1-p^2}.$$

Denoting $EV = H$ yields (*Theorema 5:tum*)

$$\mathrm{d}x = -\frac{Hp\mathrm{d}p}{1-p^2}, \qquad \mathrm{d}y = -\frac{Hp^2\mathrm{d}p}{(1-p^2)^{3/2}},$$

and so forth (Fig. 8.1).

Having proved geometrically all the 14 theorems with appropriate corollaries and scholia, Lexell proceeds to show how the properties of a curve can be derived from some knowledge of the radius of curvature. Among other things he was able to demonstrate that in the event of the radius of curvature R being proportional to some power of the y coordinate, such that $R : y^n = m : 1$, n and m being constants, the curve may be a cycloid, a catenary or even an elastic curve. He further investigates the curve described by a body which is attracted by a centripetal force proportional to any function depending on the distance from a fixed point, and shows that in some cases the curve in question must be a conic section.

In all, Lexell's systematic approach for finding the curve (function) by solving a differential equation of the second order was a remarkable achievement for a 22-year-old, even though the formula he used for the radius of curvature was already present in the work of Jakob Bernoulli and Leibniz. By this thesis, Lexell had proved, not least to himself, that he was capable of dealing with the subtlest mathematical problems independently and of grasping new and meaningful challenges quickly. However, the thesis is rarely listed among Lexell's works as it has possibly passed unnoticed among the many dissertations printed in Uppsala.

8.2 Analysis

After his earliest work in differential geometry, Lexell turned his attention to calculus and differential equations. The papers [Lexell 4] and [Lexell 5], which he submitted as specimens of his competence to the Petersburg Academy of Sciences in 1768 (they were thus essentially written before his arrival in Russia) concern in particular nonlinear differential equations of the second order (i.e. equations which involve the function, its first and second order derivatives, and possibly powers of these). Such equations have exact solutions in only few specific cases; the best known examples are Bernoulli's and Riccati's equations.

To solve equations of the kinds

$$
\begin{aligned}
&a^2yy'' + a^2b(y')^2 - y^2 = 0,\\
&x^2yy'' + bx^2(y')^2 + cxyy' - ay^2 = 0,\\
&y^2y''' + ayy'y'' - b(y')^3 = 0, \quad \text{and}\\
&y'''' + ay'''y' + b(y'')^2 + cy''(y')^2 - f(y')^4 = 0,
\end{aligned}
$$

where a, b, c and f are constants and y is a function of the variable x, Lexell employs a technique now known as *the integrating factor method*, where the equation is multiplied by an exponential of a solution expressed in terms of an integral. In this manner, the original equation turns into an ordinary first order differential equation with undefined coefficients, which may or may not depend on the variable x. The coefficients are then determined from an equation (or a system of equations) by comparison with the original equation. These equations admit a closed-form solution only in rare cases, when the coefficients are of a very specific form. Papers [Lexell 4] and [Lexell 5] contain lengthy exercises in formal manipulation and present a highly algorithmic way of solving equations of this kind.

The news of Lexell's success in solving this kind of equations quickly reached his colleagues in Finland and Sweden. In a postscript, inserted vertically in the margin of a letter to Anders Planman in Åbo (HUB Ms. Coll. 171) dated 25 June 1770 in Saint Petersburg, Lexell writes about his investigations:

> Here is a little differential equation for Professor Wallenius, which I have found a way to integrate[4]:
>
> $$\mathrm{d}y\,\mathrm{d}^3y + a\mathrm{dd}y^2 + b\mathrm{d}y^2\mathrm{dd}y + c\mathrm{d}y^4 = 0.$$

Although Lexell does not give a solution, he apparently manages to arouse the curiosity of Planman and Johan Henrik Lindquist, a student of Wallenius. Also Daniel Melander in Uppsala was kept informed, given that he wrote a short note[5] on its solution in the proceedings of the Swedish Academy of Sciences. On 4 December 1771, Lexell comments on Planman's and Lindquist's solutions (KVAC, Berg. brevs. Vol. 15, pp. 691ff.):

> The integrals that you, Herr Professor, as well as *Magister* Lindquist have found, are quite correct. The form that you, Herr Professor, have found is quite similar to mine, if only one puts $E = (2+a)cA/f$ and $C = (2+a)cB/f$. That $\mathrm{d}y$ receives a coefficient follows from my method, which is direct and not based on any substitution. Consider the integral
>
> $$\frac{\mathrm{dd}y + a'\mathrm{d}y^2}{\mathrm{d}y^n} + b'e^{\lambda' yg}\mathrm{d}x^{n+2} = 0,$$
>
> and obtain the values of a', b', λ' etc.

Lindquist's solution to the equation (which appeared in Lexell's letter cited above)

$$y'''y' + a(y'')^2 + by''(y')^2 + c(y')^4 = 0,$$

was considered in a thesis he presented in Åbo in May 1774, entitled *Methodus integrandi aequationes quasdam differentiales tertii ordinis.*[6] Key to the solution is the substitution $\mathrm{d}^2y = p\mathrm{d}y^2$, in other words $y'' = p(y')^2$, by means of which the original differential equation is reduced to $\mathrm{d}y((a+2)p^2 + bp + c) + \mathrm{d}p = 0$. After separation of variables the equation reads

$$y' = \frac{-p'}{(a+2)p^2 + bp + c},$$

---

[4]In the quotation, we have retained the original notation of the equations. Thus, we note for clarity that the forms dd and d^2 are equivalent.

[5]*Probleme, at integrera differential-æquationen* $\mathrm{d}^3y\mathrm{d}y + a\mathrm{dd}y\mathrm{dd}y + b\mathrm{dd}y\mathrm{d}y^2 + c\mathrm{d}y^4 = 0$, *hvilket nyligen af någon blifvit proponeradt*, KVAH, 1772, pp. 92–93.

[6]"Method of integrating certain differential equations of the third order". The thesis was defended *pro exercitio* by a student named Johan Tennberg (who, incidentally, would later marry a niece of Lexell's).

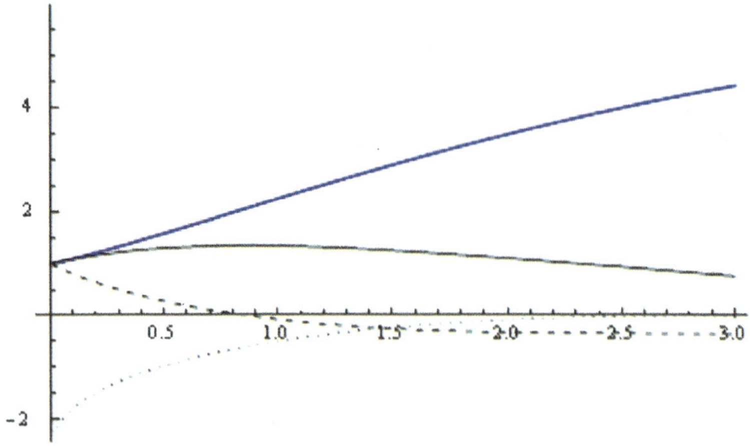

Fig. 8.2 Plot of a numerical solution to Lexell's differential equation when the constants are $a = b = 1$ and $c = 1/4$, and the initial values $y(0) = y'(0) = y''(0) = 1$. The thick line represents y, the thin line y', the dashed line y'' and the dotted line y'''

which expression is integrable in terms of inverse tangent functions and logarithms. Using the value(s) for p thus obtained and the substitution used above, the solution to the differential equation can be obtained via the equation $\log(y') = \int p dy$. In the thesis, the author (Lindquist) claims that Lexell was the first to have solved equations of this type, referring to his publications [Lexell 4] and [Lexell 5] (Fig. 8.2).

It is likely that these investigations incited Lexell to consider integrability of differential equations more generally, as we will see next. The method of reducing the order of the differential equation in the manner described above was already known to Euler and Lagrange at the time, but Lexell applied it systematically and with great skill to equations of higher order ($n = 3, 4$). The theory is also applicable to linear differential equations as shown in his paper [Lexell 5], which concerned a general and systematic method of solving linear differential equations with constant coefficients.

Another example of Lexell's algebraic capability can be found in the treatise [Lexell 7]—apparently the first paper he wrote entirely in Saint Petersburg—whose title reads in English: "Solution of an algebraic problem, the investigation of numbers in continued proportion,[7] whose sum a and sum of their squares b are prescribed". His interest in the problem was prompted by a problem given in Nicholas Saunderson's *The elements of algebra* (London, 1740, pp. 263ff.) originally proposed by Abraham de Moivre in the form:

[7] A sequence of numbers $a_1, a_2, a_3, \ldots$, where the consecutive proportions a_1/a_2, a_2/a_3 etc. are equal.

The sum of four continual proportionals being given, as also the sum of their squares, to find the proportionals. Let the four proportionals be x^3, x^2y, xy^2, y^3, their sum $= a$ and the sum of their squares $= b$.

In his book, Saunderson showed how to find a solution to a problem with four and five terms, but Lexell pursued the problem systematically for more general systems of equations of the kind

$$x^m + x^{m-1}y + x^{m-2}y^2 + \ldots + x^2y^{m-2} + xy^{m-1} + y^m = a,$$
$$x^{2m} + x^{2m-2}y^2 + x^{2m-4}y^4 + \ldots + x^4y^{2m-4} + x^2y^{2m-2} + y^{2m} = b.$$

The cases where the number of terms $m + 1$ is either even or odd were considered separately.

8.2.1 *Integrability of Differential Formulae*

When Euler was writing his foundations of the variational calculus in the appendix of the third volume of his Integral Calculus,[8] he had investigated the criteria for the integrability of higher-order differentials. He asked the question on what conditions the expression $\int Z\,\mathrm{d}x$ is a maximum or a minimum, and indeed managed to find a condition.[9] Meanwhile, the young Marquis de Condorcet (1743–1794) had independently of Euler dealt with the integrability of differential expressions involving two variables in his book *Du calcul intégral* (Paris, 1765). The historical development of the theorem is covered by Lexell in a letter to Johann Heinrich Lambert, which is given in the Appendix section "Letters".

Having already acquainted himself with Euler's and Condorcet's contributions, Lexell, in papers [Lexell 8] and [Lexell 15] set forth another approach to find out whether an arbitrary differential formula is integrable or not. Aiming at confirming Euler's condition by pure analysis without having recourse to the calculus of variations [23, pp. 904–907], he managed in these papers to prove the criterion for integrability of the expressions $\mathrm{d}x \int V\mathrm{d}x$, $\mathrm{d}x \int \mathrm{d}x \int V\mathrm{d}x$, up to $\mathrm{d}x \int \mathrm{d}x \int \mathrm{d}x \int V\mathrm{d}x$; the more integrations, the more complicated the proof will be.

To take the simplest case, let V be a function of $x, y, p, q, r, \ldots$ with $\mathrm{d}y = p\mathrm{d}x$, $\mathrm{d}p = q\mathrm{d}x$, $\mathrm{d}q = r\mathrm{d}x, \ldots, \mathrm{d}t = u\mathrm{d}x$ and

$$\mathrm{d}V = M\mathrm{d}x + N\mathrm{d}y + P\mathrm{d}p + Q\mathrm{d}q + \ldots + U\mathrm{d}u.$$

[8] *Institutionum calculi integralis volumen tertium*, St. Petersburg, 1770 (reprinted in: OO Ser. I, Vol. 13. E385).

[9] Appendix *De calculo variationum*, §§92, 129, 131.

Now, following Lexell, we can in every case put

$$V\mathrm{d}x = \mu\mathrm{d}x + \nu\mathrm{d}y + \pi\mathrm{d}p + \ldots + \tau\mathrm{d}t,$$

and therefore

$$V = \mu + \nu p + \pi q + \ldots + \tau u,$$

from which one obtains by Leibniz's chain rule

$$\begin{aligned}\mathrm{d}V &= \mathrm{d}\mu + p\mathrm{d}\nu + q\mathrm{d}\pi + \ldots + u\mathrm{d}\tau \\ &\quad + \nu\mathrm{d}p + \pi\mathrm{d}q + \ldots + \tau\mathrm{d}u.\end{aligned}$$

By comparison with the original expression, the following system of equations results:

$$\begin{aligned}M &= \left(\frac{\mathrm{d}\mu}{\mathrm{d}x}\right) + p\left(\frac{\mathrm{d}\nu}{\mathrm{d}x}\right) + q\left(\frac{\mathrm{d}\pi}{\mathrm{d}x}\right) + \ldots, \\ N &= \left(\frac{\mathrm{d}\mu}{\mathrm{d}y}\right) + p\left(\frac{\mathrm{d}\nu}{\mathrm{d}y}\right) + q\left(\frac{\mathrm{d}\pi}{\mathrm{d}y}\right) + \ldots, \\ P &= \left(\frac{\mathrm{d}\mu}{\mathrm{d}p}\right) + p\left(\frac{\mathrm{d}\nu}{\mathrm{d}p}\right) + q\left(\frac{\mathrm{d}\pi}{\mathrm{d}p}\right) + \ldots + \nu, \\ Q &= \left(\frac{\mathrm{d}\mu}{\mathrm{d}q}\right) + p\left(\frac{\mathrm{d}\nu}{\mathrm{d}q}\right) + q\left(\frac{\mathrm{d}\pi}{\mathrm{d}q}\right) + \ldots + \pi, \text{etc.},\end{aligned}$$

where the variables $x, y, p, q, \ldots, u$ are independent of each other, such that the expressions $\mathrm{d}p/\mathrm{d}x$ and the like are zero. Thus Lexell contended that, for $V\mathrm{d}x$ to be an exact differential,[10] the integrability condition will be the equation

$$0 = N - \frac{\mathrm{d}P}{\mathrm{d}x} + \frac{\mathrm{d}^2 Q}{\mathrm{d}x^2} - \frac{\mathrm{d}^3 R}{\mathrm{d}x^3} \ldots.$$

Analogously, for V being a function of four variables x, y, z and w,

$$\begin{aligned}\mathrm{d}V = M\mathrm{d}x &+ N\mathrm{d}y + P\mathrm{d}p + Q\mathrm{d}q + R\mathrm{d}r + \ldots \\ &+ N'\mathrm{d}z + P'\mathrm{d}p' + Q'\mathrm{d}q' + R'\mathrm{d}r' + \ldots \\ &+ N''\mathrm{d}w + P''\mathrm{d}p'' + Q''\mathrm{d}q'' + R''\mathrm{d}r'' + \ldots,\end{aligned}$$

[10]An expression of two variables $A(x, y)\mathrm{d}x + B(x, y)\mathrm{d}y$ is said to be an *exact differential* if there exists a function $f(x, y)$ such that $\mathrm{d}f = \left(\frac{\partial f}{\partial x}\right)_y \mathrm{d}x + \left(\frac{\partial f}{\partial y}\right)_x \mathrm{d}y = A\mathrm{d}x + B\mathrm{d}y$.

the conditions of integrability are three equations:

$$0 = N - \frac{\mathrm{d}P}{\mathrm{d}x} + \ldots, \qquad 0 = N' - \frac{\mathrm{d}P'}{\mathrm{d}x} + \ldots, \qquad 0 = N'' - \frac{\mathrm{d}P''}{\mathrm{d}x} + \ldots.$$

Lexell continued to elaborate his proofs, also giving examples, in a second dissertation [Lexell 15], realising that his original derivations were not quite watertight. In the years to come he extended his method into multivariate calculus in the papers [Lexell 53] and [Lexell 73], beginning with systems of linear equations of the kind

$$\mathrm{d}^2x + \alpha\mathrm{d}x + \beta\mathrm{d}y + \gamma x + \delta y = 0,$$
$$\mathrm{d}^2y + \alpha'\mathrm{d}x + \beta'\mathrm{d}y + \gamma' x + \delta' y = 0,$$

which by way of differentiation and a set of new expansion coefficients can be combined into an equation of the fourth order

$$\mathrm{d}^4x + (\alpha + \beta')\mathrm{d}^3x + (\alpha\beta' - \alpha'\beta + \gamma + \delta')\mathrm{d}^2x + (\alpha\delta' - \alpha'\delta + \beta\gamma' - \beta'\gamma)\mathrm{d}x + (\delta'\gamma - \delta\gamma')x = 0.$$

Lexell further considered the criteria for the integrability of equations of higher order and even several variables and managed to derive conditions for integrability analogous to the conclusions of his original investigations, see especially [23, pp. 1031–1032].

8.2.2 *Trigonometric Series*

The theory of series was a topic of great importance in eighteenth-century calculus. In particular, questions regarding the convergence of sums of infinite progressions and the development of mathematical expressions in infinite series were the subject of lively debate. The theory of trigonometric series and the orthogonality of trigonometric functions was still in its infancy when a famous controversy developed between Euler, d'Alembert and Daniel Bernoulli [79], concerning the possible solutions to the wave equation

$$c^{-2}\frac{\partial^2 y}{\partial t^2} = \frac{\partial^2 y}{\partial x^2},$$

for the shape of a taut string $y(x,t)$ (where y is the transversal displacement, x the position, t the time and c a constant), see e.g. [161]. More specifically, the question was about the class of functions which could be represented by trigonometric series, and whether it was possible to develop a discontinuous "function", or a function whose derivatives are discontinuous or even singular, into a trigonometric

(Fourier) series. In this debate, which took place in the 1740s and 1750s, Euler and, even more strongly, Bernoulli anticipated a broader concept of functions, which was systematically developed into the theory of generalised functions only a century later, whereas d'Alembert stuck to a narrower definition of "acceptable functions", see e.g. [79, 162].

The controversy regarding the vibrating string still echoed in two papers by Bernoulli and Euler, which had been incited by an article[11] by the French Abbé Charles Bossut (1730–1814). Bossut's article concerned finite trigonometric sums such as

$$S_n = \sin z + \sin 2z + \sin 3z + \ldots + \sin nz,$$

to which the author had discovered closed-form expressions. In Daniel Bernoulli's paper,[12] the more difficult question was tackled as to whether the sum for infinite n also converged. In particular, in order to arrive at closed form results such as

$$S_\infty = \sin z + \sin 2z + \ldots = \frac{\frac{1}{2}\sin z}{1 - \cos z},$$

Bernoulli questioned how one should deal with terms like $\cos \infty z$. Since Bossut's expression for S_n is composed of trigonometric expressions, it follows that the sum oscillates seemingly without a limit when $n \to \infty$. Hence, Bernoulli was compelled to introduce some "metaphysical" restrictions (as Cantor called them) in order to achieve convergence (assuming that the terms approach zero). Bernoulli's paper published in the *Novi Commentarii* of Saint Petersburg was followed by two articles, one by Euler[13] and another one by Lexell [Lexell 28].

Here, as usual, Euler's paper threw significant light on the question. Without touching Bernoulli's "metaphysical" arguments, Euler proposed that if the sum of the power expansion

$$az + bz^2 + cz^3 + \ldots + hz^n,$$

whether extending to infinity or not, is known to converge, then by introducing the formula $z = re^{\sqrt{-1}\varphi}$ and the celebrated relation $e^{\sqrt{-1}\varphi} = \cos\varphi + \sqrt{-1}\sin\varphi$, the series is transformed into a series of sines and cosines containing a progression

---

[11] "Manière de sommer les suites dont les termes sont des puissances semblables de sinus ou de cosinus d'arcs qui forment une progression arithmétique", *Mémoires de l'Académie Royale des Sciences*, 1769 (1772), pp. 453–466.

[12] *Theoria elementaris serierum, ex sinibus atque cosinibus arcuum arithmetice progredientium diversimode compositarum, dilucidata*, NCASIP XVIII, pp. 3–23, 1774 (item St.66 in Straub's catalogue [55]).

[13] *Summatio progressionum* $\sin \phi^\lambda + \ldots$ NCASIP XVIII, pp. 24–36, 1774 (OO Ser. I, Vol. 15, pp. 168–184. E447).

of the argument. Assuming plainly the position of Euler, Lexell considered further various sums of sequences of trigonometric functions, such as

$$\sin z + \sin(z + \nu) + \sin(z + 2\nu) + \sin(z + 3\nu) + \ldots + \sin(z + n\nu),$$

$$\cos z + \cos(z + \nu) + \cos(z + 2\nu) + \cos(z + 3\nu) + \ldots + \cos(z + n\nu),$$

and derived a number of closed-form results when n is either finite or infinite. As we know now, the ambiguities related to the convergence of so-called Fourier series were resolved only in the first part of the nineteenth century and in its most general sense in the late twentieth century (theorem by Lennart Carleson, 1966).

8.2.3 Elliptic Integrals

Elliptic integrals form a class of special functions that occur frequently in physics and astronomy. The history of their research goes back to 1655, when John Wallis (1616–1703) considered the lengths of elliptical arcs in terms of power series (hence the name elliptic integral). The present tripartite classification of elliptic integrals (classes E, F and Π) was proposed by Adrien-Marie Legendre (1752–1833). His first two papers on elliptic integrals concerned the properties of elliptical arcs and were published in 1786, and his great treatise on elliptic integrals *Traité des fonctions elliptiques* began to appear in 1825 [55].

However, long before Legendre's contribution, Euler had put forward a classification of elliptic integrals[14] according to their relation to the arc lengths of ellipses and hyperbolas, or, to conic sections in general. In the publication [Lexell 61] (which, incidentally, bore the same title as Euler's article), Lexell considered indefinite integrals of the form

$$\int \sqrt{\frac{1 + mz^2}{nz^2 + 1}} \mathrm{d}z,$$

which, by way of substitutions, he managed to reduce into a form involving trigonometric functions (type E standard elliptic integral), and further identified 12 forms of integrals that could be treated in such a way. Other similar formulas such as

$$\int \frac{\mathrm{d}z}{\sqrt{(1 + mz^2)(1 \pm nz^2)}}, \qquad \int \frac{\mathrm{d}z}{\sqrt{(1 \pm mz^2)(nz^2 - 1)}},$$

[14] *De reductione formularum integralium ad rectificationem ellipsis ac hyperbolae*, NCASIP X, pp. 3–50, 1766 (OO Ser. I, Vol. 20, pp. 256–301. E295).

were considered in [Lexell 65]. In the same paper, Lexell gives further evidence for the solidness of Euler's classification. Apparently, Lexell's two memoirs made no lasting impression, as they are seldom mentioned in the standard works on the history of elliptic integrals [40, 70]. It was only after Legendre's classification of elliptic integrals that a more complete theory of elliptic integrals, their inverses (i.e. the elliptic functions) and addition theorems for them could be developed by Carl Gustav Jacob Jacobi (1804–1851).

The related articles [Lexell 91] and [Lexell 107] deal with indefinite integrals containing fourth roots, in particular

$$\int \frac{\mathrm{d}x}{(1+x)\sqrt[4]{2x^2-1}},$$

which, rather surprisingly, can be done analytically in terms of elementary functions. In fact, Lexell gives two alternative solutions, both equally elegant, as is shown below [Lexell 107]: First, the substitution $\tan z = (2x^2-1)^{1/4}$ yields $2x^2 = 1 + \tan^4 z$ and further

$$x^2 = \frac{1+\cos^2 2z}{4\cos^4 z}.$$

Thus, one finds

$$\frac{\mathrm{d}x}{(1+x)\sqrt[4]{2x^2-1}} = \frac{\mathrm{d}z \tan^2 z}{x(1+x)\cos^2 z},$$

where

$$x(1+x) = \frac{\sqrt{1+\cos^2 2z}}{4\cos^2 z}(2\cos^2 z + \sqrt{1+\cos^2 2z}).$$

After the substitution of $\cos 2u$ for $\cos^2 2z$ and the ensuing differentiations and manipulations, the expression becomes integrable.

Alternatively, by expansion of the numerator and denominator by $1-x$, the original integral expression breaks down into two terms. By using the substitution $y = (2x^2-1)^{1/4}$ in one of the expressions and $v = (2x^2-1)^{1/4}/x$ in the other, Lexell shows that both expressions are integrable. In particular, he obtains the results

$$\int \frac{-x\mathrm{d}x}{(1-x^2)\sqrt[4]{2x^2-1}} = \int \frac{\mathrm{d}y}{1+y^2} - \int \frac{\mathrm{d}y}{1-y^2}, \qquad \text{and}$$

$$\int \frac{\mathrm{d}x}{(1-x^2)\sqrt[4]{2x^2-1}} = -\frac{1}{2}\int \frac{\mathrm{d}v}{1+v^2} + \frac{1}{2}\int \frac{\mathrm{d}v}{1-v^2},$$

both of which may be expressed in terms of inverse trigonometric and logarithm functions.

8.2.4 Lagrange's Problem

In October 1766, the 30-year-old Joseph Louis Lagrange had arrived in Berlin from his native town of Turin to succeed Leonhard Euler as Professor of Mathematics at the Royal Prussian Academy of Sciences. In 1787, he moved on to Paris, where he published his famous *Mécanique analytique* in 1788 and where he remained for the rest of his life. In 1770, while still in Berlin, he presented an article for the Royal Prussian Academy of Sciences entitled "A new method to solve equations with literal coefficients in terms of series",[15] which accordingly dealt with the solution of equations by means of power series. More specifically, if the equation is $y - x + \varphi(x) = 0$, the objective is so to speak to "invert" the unknown x in terms of another series. According to the theorem, for a given function ψ, we have

$$\psi(x) = \psi(y) + \frac{\mathrm{d}\psi(y)}{\mathrm{d}y}\varphi(y) + \frac{\mathrm{d}\frac{\mathrm{d}\psi(y)}{\mathrm{d}y}\big(\varphi(y)\big)^2}{2\cdot \mathrm{d}y} + \frac{\mathrm{d}^2\frac{\mathrm{d}\psi(y)}{\mathrm{d}y}\big(\varphi(y)\big)^3}{2\cdot 3\cdot \mathrm{d}y^2} + \ldots.$$

Lagrange used the method to solve algebraic and transcendental equations, such as those arising in the solution of Kepler's problem involving two bodies (i.e. for the equation $t = x - e\sin x$). His approach was quite different from all earlier attempts and attracted much interest among contemporary mathematicians (for a review of the development of Lagrange's inversion theorem and its demonstrations, see [105]). Even much later, Lagrange's theorem would play an important part in Laplace's five-volume *Traité de mécanique céleste* (1799–1825) and in the calculus developed by Cauchy.

Finding a proof of the theorem which Lagrange had communicated to Euler also engaged Lexell. In a letter to Anders Planman (KVAC, Berg. brevs. Vol. 16, pp. 691ff.) dated 4 December 1771, Lexell describes his encounter with the problem in the following words:

> I have searched for a proof for the following rather curious theorem discovered by la Grange, which Euler told me about, since la Grange had only used induction and that of a very particular kind. If $t - x + \varphi x = 0$, then
>
> $$\psi x = \psi t + \varphi t \psi' t + \frac{\mathrm{d}(\varphi t)^2 \psi' t}{1\cdot 2\cdot \mathrm{d}t} + \frac{\mathrm{dd}(\varphi t)^3 \psi' t}{1\cdot 2\cdot 3\cdot \mathrm{d}t^2} + \frac{\mathrm{d}^3(\varphi t)^4 \psi' t}{1\cdot 2\cdot 3\cdot 4\cdot \mathrm{d}t^3} + \text{etc.}$$
>
> Even more general is the following theorem. If $t = x - P$ when P is a function of both x and t, and P' is the value of that function when instead of x we write t and instead of t, a, one obtains
>
> $$\psi x = \psi t + P' \psi' t + \frac{\mathrm{d}(P')^2 \psi' t}{1\cdot 2\cdot \mathrm{d}t} + \frac{\mathrm{dd}(P')^3 \psi' t}{1\cdot 2\cdot 3\cdot \mathrm{d}t^2} + \frac{\mathrm{d}^3(P')^4 \psi' t}{1\cdot 2\cdot 3\cdot 4\cdot \mathrm{d}t^3} + \text{etc.},$$

[15] "Nouvelle méthode pour résoudre les équations littérales par le moyen des séries", *Histoire de l'Académie Royale des Sciences et Belles-Lettres de Berlin*, 1770, pp. 251–328; also in *Œuvres* [85, Vol. III, pp. 5–73].

> when after the differentiation a is in each case replaced by t, NB $\psi' t = d\psi/dt$. Can anything more general be invented in analysis? I also found a demonstration. Later on old Euler found several proofs; he is in the habit of turning things around from all sides, but I am happy with the one I have found, as it seems to be rather direct. La Grange has used this theorem for many things, such as to search for *radices aequationum per approximationem* [roots of equations by means of approximations] to solve the well-known *Problema Kepplerianum*. For this purpose it is not suitable [...] unless the eccentricity is rather small. Yet there may be given a way to find a rapidly converging series, since this theorem is rather general, but so far I have not found anything which pleases me.

Although Euler had found many ways to prove the theorem, he encouraged Lexell to communicate to Lagrange his own demonstration, which unlike Lagrange's did not depend on induction but was more complete, as it was based on recursion. Three demonstrations of the theorem are given in Lexell's letter of 5 March 1772 [85, Vol. XIV, pp. 228–235], two of them being due to Euler and one due to Lexell.

In a preliminary lemma in his paper [Lexell 16] (in English: "Proof of a theorem by Lagrange"), Lexell carefully showed that for two functions y and z, if the relation

$$\sum_{k=0}^{m}(-1)^k y^{m-k} C_m^k d^{m-1}(y^k z) = 0$$

holds, then it must hold also for a sum extended to $m + 1$. Obviously the relation holds for the order 1, because it gives $yd^0 y - y^0 d^0(yz) = yz - yz = 0$. Lagrange's formula can now be deduced by using Taylor's formula by the following steps: Let the equation to be solved be $t = x - \varphi(x)$. Then, for any function ψ, one has to calculate $\psi(x)$ as a function of t, and thus,

$$\psi(t) = \psi(x - \varphi(x)) = \sum_{n=0}^{\infty}(-1)^n \frac{1}{n!}\frac{d^n(\psi(x))}{dx^n}(\varphi(x))^n.$$

Substituting for $\psi(t)$ each of the terms appearing in the series expansion of Lagrange, viz.

$$\frac{1}{k!}\frac{d^{k-1}}{dt^{k-1}}\left[(\varphi(x))^k \psi'(t)\right],$$

Lexell develops the sum of different series for $k \geq 0$. He concludes his demonstration of Lagrange's theorem by changing the order of summation and taking advantage of the first sum expression given in the preliminary lemma. A similar proof was later given by Johann Friedrich Pfaff.[16]

---

[16] "Analysis einer wichtigen Aufgabe des Herrn de la Grange". *Archiv der reinen und angewandten Mathematik*, 1795, pp. 81–88.

8.3 Geometry

8.3.1 Science of the Spheres

Owing to the rapid development of astronomy, geography and geodesy since the sixteenth century, spherical geometry, or the science of geometric figures drawn on the surface of a sphere, attracted the attention of mathematicians. The interest was mainly due to astronomy: the clear night sky may give the impression of the celestial objects appearing as bright dots attached to a vast sphere of arbitrary radius—the celestial sphere—in the midpoint of which the observer is situated. Only the direction (and movement) of the objects, but not their distance, can be observed. Before Euler and Lexell, only a few theorems dealing with spherical triangles were known, one of them due to Menelaus of Alexandria (ca. 70–130). Key to the research in spherical geometry for Lexell were a theorem by Albert Girard (1595–1632) and the compendium of trigonometric formulas in the *Canon mathematicus, seu ad triangula* (1579) of François Viète (1540–1603). Euler had proved and elaborated some of the theorems in a series of papers.[17] However, some questions of spherical geometry extending the results of plane geometry still waited for research [23].

Lexell's work on spherical geometry is summarised in the following publications, the titles of which are given here in English:

—"On epicycloids described on spherical surfaces", 1779 (published in 1781) [Lexell 69],
—"Solution to a problem of geometry in the theory of the sphere", 1781 (published in 1784) [Lexell 88],
—"On the properties of circles described upon a spherical surface", 1782 (published in 1786) [Lexell 95],
—"Demonstration of some theorems in the theory of the sphere", 1782 (published in 1786) [Lexell 98].

The first of these papers was clearly prompted by an astronomical problem: the apparent motion of comets and planets as viewed from the Earth. In the second paper [Lexell 88], Lexell presents a theorem on spherical triangles, which is now known by his name, as follows:

Theorem [Lexell 88] (1781/1784)
On the surface of the sphere, the line in which are situated the vertices of all the triangles having the same base and the same surface area is a small circle of the sphere.

A spherical triangle is a triangle drawn upon a spherical surface, whose every side is an arc of a great circle. The theorem thus asserts that the locus of the vertices of the triangles which have the same base and the same area is itself a small circle on the sphere. This circle is known as the "Lexell circle".

[17]On Euler's work in spherical trigonometry, see: OO Ser. III, Vol. 10.

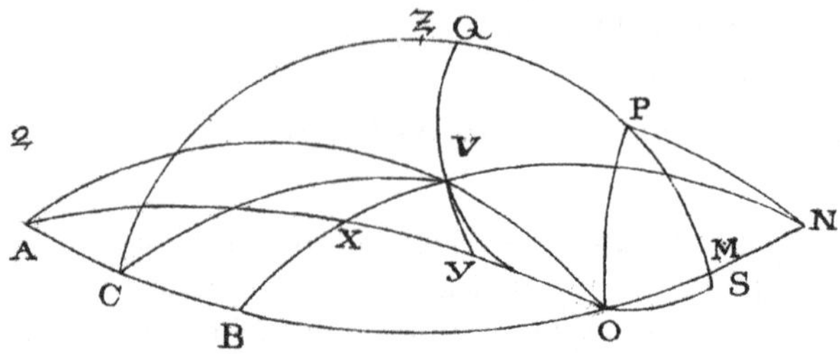

Fig. 8.3 Lexell's construction of the theorem in [Lexell 88]

In his paper [Lexell 88], which was published in the Academy's *Acta* in 1784, but which was written no later than 1778, Lexell proves the theorem first analytically, using the rules of spherical trigonometry, and finally *more geometrico*, which is traditionally regarded as the most persuasive mode of demonstration. On the other hand, Euler also gave a derivation of this theorem, using both analysis (surface differentials) and geometry, in a paper which was read in 1778, but not published until 1797.[18] The construction of Lexell's own demonstration is the following [Lexell 88; §5] (cf. the spherical triangle ABV as in Fig. 8.3):

Draw a great circle ZC through the point C normally on AB and join CV. Then, [by appropriately denoting the angles and arcs involved] ... the equation [given earlier in §4] shows that the curve obtained must be a small circle, the construction of which proceeds as follows: Construct a great circle CZ such that it meets with the great circle ABO for the second time in the point M and intersects it so that MO = CB. Next, draw PO which with the arc MO forms the angle POM = $90^\circ - \delta$; now if one describes around the pole P a small circle OVQ whose radius is the arc PO, this circle is the locus of those points V for which, if from the given points A and B the arcs of a great circle AV, BV, respectively, be drawn (to an arbitrary point V) the triangle BVA is of a given size, the sum of the angles is $= 180^\circ + 2\delta$. For, from

$$\mathrm{POM} = 90^\circ - \delta, \qquad \tan \mathrm{POM} = \cot\delta = -\frac{\tan\varepsilon}{\sin a},$$

it follows that

$$\tan \mathrm{POM} = \frac{\tan \mathrm{PM}}{\sin \mathrm{MO}}$$

and so $\tan \mathrm{PM} = -\tan\varepsilon$, and $\mathrm{PM} = 180^\circ - \varepsilon$, because the arc $\mathrm{CZP} = \varepsilon$. Because $\cos \mathrm{PO} = \cos \mathrm{OM} \cos \mathrm{PM}$, and therefore $\cos \mathrm{PO} = -\cos a \cos\varepsilon$, from $\mathrm{OM} = \mathrm{CB} = a$ we obtain PO = γ; On the other hand, it is clear that for a spherical triangle PCV

$$\cos \mathrm{PV} = \cos \mathrm{PO} = \cos \mathrm{PC} \cos \mathrm{VC} + \sin \mathrm{PC} \sin \mathrm{VC} \cos \mathrm{PCV},$$

that is, $\cos\gamma = \cos\varepsilon\cos z + \sin\varepsilon\sin z\cos\varphi$, which is the equation found [in §4]. Because it has already been shown that if the arc $\mathrm{CP} = 180^\circ - \mathrm{PM} = \varepsilon$, and therefore $\tan \mathrm{CP} = \tan\varepsilon = -\cot\delta \sin a$, then

$$\mathrm{CQ} = \mathrm{CP} - \mathrm{PQ} = \mathrm{CP} - \mathrm{PO} = \varepsilon - \gamma \quad \text{and} \quad \mathrm{CS} = \mathrm{CP} + \mathrm{PO} = \varepsilon + \gamma,$$

[18] *Variae speculationes super area triangulorum sphaericorum*, NAASIP X, pp. 47–62, 1797 (OO Ser. I, Vol. 29, pp. 253–266. E698). In that paper, Euler gives full credit for the discovery of the theorem to Lexell.

which rule determines the two points in which the small circle QVO meets the great circle CZP. [19]

Hence the geometric proof runs as follows [Lexell 88; §10]:

Take a point V anywhere on this small circle such that it is located on the arc OVQ, which is above COM, and draw the great semicircles AVO, BVN, as well as the arcs of a great circle PV, PN. Now, on account of the arcs BON = CBM = ACO, it follows that CB = MO = MN, and thus PN = PO = PV, hence in the isosceles triangle PVN, the angle PVN = PNV and in the isosceles Δ PVO, the angle PVO = POV. Now, on account of the angle PON = PNO = PNV + VNO, and VNO = VBO, the angle PON = 180°− POB = PVN + VBO, whereby

$$180^\circ - \mathrm{VOB} - \mathrm{POV} = 180^\circ - \mathrm{VOB} - \mathrm{PVO} = \mathrm{PVN} + \mathrm{VBO}.$$

Adding the angle BVO on both sides,

$$180^\circ + \mathrm{BVO} - \mathrm{VOB} - \mathrm{PVO} = \mathrm{BVO} + \mathrm{VBO} + \mathrm{PVN},$$

where, since PVN = PVO − VNO = PVO − AVB, it follows that

$$180^\circ + \mathrm{BVO} - \mathrm{VOB} - \mathrm{PVO} = \mathrm{BVO} + \mathrm{VBO} + \mathrm{PVO} - \mathrm{AVB},$$

so that

$$360^\circ - 2\,\mathrm{POV} = \mathrm{BVO} + \mathrm{VBO} + \mathrm{BOV}.$$

Since the angle VOX = 90°− POV, then 180°− 2 POV = 2 VOX, and so

$$180^\circ + 2\,\mathrm{VOX} = \mathrm{BVO} + \mathrm{VBO} + \mathrm{BOV},$$

---

[19]Citing the original Latin text in [Lexell 88; §5]: "Per punctum C ducatur circulus maximus ZC normalis ad AB et iungatur CV, tum vero dicatur CV=z et angulus ZCV=ϕ, eritque ob $\cos \mathrm{CR} \cos \mathrm{VR} = \cos \mathrm{VC}$, et $\sin \mathrm{VR} = \sin \mathrm{VC} \sin \mathrm{VCR} = \sin \mathrm{VC} \cos \mathrm{ZCV}$, $\cos x \cos y = \cos z$ et $\sin y = \sin z \cos \phi$, his igitur valoribus in aequatione allata substitutis, fiet $\cot \delta \sin a \sin z \cos \phi = \cos z + \cos a$, quae iam facile ad huiusmodi formam reducitur: $\cos \gamma = \cos z \cos \varepsilon + \sin z \sin \varepsilon \cos \phi$, ponendo $\frac{\cos \gamma}{\cos \varepsilon} = -\cos a$, et $-\tan \varepsilon = \cot \delta \sin a$, ex qua aequatione manifesto liquet curvam istam quaesitam esse circulum minorem, cuius constructio hunc in modum adornatur: Concipiatur circulum maximum CZ productum iterum occurrere circulo maximo ABO in puncto M, tumque resecetur MO = CB, et ducatur PO, qui cum arcu MO facit angulum POM = $90^\circ - \delta$; iam si polo P intervallo arcu PO describatur circulus minor OVQ, erit hic circulus locus istorum punctorum V, ita sitorum, ut si ex datis punctis A, B ad punctum quodpiam V ducantur arcus circulorum maximorum AV, BV, erit triangulum BVA datae magnitudinis, summa angulorum existente = $180^\circ + 2\delta$. Nam ob $\mathrm{POM} = 90^\circ - \delta$, erit $\tan \mathrm{POM} = \cot \delta = -\frac{\tan \varepsilon}{\sin a}$, est vero $\tan \mathrm{POM} = \frac{\tan \mathrm{PM}}{\sin \mathrm{MO}}$, eritque igitur $\tan \mathrm{PM} = -\tan \varepsilon$, et $\mathrm{PM} = 180^\circ - \varepsilon$, quare arcus CZP=$\varepsilon$, tum vero sit $\cos \mathrm{PO} = \cos \mathrm{OM} \cos \mathrm{PM}$, ideoque $\cos \mathrm{PO} = -\cos a \cos \varepsilon$, ob $\mathrm{OM} = \mathrm{CB} = a$, hinc obtinemus PO=$\gamma$; sponte autem liquet esse pro triangulo sphaerico PCV; $\cos \mathrm{PV} = \cos \mathrm{PO} = \cos \mathrm{PC} \cos \mathrm{VC} + \sin \mathrm{PC} \sin \mathrm{VC} \cos \mathrm{PCV}$, hoc est $\cos \gamma = \cos \varepsilon \cos z + \sin \varepsilon \sin z \cos \phi$, quae est aequatio supra inventa. Quia uti iam observavimus, sit arcus $\mathrm{CP} = 180^\circ - \mathrm{PM} = \varepsilon$, ideoque $\tan \mathrm{CP} = \tan \varepsilon = -\cot \delta \sin a$, erit $\mathrm{CQ} = \mathrm{CP} - \mathrm{PQ} = \mathrm{CP} - \mathrm{PO} = \varepsilon - \gamma$ et $\mathrm{CS} = \mathrm{CP} + \mathrm{PO} = \varepsilon + \gamma$, qua ratione bina puncta definiuntur, in quibus circulus minor QVO circulo maximo CZP occurrit."

from which it can be concluded that the spherical triangle BVO equals the spherical segment AVOXA, and if from both members VOX be subtracted, the triangle AVX = BOX, and if to both members the triangle AXB be added, the segment AXOBA = Δ AVB. [20]

In order to give a modern proof of Lexell's theorem we follow Fejes Tóth's geometric construction [42] closely and utilise the concept of a polar triangle, defined as follows: if A, B and C are the vertices defining the triangle Δ =ABC, then the polar triangle Δ' =A′B′C′ is the dual of Δ such that A′ is the pole of the great circle BC closest to A (meaning the pole whose spherical distance to A is $< \pi/2$), B′ is the pole of the great circle AC closest to B, and C′ is the pole of the great circle AB closest to C. The polar triangle of the polar triangle of Δ is the triangle itself, i.e. Δ. (An obvious corollary is that there exists a certain spherical triangle which coincides with its own polar triangle) (Fig. 8.4).

We also use Girard's theorem, stating that

$$\alpha + \beta + \gamma - \pi = \text{Area of } \Delta,$$

where the angles of the triangle Δ are denoted by α, β and γ.

Proof. From the point A on the sphere, let two great circles G_1 and G_2 be drawn, which intercept at $A^\star$ (the antipode of A). Inside the figure, closer to the pole $A^\star$, a small circle is drawn at a tangent to both the great circles G_1 and G_2. Consider the arc S of this circle K, whose end points are the two points T_1 and T_2, where the lines touch the circle, and which is located closer to A. The segment from A to the centre of S bisects the triangle formed by A, T_1 and T_2. Let the points B and C be two points on the great circles G_1 and G_2, respectively, between A and the respective points T_1 and T_2. If one of the intersecting points, say B, is given, the other, say C, is determined such that a great circle drawn from B is tangent to the arc of the small circle at T_3. Let the sides of Δ be denoted by a, b and c. Then it is easy to see from the construction that the perimeter of Δ is constant and independent of the tangent point T_3, viz. $a+b+c = p$, where $p/2$ is the distance from A to the tangent points T_1 and T_2.

---

[20] Citing the original §10: "Sumatur punctum V ubicunque in hoc circulo minore, modo in arcu OVQ supra COM elevato reperiatur, et ducantur semicirculi maximi AVO, BVN, tumque arcus circulorum maximorum PV, PN. Iam ob arcum BON = CBM = ACO, fiet CB = MO = MN, ideoque PN = PO = PV, hinc in triangulo PVN aequicruro, erit ang. PVN = PNV et in Δ PVO aequicruro, ang. PVO = POV. Nunc vero ob ang. PON = PNO = PNV + VNO, et VNO = VBO, fiet ang. PON = 180°− POB = PVN + VBO, ideoque 180°− VOB − POV = 180°− VOB − PVO = PVN + VBO, addatur utrinque angulus BVO, eritque 180° + BVO − VOB − PVO = BVO + VBO + PVN, unde ob PVN = PVO − VNO = PVO − AVB, fiet: 180° + BVO − VOB − PVO = BVO + VBO + PVO − AVB, unde 360°− 2 POV = BVO + VBO + BOV, quum igitur sit angulus VOX = 90°− POV, erit 180°− 2 POV = 2 VOX, hincque 180° + 2 VOX = BVO + VBO + BOV, ex quo iam constat triangulum sphaericum BVO aequari segmento sphaerae AVOXA, hincque demto utrinque triangulo communi VOX, fiet triangulum AVX = BOX et addito denuo utrinque triangulo AXB fiet segmentum AXOBA=Δ AVB."

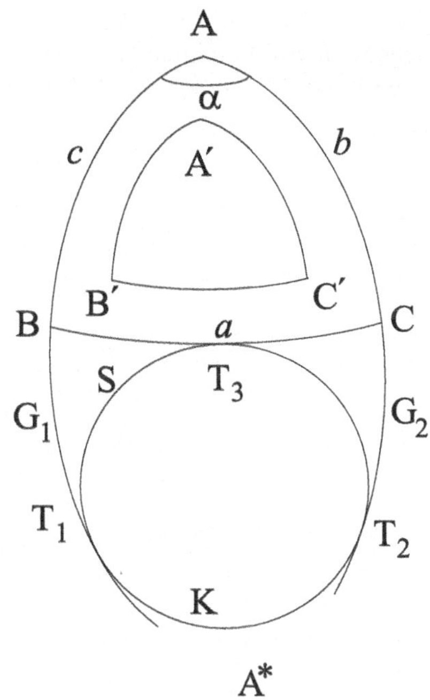

Fig. 8.4 The triangle and its polar triangle in Fejes Tóth's proof of Lexell's theorem

Analogously, let the angles of the polar triangle Δ' be α', β' and γ', and its sides a', b' and c'. From the properties of polar triangles, it is obvious that $\alpha' + a = \pi$, $\beta' + b = \pi$, and $\gamma' + c = \pi$. Thus, according to Girard's theorem

$$\pi - a + \pi - b + \pi - c - \pi = 2\pi - p = \text{Area of } \Delta'.$$

Thus, whereas in the triangle Δ the angle α remains constant (so that the triangle Δ may vary while its perimeter is unchanged), in the triangle Δ' the area is constant when the length of its base B′C′ is constant viz. $a' = \pi - \alpha$. Now, if the base BC forms a tangent to the circle K, the closest pole A′ must also lie on an arc of the circle, called K′, which is located at all points at a distance $\pi/2$ from K, and bounded by $B'^{\star}$ and $C'^{\star}$. □

In the paper [Lexell 95] Lexell pursues his investigations by proving many theorems pertaining to spherical triangles and spherical quadrilaterals. Among other things he notes that "if on a sphere the vertices of a spherical quadrilateral lie on a small circle, the sums of the pairs of opposite angles are equal" (§33), which is a generalisation of the corresponding theorem for cyclic quadrilaterals in Euclidean geometry. Furthermore, he shows that the sums of the lengths of the opposing sides of a spherical quadrilateral circumscribed around a circle are equal (§40), which is the dual theorem of the preceding one. Lexell also derives various expressions for the relationship between the angles and the sides of spherical triangles and quadrilaterals. In particular, when a, b, c are the lengths of the sides of the spherical triangle and

$$s = (a + b + c)/2,$$

he shows that the value of d given by the expression [19]

$$d = 2\sqrt{\sin s \sin(s - a) \sin(s - b) \sin(s - c)}$$

is a constant. Likewise, when A, B, C are the angles corresponding to the sides a, b, c and $s = (A + B + C)/2$, he shows that the expression

$$\delta = 2\sqrt{-\cos S \sin(S - A) \sin(S - B) \sin(S - B)}$$

remains constant. By means of this theorem, Lexell also found expressions for the radii of the circles which can be inscribed in and circumscribed around the triangle.

Analogous theorems were next given for spherical quadrilaterals. In particular, denoting by a, b, c and d the length of the sides, and assigning $s = (a+b+c+d)/2$, then the excess angle ε (over 2π) of the sum of the angles of a spherical quadrilateral which can be inscribed in a small circle

$$\tan\frac{\varepsilon}{4} = \sqrt{\tan\frac{s-a}{2}\tan\frac{s-b}{2}\tan\frac{s-c}{2}\tan\frac{s-d}{2}}$$

follows from the equations on page 88 of [Lexell 95] [109], although the formula is only implicitly present (contrary to the claim of [145, p. 34]). By applying Girard's theorem for the area of spherical triangles, it also follows that the angle ε will be proportional to the surface area of the spherical quadrilateral. This is the analogue of Brahmagupta's formula for the area of a quadrilateral inscribed in a circle in the plane:

$$S = \sqrt{(s - a)(s - b)(s - c)(s - d)}.$$

Furthermore, in the event that either a, b, c, or $d = 0$, it is analogous to the formula originated by Archimedes and Heron of Alexandria for the surface area of a triangle. Lexell's formula is also more general than a later theorem by L'Huilier[21] for a spherical triangle.

[21] Simon Antoine Jean L'Huilier (1750–1840), Genevan mathematician [55, 175].

8.3.2 Euler's Theorem on Rotation

Euler's rotation theorem (1775) [1, 23, 39, 133, 172] and the Euler angles are today commonplace in geometry and many branches of mathematical physics. The theorem states that any displacement of a rigid body such that a point on the body remains fixed is equivalent to a rotation about a fixed axis through this point. In connection to the theorem, Euler also showed how to perform the rotation systematically by using three angles. The theorem came into being while Euler was reorganising the laws of motion, having realised that the law of linear momentum is not sufficient to describe the rotational motion of a rigid body and that it must be appropriately complemented with the law of conservation of angular momentum [161].

The rotation theorem was presented in the *Novi Commentarii* for the year 1775 in a Latin memoir entitled in English "General formulae for the arbitrary translation of a rigid body".[22] Knowing that his algebraic proof was incomplete, Euler put forward this problem for other "geometers" to test their skills. In an article published in the same volume of the *Novi Commentarii* immediately after Euler's, Lexell took up the challenge and further elaborated the procedure in his Latin paper [Lexell 45] entitled in English "Some general theorems concerning the translation of a rigid body". More specifically, he proved by theorems of spherical geometry that an expression, which today is easily identified as a determinant of a 3×3 eigenvalue equation, which Euler had derived, vanishes identically in every rotation of a rigid body. Adopting Lexell's notation, if a sphere be turned around its fixed centre so that the points A, B, C on mutually orthogonal axes (the "old" coordinate system), arrive after rotation at the points a, b, c (the "new" coordinate system), then any point Z will be transferred to z so that (Fig. 8.5)

$$\begin{aligned}\cos \mathrm{zA} &= \cos \mathrm{ZA} \cos \mathrm{Aa} + \cos \mathrm{ZB} \cos \mathrm{Ab} + \cos \mathrm{ZC} \cos \mathrm{Ac},\\ \cos \mathrm{zB} &= \cos \mathrm{ZA} \cos \mathrm{Ba} + \cos \mathrm{ZB} \cos \mathrm{Bb} + \cos \mathrm{ZC} \cos \mathrm{Bc},\\ \cos \mathrm{zC} &= \cos \mathrm{ZA} \cos \mathrm{Ca} + \cos \mathrm{ZB} \cos \mathrm{Cb} + \cos \mathrm{ZC} \cos \mathrm{Cc}.\end{aligned}$$

Since the point z is situated in the same manner with respect to the points a, b, c as the point Z with respect to the points A, B, C, it follows that the arcs za = ZA, zb = ZB, and zc = ZC. Thus,

$$\begin{aligned}\cos \mathrm{zA} &= \cos \mathrm{za} \cos \mathrm{Aa} + \cos \mathrm{zb} \cos \mathrm{Ab} + \cos \mathrm{zc} \cos \mathrm{Ac},\\ \cos \mathrm{zB} &= \cos \mathrm{za} \cos \mathrm{Ba} + \cos \mathrm{zb} \cos \mathrm{Bb} + \cos \mathrm{zc} \cos \mathrm{Bc},\\ \cos \mathrm{zC} &= \cos \mathrm{za} \cos \mathrm{Ca} + \cos \mathrm{zb} \cos \mathrm{Cb} + \cos \mathrm{zc} \cos \mathrm{Cc},\end{aligned}$$

---

[22] *Formulae generales pro translatione quacunque corporum rigidorum*, NCASIP XX, pp. 189–207, 1776 (OO Ser. II, Vol. 9, pp. 84–98. E478). However, the Euler angles of coordinate transformation appear for the first time in an algebraic context in the paper *Problema algebraicum ob affectiones prorsus singulares memorabile*, NCASIP XV, 1770, pp. 75–106 (OO Ser. I, Vol. 6, pp. 287–315. E407).

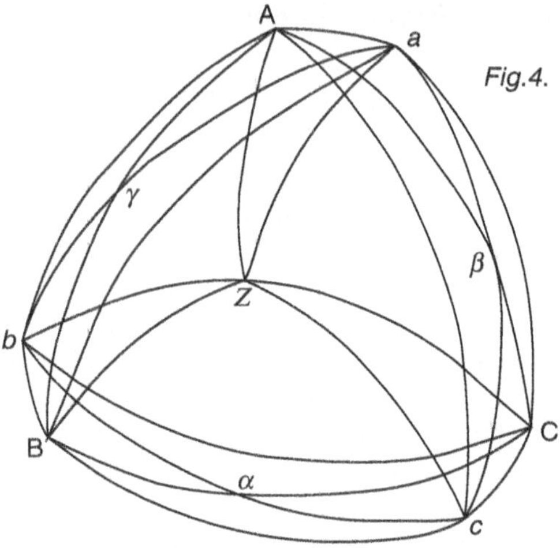

Fig. 8.5 Illustration referring to Lexell's investigation of rotation [Lexell 45]

and in order for z to coincide with Z after the rotation,

$$\cos \mathrm{ZA}(\cos \mathrm{Aa} - 1) + \cos \mathrm{ZB} \cos \mathrm{Ab} + \cos \mathrm{ZC} \cos \mathrm{Ac} = 0$$

$$\cos \mathrm{ZA} \cos \mathrm{Ba} + \cos \mathrm{ZB}(\cos \mathrm{Bb} - 1) + \cos \mathrm{ZC} \cos \mathrm{Bc} = 0$$

$$\cos \mathrm{ZA} \cos \mathrm{Ca} + \cos \mathrm{ZB} \cos \mathrm{Cb} + \cos \mathrm{ZC}(\cos \mathrm{Cc} - 1) = 0.$$

Here, the modern reader may recognise the eigenvalue equation $(\bar{\bar{\mathrm{R}}} - \lambda \bar{\bar{\mathrm{I}}}) \cdot \mathbf{r} = 0$, or explicitly

$$\begin{pmatrix} \alpha & \beta & \gamma \\ \alpha' & \beta' & \gamma' \\ \alpha'' & \beta'' & \gamma'' \end{pmatrix} \begin{pmatrix} x \\ y \\ z \end{pmatrix} = \begin{pmatrix} 0 \\ 0 \\ 0 \end{pmatrix}$$

where $x = \cos \mathrm{AZ}$, $y = \cos \mathrm{BZ}$, $z = \cos \mathrm{CZ}$, $\alpha = \cos \mathrm{Aa} - \lambda$, $\beta = \cos \mathrm{Ab}$, $\gamma = \cos \mathrm{Ac}$, $\alpha' = \cos \mathrm{Ba}$, $\beta' = \cos \mathrm{Bb} - \lambda$, $\gamma' = \cos \mathrm{Bc}$, $\alpha'' = \cos \mathrm{Ca}$, $\beta'' = \cos \mathrm{Cb}$, $\gamma'' = \cos \mathrm{Cc} - \lambda$. The eigenvalue is in this case $\lambda = 1$.

To solve the equation, Lexell combined the first and the second row and, correspondingly, the first and the third row, and by solving for y/z from both equations, he obtained

$$\alpha\beta'\gamma'' - \alpha\gamma'\beta'' - \alpha'\beta\gamma'' - \alpha''\beta'\gamma + \alpha''\beta\gamma' + \alpha'\beta''\gamma = 0.$$

This is what would be called the characteristic equation for the problem. With the values of $\alpha, \beta, \ldots$ inserted, it reads

$$\begin{aligned} &(\cos \mathrm{Aa} - 1)(\cos \mathrm{Bb} - 1)(\cos \mathrm{Cc} - 1) + \cos \mathrm{Ca} \cos \mathrm{Ab} \cos \mathrm{Bc} \\ &+ \cos \mathrm{Ba} \cos \mathrm{Ac} \cos \mathrm{Cb} - \cos \mathrm{Bc} \cos \mathrm{Cb}(\cos \mathrm{Aa} - 1) \\ &- \cos \mathrm{Ba} \cos \mathrm{Ab}(\cos \mathrm{Cc} - 1) - \cos \mathrm{Ca} \cos \mathrm{Ac}(\cos \mathrm{Bb} - 1) = 0, \end{aligned}$$

which, Lexell maintained, is exactly the expression Euler had derived in his aforementioned paper. Further, he showed that Euler's three necessary conditions for the rotation to be unique can be written in terms of the *direction cosines* as

$$\cos^2 \mathrm{Aa} + \cos^2 \mathrm{Ba} + \cos^2 \mathrm{Ca} = 1,$$
$$\cos^2 \mathrm{Ab} + \cos^2 \mathrm{Bb} + \cos^2 \mathrm{Cb} = 1,$$
$$\cos^2 \mathrm{Ac} + \cos^2 \mathrm{Bc} + \cos^2 \mathrm{Cc} = 1,$$

in addition to which the conditions

$$\cos \mathrm{Aa} \cos \mathrm{Ab} + \cos \mathrm{Ba} \cos \mathrm{Bb} + \cos \mathrm{Ca} \cos \mathrm{Cb} = \cos \mathrm{ab} = 0,$$
$$\cos \mathrm{Aa} \cos \mathrm{Ac} + \cos \mathrm{Ba} \cos \mathrm{Bc} + \cos \mathrm{Ca} \cos \mathrm{Cc} = \cos \mathrm{ac} = 0,$$
$$\cos \mathrm{Ab} \cos \mathrm{Ac} + \cos \mathrm{Bb} \cos \mathrm{Bc} + \cos \mathrm{Cb} \cos \mathrm{Cc} = \cos \mathrm{bc} = 0,$$

must be satisfied. In the remaining part of the paper, Lexell proceeds to prove by means of spherical trigonometry that these expressions hold true for all kinds of rotations.

Today, Lexell's contribution is seldom associated with the investigation of rotation (see e.g. [172]), but he was nevertheless among the first to confirm and to further investigate Euler's solution. Almost a century later, Olinde Rodrigues (1795–1851) completed the theory in a more symmetrical form. In vector form, any rotation of a vector **r** around the unit vector **u** by the angle ϕ can be described by means of the formula [1]

$$\overline{\overline{\mathrm{R}}} \cdot \mathbf{r} = \cos\phi\ \mathbf{r} + 2\sin^2(\frac{\phi}{2})\ \mathbf{uu} \cdot \mathbf{r} + \sin\phi\ \mathbf{u} \times \mathbf{r},$$

which with complete justification could be called the "Euler-Lexell-Rodrigues" formula of rotation.

8.3.3 Polygonometry

Through the rapid development of analysis in the eighteenth century, trigonometric calculations and manipulations had become easy to perform. The question soon arose—possibly in connection with surveying and astronomy—whether a method could be found by which the properties of an arbitrary polygon (triangle, quadrangle or n-angle) could be determined without first breaking down the object into triangles. The first to investigate this idea systematically seems to have been

Johann Heinrich Lambert[23] in 1770, followed by Johann Tobias Mayer (1752–1830; son of the astronomer Tobias Mayer of Göttingen) in 1773 [30, 55, 59], and the Danish mathematician Stephan Bjørnsen[24] in 1780 [19, 23]. However, to a certain degree, Lexell can be regarded as the founder of general polygonometry, since in a letter to the Royal Society dated 14 June 1774 and published in the *Philosophical Transactions* in 1775 [Lexell 40] (Fig. 8.6), he reported his discovery of the laws governing general polygons. The first announcement of the discovery can nevertheless be found in a letter written on 22 December 1773 to Johann III Bernoulli (UBB Mscr. L I a 703, fo 108r–158v; see Fig. 8.7). Having first expressed the two general equations, Lexell presented the six equations for the hexagon:

> Il y a fort peu de temps que je viens de trouver deux théorèmes de géométrie fort beaux et fort simples, que je me fais un plaisir de Vous communiquer ici. Soit un polygone quelconque dont les côtés $a, b, c, [\dots f]$ et les angles externes $\alpha, \beta, \gamma, \dots \xi$, et il sera
>
> $$\begin{aligned} \text{I}^\circ.\ & a \sin\alpha + b\sin(\alpha+\beta) + c\sin(\alpha+\beta+\gamma) + \dots \\ & + f\sin(\alpha+\beta+\dots+\xi) = 0 \\ \text{II}^\circ.\ & a\cos\alpha + b\cos(\alpha+\beta) + c\cos(\alpha+\beta+\gamma) + \dots \\ & + f\cos(\alpha+\beta+\dots+\xi) = 0 \end{aligned}$$
>
> Parce que $\alpha+\beta+\gamma+\dots+\xi = 360^\circ$, son sinus sera =0 et le cosinus =1, on auroit donc pû dans la première formule négliger le dernier terme et dans le second au lieu de $f\cos(\alpha+\beta+\gamma+\dots+\xi)$ écrire f, mais pour mieux conserver l'uniformité, je les ai plustôt représenté comme ci dessus. En faisant différentes combinaisons de ces deux équations, on trouvera aisément toutes celles, qui servent pour la résolution trigonométrique des polygones quelconques. Ainsi on aura pour l'hexagone ces six équations
>
> $$\begin{aligned} \text{I.}\ \ ff = {} & a^2+b^2+c^2+d^2+e^2+2ab\cos\beta \\ & + 2ac\cos(\beta+\gamma) + 2ad\cos(\beta+\gamma+\delta) + 2ae\cos(\beta+\gamma+\delta+\varepsilon) \\ & + 2bc\cos\gamma + 2bd\cos(\gamma+\delta) + 2be\cos(\gamma+\delta+\varepsilon) \\ & + 2cd\cos\delta + 2ce\cos(\delta+\varepsilon) \\ & + 2de\cos\varepsilon \end{aligned}$$

[23] Johann Heinrich Lambert (1728–1777), mathematician, physicist and philosopher, born in Mülhausen (today: Mulhouse) but working at different places in Europe; from 1764 to his death he was with the Prussian Academy of Sciences in Berlin. His works include a systematic treatise on heat (*Pyrometrie*, 1779), a work on orbit determination (*Insigniores orbitae cometarum proprietates*, 1761) and the law of exponential absorption of light now bearing his name (*Photometria*, 1760). He also proved rigorously that π is irrational using a continued fraction expression for the tangent function (*Mémoires de l'Académie Royale des Sciences et des Belles-Lettres de Berlin* Vol. XVII, 1768, pp. 265–322).

[24] Bjørnsen is also known by his Icelandic name Stefán Bjarnarson (1730–1798). He moved to Copenhagen, where, besides mathematics, he devoted himself to editing such works as the ancient Icelandic *Rimbegla*, parts of which deal with chronology and computation.

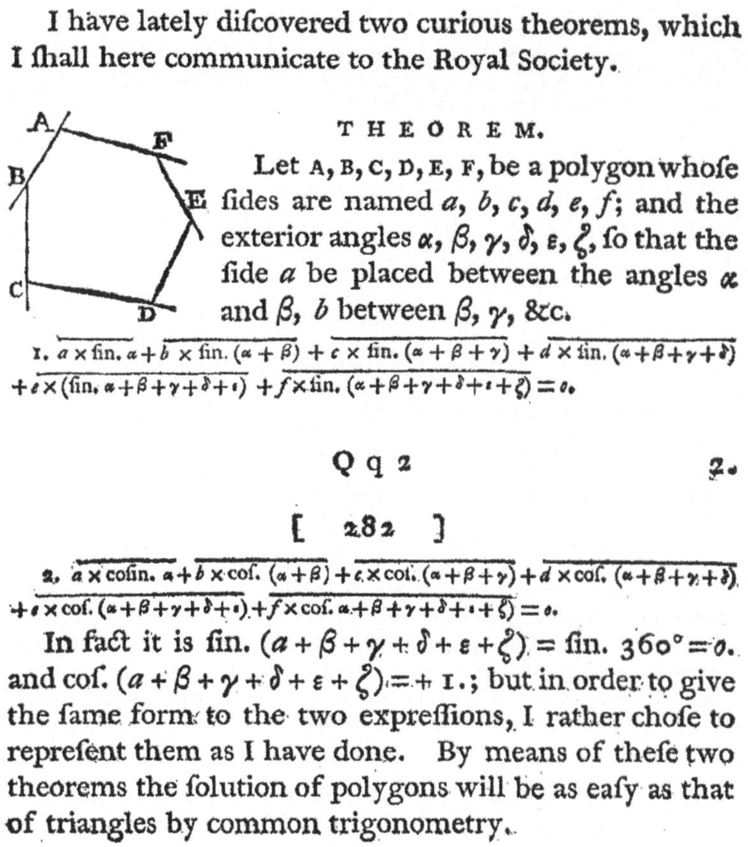

I have lately diſcovered two curious theorems, which I ſhall here communicate to the Royal Society.

THEOREM.

Let A, B, C, D, E, F, be a polygon whoſe ſides are named a, b, c, d, e, f; and the exterior angles α, β, γ, δ, ε, ζ, ſo that the ſide a be placed between the angles α and β, b between β, γ, &c.

$$1.\ \overline{a\times \text{ſin.}\,\alpha}+\overline{b\times \text{ſin.}\,(\alpha+\beta)}+\overline{c\times \text{ſin.}\,(\alpha+\beta+\gamma)}+\overline{d\times \text{ſin.}\,(\alpha+\beta+\gamma+\delta)}$$
$$+\overline{e\times(\text{ſin.}\,\alpha+\beta+\gamma+\delta+\varepsilon)}+\overline{f\times \text{ſin.}\,(\alpha+\beta+\gamma+\delta+\varepsilon+\zeta)}=0.$$

Qq 2 2.

[282]

$$2.\ \overline{a\times \text{coſin.}\,\alpha}+\overline{b\times \text{coſ.}\,(\alpha+\beta)}+\overline{c\times \text{coſ.}\,(\alpha+\beta+\gamma)}+\overline{d\times \text{coſ.}\,(\alpha+\beta+\gamma+\delta)}$$
$$+\overline{e\times \text{coſ.}\,(\alpha+\beta+\gamma+\delta+\varepsilon)}+\overline{f\times \text{coſ.}\,\alpha+\beta+\gamma+\delta+\varepsilon+\zeta}=0.$$

In fact it is ſin. $(\alpha+\beta+\gamma+\delta+\varepsilon+\zeta)$ = ſin. $360°=0$. and coſ. $(\alpha+\beta+\gamma+\delta+\varepsilon+\zeta)=+1$.; but in order to give the ſame form to the two expreſſions, I rather choſe to repreſent them as I have done. By means of theſe two theorems the ſolution of polygons will be as eaſy as that of triangles by common trigonometry.

Fig. 8.6 In a letter printed in the *Philosophical Transactions of the Royal Society* Lexell announces to its secretary, Dr. Charles Morton, "two curious theorems related to the angles of general polygons" [Lexell 40]

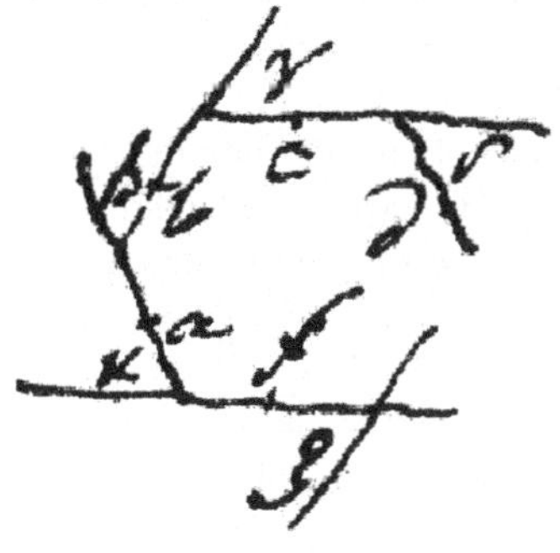

Fig. 8.7 Drawing in the letter to Johann III Bernoulli referring to the polygonometric theorem. The sides of the polygon are denoted by $a, b, c, \ldots$ and the respective external angles $\alpha, \beta, \gamma, \ldots$

$$\begin{aligned}
\text{II.}\quad & f^2 + a^2 + 2af\cos\alpha = b^2 + c^2 + d^2 + e^2 \\
& + 2bc\cos\gamma + 2bd\cos(\gamma+\delta) + 2be\cos(\gamma+\delta+\varepsilon) \\
& + 2cd\cos\delta + 2ce\cos(\delta+\varepsilon) \\
& + 2de\cos\varepsilon \\
\text{III.}\quad & f^2 + a^2 + b^2 + 2af\cos\alpha + 2bf\cos(\alpha+\beta) = c^2 + d^2 + e^2 \\
& + 2cd\cos\delta + 2cd\cos(\delta+\varepsilon) \\
& + 2de\cos\varepsilon \\
\text{IV.}\quad & a\sin\alpha + b\sin(\alpha+\beta) + c\sin(\alpha+\beta+\gamma) \\
& + d\sin(\alpha+\beta+\gamma+\delta) + e\sin(\alpha+\beta+\gamma+\delta+\varepsilon) = 0 \\
\text{V.}\quad & a\sin\alpha + b\sin(\alpha+\beta) + c\sin(\alpha+\beta+\gamma) \\
& + d\sin(\alpha+\beta+\gamma+\delta) - e\sin\zeta = 0 \\
\text{VI.}\quad & a\sin\alpha + b\sin(\alpha+\beta) + c\sin(\alpha+\beta+\gamma) \\
& - e\sin(\varepsilon+\zeta) - e\sin\zeta = 0
\end{aligned}$$

Ces deux théorèmes étant fort faciles à démontrer il est très singulier qu'aucun géomètre n'y a fait attention jusqu'à présent, au moins tant que je sçai. Si Vous voudriez bien Monsieur les communiquer à M[rs] la Grange et Lambert, Vous me feriez un plaisir fort sensible. Je ne les donne pas pour une découverte importante, mais seulement à cause de leur grande généralité et de l'usage singulier qu'on en peut tirer. On pourroit encore les généraliser un peu plus en introduisant au lieu de l'angle α, un angle φ que fait le côté a avec une ligne droite donnée de position.

In other words, Lexell finally suggested a generalisation involving the angle φ (instead of α) between the side a and an arbitrary straight line. The generalisation was presented in §8 of [Lexell 35].

In fact, in the two-part Latin paper [Lexell 35] and [Lexell 44], entitled in English "On the resolution of plane polygons" (parts 1 and 2), the subject was given a systematic and complete treatment. In the former paper [Lexell 35], Lexell provided a geometric construction of n-gons when $2n - 3$ sides and angles are prescribed (at least $n - 2$ of these must be sides). His solutions of all the problems were based on two sets of equations obtained when the sides of a polygon are projected orthogonally onto two mutually perpendicular lines, one of which must coincide with a side of the polygon. At the same time, he also dealt with the question of the number of diagonals in an n-gon (which he gave as $n(n-3)/2$) and related issues of combinatorics. In this way, Lexell derived formulas for the resolution of triangles and quadrangles, pentagons, hexagons, and heptagons. In the latter paper [Lexell 44], he put forward a complete classification for the solution of problems of this kind.

The book *Polygonometrie* (1784) [Lexell 106] is essentially a compilation of the aforementioned two articles translated into German by Johann Friedrich Lempe, a teacher of mathematics and surveyor at the Freiberg mining school. Since the work does not appear in the list of Lexell's books [47], it is quite likely that he was himself unaware of its publication. The book was not known even to Moritz Cantor, who claims that the first book in polygonometry was Simon l'Huilier's major work *Polygonométrie*[25] published in 1789 [23, p. 432].

8.3.4 Polyhedrometry

Possibly as a digression from the study of polygonometry, the theory of polyhedra was advanced by Lexell in the case of a tetrahedron in the short paper [Lexell 60] in the proceedings of the Royal Swedish Academy of Sciences, the title of which title translates as "A remarkable theorem concerning the angles of planes in triangular pyramids". In this succinct paper, which seems to have passed unnoticed in the general history of polyhedrometry (usually attributed to l'Huilier, see e.g. [175, Pt. 1]), Lexell writes:

> It is easily understood that in triangular pyramids, or polyhdera enclosed by triangles, the six angles which the triangles form mutually [the so-called dihedral angles] are in such a way connected to each other that as soon as five of them are given, the sixth one may be thereby determined; but as far as I know, no one has hitherto given a formula to express this connection between the aforementioned angles. When I some time ago was occupied with a number of investigations of the properties of triangular pyramids, I discovered the following theorem, which I consider worthy of attention, the more so as the whole doctrine of bodies with planar faces is based on the proofs that can be given for triangular pyramids; just like the properties of polygons in the plane depend on the properties of triangles.
> If, in a triangular pyramid ABCD [...], the angles between the plane BCD and the planes ACD, ABD, ABC are denoted by A, B, C, respectively, and the angles between the plane ABC and ABD, ACD by A', B', respectively, and finally the angle between the planes ABD, ACD by C', then

$$
\begin{aligned}
0 = 1 &- \cos^2 A - \cos^2 B - \cos^2 C - \cos^2 A' - \cos^2 B' - \cos^2 C' \\
&+ \cos^2 A \cos^2 A' + \cos^2 B \cos^2 B' + \cos^2 C \cos^2 C' \\
&- 2 \cos A \cos B \cos C' - 2 \cos A \cos B' \cos C \\
&- 2 \cos A' \cos B \cos C - 2 \cos A' \cos B' \cos C' \\
&- 2 \cos A \cos A' \cos B \cos B' - 2 \cos B \cos B' \cos C \cos C' \\
&- 2 \cos A \cos A' \cos C \cos C'.
\end{aligned}
$$

[25] *Polygonométrie ou de la mesure des figures rectilignes. Et abrégé d'isopérimétrie élémentaire ou de la dependance mutuelle des grandeurs et des limites des figures.* Genève, 1789. In the Introduction, l'Huilier says that he had seen Lexell's *Polygonometrie* [Lexell 106] and thought that it was inferior to the original memoirs.

To prove the theorem Lexell introduces a property of the angles defined at a corner:

> In the triangular pyramid, consider the corner at B, which is enclosed by the plane angles ABC, ABD, CBD, and associated with the plane angles A', B, C, then one has:
>
> $$\cos C = \sin A' \sin C \cos ABC - \cos A' \cos C,$$
>
> or if $\angle ABC$ is denoted by L,
>
> $$\cos L = \frac{\cos C + \cos A' \cos C}{\sin A' \sin C}$$

Lexell admits that the formula could be derived using the formula for the angle between two intersecting planes. However, since the derivation of the formula in this way is tedious, he prefers to use an easier method based on a theorem for spherical triangles. Thus, let us imagine a spherical triangle EFG such that the arc EF corresponds to the angle $\angle ABD$ in the triangular pyramid, and similarly the arcs EG, FG to the angles $\angle ABC$, $\angle CBD$. Then the angles $\angle A'$, $\angle B$, $\angle C$ at the corner B of the pyramid will be equal to the angles $\angle E$, $\angle F$, $\angle G$, respectively, in the spherical triangle. Now, since it is known that for spherical triangles

$$\cos \mathrm{EG} = \frac{\cos F + \cos E \cos G}{\sin E \sin G},$$

the substitution of L, A, B, C for EG and E, F, G yields the desired formula. Hence, by applying this expression to each of the corners of the tetrahedron, the proof of the original theorem proceeds rather straightforwardly.

8.3.5 Castillon's Problem

The Castillon problem, or the Cramer-Castillon problem, is one of the more remarkable problems of ancient Greek origin, which occupied a number of eighteenth-century mathematicians. It is a generalisation of proposition N° 117 of the seventh book of the *Collection* of Pappus of Alexandria, originally formulated by the Genevan mathematician Gabriel Cramer[26] to his friend Johann Castillon,[27] who expressed it as follows:

> *Given a circle and three points, to inscribe in the circle a triangle whose sides (extended if necessary) each pass through these points.*

---

[26]Gabriel Cramer (1704–1752) was a student of Johann I Bernoulli and the editor of his *Opera Omnia* [32]. Cramer is known for the rule in linear algebra bearing his name.

[27]Johann (Giovanni) Francesco Melchiore Salvemini Castillon (1704–1791), Italian mathematician who after teaching in Lausanne and Utrecht was employed at the Prussian Academy of Sciences, where he became Astronomer Royal. He published the correspondence of Leibniz and Johann I Bernoulli in 1745 and, in 1761, a commentary on Newton's *Arithmetica universalis*.

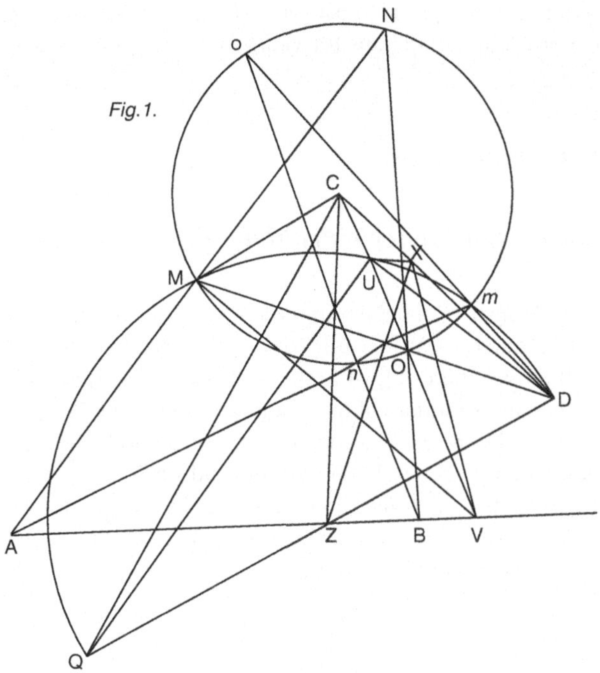

Fig. 8.8 Diagram illustrating Lexell's construction in the solution of the Cramer-Castillon problem

In 1776 Castillon published the problem along with his geometrical solution of it in a paper entitled "On a problem in plane geometry which has been regarded as very difficult" in the proceedings of the Berlin Academy of Sciences[28]: Joseph Louis Lagrange quickly submitted to the same journal his elegant algebraic solution, reported by Castillon in a paper in the same volume of the Berlin proceedings, entitled "On a new property of conical sections".[29] However, in the *Acta* of Saint Petersburg for the year 1780 (printed only in 1783–1784), the respective solutions of three academicians from Saint Petersburg, namely Euler,[30] Fuss and Lexell were published. In his article [Lexell 84], Lexell reviewed the solutions that had been published so far and made some remarks as to the number of possible solutions of the problem in general, given that the analysis leads to a quadratic equation, which may have either zero, one or two solutions. He also managed to give a geometric construction for Lagrange's solution, as shown in Fig. 8.8. Yet this

[28]"Sur un problème de géométrie plane, qu'on regarde comme fort difficile", *Nouveaux Mémoires de l'Académie Royale des Sciences et Belles-Lettres de Berlin*, 1776, pp. 265–286.

[29]"Sur une nouvelle propriété des sections coniques", *Nouv. Mémoires* (Berlin), 1776, pp. 284–311. See [85, Vol. IV, pp. 335–339]

[30]*Problematis cuiusdam Pappi Alexandrini constructio*, AASIP 1780 (printed 1783) Pt. I, pp. 91–96 (OO Ser. I, Vol. 26, pp. 237–242. E543). Fuss's article "Solutio problematis geometrici Pappi Alexandrini" followed immediately after Euler's.

seemingly simple proposition, which is a classic of geometry [23], continued to occupy mathematicians for years to come.

8.4 Mechanics

8.4.1 On Kepler's Problem and the Three-Body Problem

In the paper in Swedish entitled "Solution of the inverse problem of the so-called centripetal forces" [Lexell 59], Lexell proposes finding the curve along which a body will move when subject to the attraction of a fixed body with the force proportional to the inverse square of the distance to the point. Admitting that this classic problem, attributed to Kepler, has been solved by a number of geometers since Newton, Lexell nevertheless argues that his simple and straightforward solution by means of integral calculus has a definite didactic value. In the paper, a step by step integration of the equation of motion in the plane is performed using polar coordinates, to finally yield the locus of a conic section:

$$r = \frac{b}{1 + e \cos \varphi},$$

where b is the parameter, and e the eccentricity of the conic section (for $e = 1$ the orbit is a parabola, $e < 1$ gives an ellipse and $e > 1$ a hyperbola).

Another paper by Lexell, published in the proceedings of the Swedish Academy of Sciences and entitled "Solution of an astronomical problem" [Lexell 39], concerns the determination of the orbit of a celestial body presumed to move in a conic section. If AS be the smallest distance of the body to the focus S of the section, and two other distances MS and NS are also known, together with the angle $\angle MSN$; the task is to find the angles $\angle ASM$ and $\angle ASN$ or the true *anomaly* of the body and the eccentricity of the conic section, or the distance between the focus S of the section and its centre. Applying this description to a comet, Lexell shows that the motion of the body can be predicted at any instant of time provided that the following information is available:

1. The locus of the intersection between the ecliptic and the comet's orbit,
2. The inclination of the orbital plane with the ecliptic,
3. The angle between the node line and the axis of the orbit,
4. The shortest distance of the orbit to the Sun,
5. The eccentricity of the orbit and
6. The moment of perihelion.

The solution Lexell provides is directly applicable to the analysis of the motion of the planets and comets, and he further discusses the sensitivity of the solution to errors in the different parameters with special application to the comet of 1769.

In one of his last papers [Lexell 105], written and presented in 1784, Lexell considered a number of theorems given by Johann Heinrich Lambert in his book on orbit calculation *Insigniores orbitae cometarum proprietates* (1761). Of particular interest was to determine the orbit of a comet (moving along a conic section) when its position is only known at two instants of time: this important boundary value problem is today known as Lambert's problem. Related problems had been studied by Euler in 1743–1744 for parabolic orbits.[31] However, as Lambert's problem was originally presented in a geometric form, an analytical formulation still had to be developed. Prompted by a recent treatise on the subject by Lagrange "Sur le problème de la détermination des orbites des comètes d'après trois observations",[32] Lexell in [Lexell 105] obtained a solution of Lambert's problem *inter alia* for hyperbolic orbits.

Most of the astronomical problems with which Lexell was occupied during his career as an astronomer were concerned with the gravitational interaction of more than two bodies, one or two of the bodies being considerably more massive than the third. In general, the three-body problem is known to have no closed-form solution [166, 172], a fact which even Euler and Lagrange ignored at the time. Thus, Lexell's approach to the general problem of multiparticle interaction was more pragmatic, i.e.: given the positions and velocities of the bodies, compute the outcome after a limited time period. However, such an approach does not state anything about the orbits in general. In fact, it is not possible to find a closed form expression for the orbit of a body of negligible mass moving in the same plane as two massive (primary) bodies in fixed circular or elliptic orbits.[33] Such a problem is called a restricted three-body problem.[34] There are nevertheless five special solutions, in which the third body remains stationary with respect to the two primary bodies. These are the so-called Lagrangian points.

It was only in 1889 that Ernst Heinrich Bruns (1848–1919) formulated a theorem which states that there are only ten integrals which constrain the motion of n interacting bodies (i.e. three each for the centre of mass, the linear momentum and the angular momentum, and one for the energy). Thereby he was able

---

[31] In particular, in *Theoria motuum planetarum et cometarum*, Berlin, 1744 (OO Ser. II, Vol. 28, pp. 105–251. E66).

[32] *Nouveaux Mémoires de l'Académie Royale des Sciences et Belles-Lettres de Berlin*, 1778 (1780), pp. 111–161.

[33] Four integration coefficients would be needed to appropriately specify the orbit of the "massless" body (that is, the body whose mass is too small to influence the others), but only one of them—the so-called Jacobian integral [172]—can be evaluated analytically. It specifies the range within which the body can move, but nothing else.

[34] Euler's first attempt to solve the three-body problem dates back to 1725–1727 [168]. In a paper published in 1765, *De motu rectilineo trium corporum se mutuo attrahentium*, NCASIP XI, pp. 144–151, 1767 (OO Ser. II, Vol. 25, pp. 281–289. E327) Euler considered the problem of three bodies constrained to move along a line. Lagrange's solution to the restricted three-body problem was given in *Essai sur le problème des trois corps* [85, Vol. VI], which was awarded the prize of the *Académie Royale des Sciences* in 1772. The paper concerned periodic solutions where the bodies occupy the vertices of a rotating equilateral triangle.

to prove the impossibility of solving algebraically the most general multi-body problem. A few years later, Henri Poincaré (1854–1912) managed to generalise Bruns's theorem, and building on Poincaré's work, the Finnish astronomer Karl Frithiof Sundman (1873–1949) succeeded in regularising the three-body problem by analytical continuation and elimination of singularities (collisions), and finally discovering a solution in terms of a slowly converging infinite series [9].

Except for the theoretical article [Lexell 96] concerning the motion of a body in the presence of two fixed centres of attraction,[35] the three-body problems Lexell dealt with concerned either comets, planets or the Moon interacting with each other or with the gravitational field of the Earth, the Sun being the principal stationary attractor. Lexell's theory of the comet of 1770 (discussed previously in Sect. 5.5) is an example of a restricted three-body problem manifesting a particular kind of resonance. In the following, we elucidate briefly his research, performed under Euler's direction, concerning the perturbation caused by Venus in the motion of the Earth. One of Euler's many contributions in this vein is a paper entitled *De perturbatione motus Terrae ab actione Veneris oriunda*,[36] which Lexell on behalf of Euler presented to the Petersburg Academy of Sciences on 25 (14) May 1772. Although this was not indicated in the paper, it transpired that Lexell was mainly responsible for the calculations and for an unfortunate sign error which was later corrected in [Lexell 75]: a thorough account of it has been given by Verdun [168].

A particular difficulty with this problem stems from the fact that the orbits of Earth and Venus are rather close to each other, the ratio between the mean radii of their orbits being around 0.72. Denoting these distances by a and a' respectively, the distance between the Earth and Venus can be expressed as [168, 174]

$$w = \sqrt{a^2 + a'^2 - 2aa' \cos \omega},$$

where ω denotes the heliocentric angle between the planets. Now, the integration of terms that are proportional to w^{-3} (the gravitational interaction decomposed in Cartesian coordinates involves such terms) leads after an expansion to integrals of the form $\int \mathrm{d}\omega (1 - g \cos \omega)^{\mu}$, where $g = \dfrac{2a'/a}{1 + (a'/a)^2}$ and $\mu = 3/2$. In the case of Venus and the Earth, the ratio $a'/a \approx 0.72$, whereby $g \approx 0.95$, which makes the series expansions (and their integrals) converge very slowly. On the other hand, in the case of the three-body system formed by the Sun, the Earth and the Moon, g is only 0.005, and hence the expansions converge fairly quickly.

---

[35]This article originated in Euler's paper with the identical title *De motu corporis ad duo centra virium fixa attracti*, NCASIP XI, 1767, pp. 152–184 (OO Ser. II, Vol. 6, E328).

[36]NCASIP XVI, pp. 426–467, 1771 (OO Ser. II, Vol. 26. E425). In his previous prize-winning essay, *Investigatio perturbationum quibus planetarum motus ob actionem eorum mutuam afficiuntur* (1756), published by the Paris Academy of Sciences (OO Ser. II, Vol. 26. E414), Euler had treated the same kind of problem analytically by the aid of approximations and series expansions [16,168]. He nevertheless indicated the need to perform numerical integrations in future investigations.

The main steps of Euler's work are the following (for details, see [168]): ignoring the relatively small orbital inclination of Venus with respect to the plane of the ecliptic, Euler first writes down the equations of motion with regard to a two-dimensional rectangular coordinate system whose origin lies at the centre of the Sun:

$$\frac{\mathrm{d}^2 x}{\alpha \mathrm{d}\tau^2} + \frac{(\odot + ♁)x}{u^3} + \frac{♀ \cos\phi}{vv} + \frac{♀(x - v\cos\phi)}{w^3} = 0,$$

$$\frac{\mathrm{d}^2 y}{\alpha \mathrm{d}\tau^2} + \frac{(\odot + ♁)y}{u^3} + \frac{♀ \sin\phi}{vv} + \frac{♀(y - v\sin\phi)}{w^3} = 0,$$

where x and y are the coordinates of the Earth, u and v are the distances of the Earth and Venus from the Sun respectively, w is the distance between the Earth and Venus, ϕ is the angular distance of Venus from the vernal equinox point and $\odot$, ♁ and ♀ are the masses of the Sun, the Earth and Venus respectively, $\mathrm{d}\tau$ being the time differential. A new set of coordinates X, Y is next introduced which define the rotation of the Earth around the Sun. These are: $x = X\cos t - Y\sin t$ and $y = Y\cos t + X\sin t$, X and Y being mean distances and t the angular variable. The necessary differentiations then lead to a coupled system of differential equations of the second order. Knowing that the ellipticity of the Earth's orbit is very small and that the disturbances are bound to change the orbit only slightly, another change of variables is performed viz. $X = a(1 + x)$ and $Y = ay$, where x and y are small quantities (and unrelated to the above mentioned coordinates). Hence, Euler first solves the set of homogeneous equations, which does not take into account the influence of Venus:

$$\frac{\mathrm{d}^2 x}{\mathrm{d}t^2} - 2\frac{\mathrm{d}y}{\mathrm{d}t} - (1 + x) + \frac{(1 + x)}{((1 + x)^2 + y^2)^{3/2}} = 0,$$

$$\frac{\mathrm{d}^2 y}{\mathrm{d}t^2} + 2\frac{\mathrm{d}x}{\mathrm{d}t} - y + \frac{y}{((1 + x)^2 + y^2)^{3/2}} = 0.$$

The final terms containing roots are next expressed as convergent series, of which the first four terms are retained. By comparing coefficients, the parameters are determined as

$$x = -\frac{1}{2}K^2 + K\cos t + \frac{1}{2}K^2\cos 2t - \frac{3}{8}K^3\cos 3t,$$

$$y = -(2K - \frac{9}{8}K^3)\sin t + \frac{1}{4}K\sin 2t - \frac{7}{24}K^3\sin 3t,$$

where K signifies the eccentricity of the Earth's orbit. Next, to solve the inhomogeneous equations, a small perturbational term is introduced with $x = X + \lambda X'$, $y = Y + \lambda Y'$, where X' and Y' express the disturbances caused by Venus to the orbit of the Earth. In the ensuing numerical integrations involving the inhomogeneous terms, Lexell commits a sign error which made the resulting values significantly depart from the reference values for Venus tabulated by Abbé La Caille.

Some time later, when the discrepancy was discovered, Lexell, after consultations with Euler, decided to correct his mistake in the paper [Lexell 75], which was presented at the conference on 14 (3) October 1782 and read on 18 (7) October [141, Vol. III, pp. 621–622]. Indeed, in the introduction he formulates his *mea culpa* as follows:

> Since in the new and unique method of determining the perturbations in the Earth's motion due to the action of Venus delivered by the illustrious Euler in the 16th volume of the *Novi Commentarii*, where I did the computation of the tables for the perturbation, large discrepancies were found with those numbers which the famous de la Caille and Tobias Mayer have found on other principles; there arose a well-founded suspicion that this discrepancy could be ascribed either to some of the methods used or an error committed in the calculations. So recently, when the illustrious French mathematician Laplace, having been asked about it by the famous Lalande, explained his thinking in private letters, at the same time inviting the most illustrious Euler and me to state our opinion; and because the principal reason for the discrepancy is ascribed to an error in the calculations committed by me, I judged it to be my role to correct the places where there is an error, and at the same time treat all the material as exactly as possible; so that there will remain no doubt in the minds of astronomers that the conclusions drawn from the diverse methods given for this intention will conform and agree with each other. [37]

In a letter dated 10 February 1783 [85, Vol. XIV, p. 122] Laplace wrote to Lagrange that he was very content having seen that Lexell's new calculations (which involved numerical integration) were in line with Clairaut's theory. He also marveled at the patience of "cet habile géomètre", referring to Lexell, for having undertaken this arduous task, only to confirm a more or less evident result [174].

8.4.2 Complements to Euler's Research in Mechanics

Lexell's other contributions to rational mechanics are few, namely [Lexell 48], [Lexell 89] and [Lexell 92]. They are completely unrelated among themselves and of

---

[37]Cum Illustris *Eulerus* in Nov. Comment. Vol. XVI. novam et singularem tradidisset Methodum pro determinanda perturbatione in motu Telluris, ab actione Veneris, meaque opera tum usus sit, ad computandam Tabulam pro ista perturbatione, quae quia valde discrepans inventa est, ab iis numeris, quos Celeb. de la *Caille* et *Tobias Mayerus* aliis principiis insistentes, invenerunt; merito suspicio oborta est, an diversitati Methodorum, utrum vero errori in calculis commisso, haec discrepantia tribuenda esset. Dum igitur nuper illustris *Galliae* Mathematicus de la *Place*, Celeb. de la *Lande* cum hac de re sciscitanti, in litteris privatis mentem exposuisset, simulque rogasset ut Illustrissimus *Eulerus* et ego de hac quaestione sententiam nostram exponeremus; quum principalis ratio huius discrepantiae errori in calculis *a me* commisso adscribenda sit, mearum partium esse iudicavi, ea in quibus tum erratum est corrigendi, simulque materiam hanc omni qua fieri potest exactitudine tractandi; quo scilicet nullum dubium animis Astronomicorum infixum resideat, quin diversis istis Methodis pro hoc instituto traditis, conclusiones inter se conformes et amicae inveniantur.

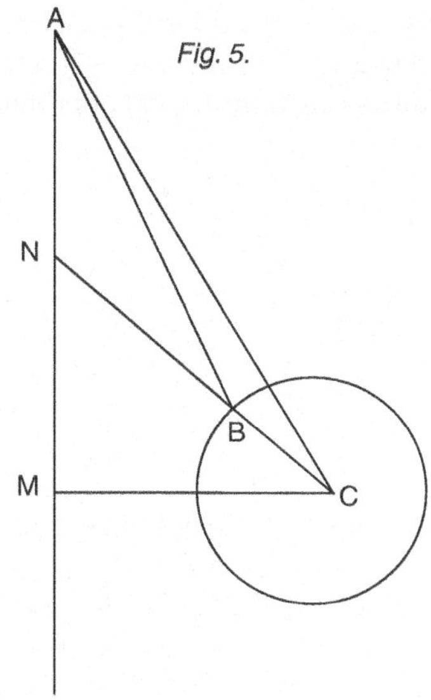

Fig. 8.9 Geometry pertaining to Lexell's analysis of the motion of a physical pendulum (Lexell 89)

the nature of completing Euler's researches. The first one is a supplement to Euler's work in naval science, *Théorie complète de la construction et de la manoeuvre des vaisseaux* (Paris, 1776 [E426]), containing a simple but exact solution to the problem of finding the maximum angle between the obliquity of the route direction of a ship and the drag force.

[Lexell 89] deals with the classical (but very difficult) problem of the motion of a physical pendulum, whose point of suspension (B) and centre of inertia (C) are distinctly separate from each other (see Fig. 8.9). Lexell begins by formulating the three governing differential equations of the second order, which must be satisfied simultaneously. Denoting by M the mass of the pendulum bob and by T the tension of the chord AB, the equations are

$$\frac{\mathrm{d}^2 x}{\alpha \mathrm{d}t^2} = -\frac{T\cos\varphi}{M}, \qquad \frac{\mathrm{d}^2 y}{\alpha \mathrm{d}t^2} = -\frac{T\sin\varphi}{M},$$

and

$$\frac{\mathrm{d}^2 \psi}{\alpha \mathrm{d}t^2} = -\frac{bT\sin(\psi-\varphi)}{Mc^2}.$$

In these equations, $x = |AM|$, $y = |MC|$, $\varphi = \angle$ MAB, $\psi = \angle$ MNB, and $b = |BC|$. The third equation involves the mass moment of inertia,[38] expressed

[38]The concept of *moment of inertia* (rotational inertia) was introduced by Euler in his monumental *Theoria motus corporum solidorum seu rigidorum*, Rostock and Greifswald, 1765 (OO Ser. II, Volumes 3 and 4. E289).

by the formula Mc^2 associated with the rotation around the point B. Using the relations $x = a\cos\varphi + b\cos\psi$ and $y = a\sin\varphi + b\sin\psi$, Lexell combines the three equations to find after integration

$$\frac{x\mathrm{d}y - y\mathrm{d}x + c^2\mathrm{d}\psi}{\mathrm{d}t} = \mathrm{L}.$$

Further, since $\mathrm{d}x = -a\mathrm{d}\varphi\sin\varphi - b\mathrm{d}\psi\sin\psi$ and $\mathrm{d}y = a\mathrm{d}\varphi\cos\varphi + b\mathrm{d}\psi\cos\psi$, he obtains

$$\frac{\mathrm{d}x^2 + \mathrm{d}y^2 + c^2\mathrm{d}\psi^2}{\mathrm{d}t^2} = \mathrm{N}.$$

In the paper, Lexell pursues the investigation by considering the solution in terms of the polar variables $v = \mathrm{AC}$ and $\theta = \angle\mathrm{MAC} = \arctan(y/x)$ and obtains a pair of equations, which nevertheless turn out to be too complicated to integrate in closed form. Without clearly defined concepts of kinetic and potential energy, which considerably facilitate the solution, Lexell's analysis remains awkward and seems to have fallen into oblivion.

The title of [Lexell 92] translates as "Thoughts on the formula by which is expressed the motion of elastic strips curved into circular rings". The latter subject was initiated by Euler's treatments of a long series of problems concerning flexible bodies, such as the vibration of strings and laminae [16]. In a comprehensive survey of Euler's contribution to the mechanics of elastic bodies [162], Truesdell has shown that while the differential equations of fourth order derived by Lexell for a rod shaped into a circle as well as the straight flexible rod were not correct—partly due to the lack of appropriate boundary conditions, such as inextensibility—their value was rather to point out the need for a systematic linearisation of the equations of finite motion. A correct theory for the vibration of rods was not to be established until a century later.

8.5 Miscellaneous Problems

In a letter to Anders Planman from Saint Petersburg dated 22 November 1778, Lexell writes about one of his latest mathematical preoccupations (HUB Ms. Coll. 171):

> ...Recently, a curious problem from the theory of combinations has engaged every mathematician here in Petersburg, without being able to make any headway, as it is probably impossible, and it goes as follows: Arrange 36 officers of 6 different regiments and of 6 different ranks into a square, 6 in every row, such that there is in every row only one of each regiment and of each rank. It is peculiar that the problem easily admits a solution for every odd number, as well as for every even number *pariter pares*, such as four, eight, twelve, but not with six, ten etc. To make this problem easier to conceive, I attach here its solution for the number seven.

Aa	Cc	Ee	Gg	Bb	Dd	Ff
De	Fg	Ab	Cd	Ef	Ga	Bc
Gb	Bd	Df	Fa	Ac	Ce	Eg
Cf	Ea	Gc	Be	Dg	Fb	Ad
Fc	Ae	Cg	Eb	Gd	Bf	Da
Bg	Db	Fd	Af	Ca	Ec	Ge
Ed	Gf	Ba	Dc	Fe	Ag	Cb

In this square, the diagonal rows also possess the desired property, but as soon as one brings on this requirement to a square of 36 [elements], it can be shown that the problem does not admit a solution.

In the following, undated, letter from Saint Petersburg, Lexell briefly returns to the problem

...The curious problem concerning a species of magic square, which I mentioned in my previous letter, was given by an officer of artillery in Berlin by the name of Tempelhof,[39] who is quite versed in mathematics. By way of a misunderstanding it was believed here that he had found a solution to it, but this turned out to be false and the more so as there is no possible solution; this is the more curious, as there is a solution to the problem for larger numbers such as 7, 8 and 9.

The theory was outlined in a frequently cited paper by Euler entitled "Researches on a new kind of magic squares".[40] One reason behind the interest in the paper and the subject generally is that it contains the germ of today's popular "sudoku" puzzles.

Another subject area where Lexell made an incidental contribution is found among Euler's posthumously published manuscripts.[41] It concerns an attempt to apply the method of infinite descent to Fermat's famous equation $a^n + b^n = c^n$ (where a, b, c, n are positive integers and $n \geq 3$) in a way which Pierre de Fermat had possibly tried to use himself in order to prove his theorem, but which "did not fit into the narrow margin". The method of infinite descent was successfully used by Fermat for many types of Diophantine equations. It consists of reducing the order of the original statement—say, that there are no integer solutions of $x^4 + y^4 = z^2$—by induction to examples which are in some sense smaller, in the end leading to a contradiction since there is no infinite sequence of decreasing natural numbers.

[39]Georg Friedrich Ludwig Tempelhoff (1737–1807), teacher of mathematics and ballistics and officer of the Prussian Army, who rose to the rank of *Generalleutnant* (Lieutenant General).

[40]"Recherches sur une nouvelle espèce de quarrés magiques". *Verhandelingen uitgegeven door het zeeuwsch Genootschap der Wetenschappen te Vlissingen* 9, Middelburg (Netherlands) 1782, pp. 85–239, also reprinted in *Commentationes arithmeticae* 2, 1849, pp. 302–361 (OO Ser. I, Vol. 7, pp. 291–392. E530).

[41]"Fragmenta arithmetica ex adversariis mathematicis deprompta". *Leonhardi Euleri Opera Posthuma*. Mathematica et Physica, 1862, Vol. 1, pp. 231–232. E806.

Now, in the proposition $a^n + b^n = c^n$, assuming n to be odd (Lexell actually considered the case $n = 5$), c to be even and a, b to be odd (with the greatest common divisor = 1), Lexell used the substitutions [143]

$$x + y = a^n, \qquad x - y = b^n, \quad \text{and} \quad z = \frac{1}{2}abc^{n-2},$$

to find

$$\frac{x^2 - y^2}{4x^2} = \left(\frac{z}{x}\right)^n,$$

and thereby $x(x^n - 4z^n) = x^{n-1}y^2$, which is a square (since n is odd). Hence, concluding that there must exist integers r, s such that

$$r^2 = \frac{x}{d} \quad \text{and} \quad s^2 = \left(\frac{x}{d}\right)^n - 4\left(\frac{z}{d}\right)^n$$

where $d = \frac{1}{2}c^{n-2}$ is the greatest common divisor of x and z, it follows by factorisation that

$$r^{2n} - s^2 = (r^n + s)(r^n - s) = 4\left(\frac{z}{d}\right)^n.$$

The greatest common divisor of both factors is 2. Thus,

$$r^n + s = 2t^n, \quad \text{and} \quad r^n - s = 2u^n,$$

whence Fermat's equation is obtained anew as $t^n + u^n = r^n$ for even r. A contradiction would indeed result if $r < c$, but since $r^2 = x/d = c^2$, we have $r = c$. Thus, the method of descent does not work. The argument was undoubtedly a tempting one, but not tenable. Thus, we cannot blame Lexell too much for proposing this erroneous argument, nor for the fact that his proof of Fermat's Last Theorem for $n = 5$ is false.

To conclude this review of Lexell's research topics in mathematics, we have seen that during his time at Euler's side in Saint Petersburg, he occupied himself with a large variety of problems of analysis, geometry, number theory, combinatorics as well as with celestial mechanics, all with great success. Yet one feels that his attention was mostly confined to problems of immediate interest—except when it came to spherical geometry and polygonometry, where his contributions were substantial. His achievements are for the most part incidental—topics which no doubt had been suggested by Leonhard Euler—and lacking in the consistency and careful preparation which would have rendered them more memorable. An obvious reason for this is the enormous workload the Saint Petersburg academicians had to cope with within a short time. The present chapter has nevertheless shown that Lexell's mathematical papers are a rich testimony to his creative genius and still worth revisiting.

Chapter 9
Academic Journey 1780–1781

At the time of writing, well over two centuries have passed since the *grand tour* of Anders Johan Lexell to the heart of the enlightened Europe. Yet, to call this scholarly journey[1] a *grand tour* may be a misnomer when compared with the educational journeys that young aristocrats traditionally undertook to complete their instruction and cultivation. As a matter of fact, Lexell's journey to the academic venues of Berlin, Göttingen, Paris, London and Oxford was the real zenith of his career. Everywhere he went, he was warmly, sometimes even overwhelmingly received by the most renowned scientists of the day. He was known not only as a close associate of Leonhard Euler (who, it may be noted in passing, never visited Paris or London), but also on the basis of his own discoveries.

This chapter contains Lexell's own description of his journey as he presented it in a series of letters addressed principally to Pehr Wargentin in Stockholm. We begin with an account of the preceding course of events, which concerned Lexell's resignation from Saint Petersburg and his approved appointment as a Professor of Mathematics at the Royal Academy of Åbo. For other accounts of Lexell's journey, based on the letters preserved in Saint Petersburg, the sources [106–108] and [15] may be recommended. The extracts they contain of Lexell's reports from his travels in Europe are, as far as we know, the only ones that have hitherto been published.

[1]Lexell's journey was termed a *Voyage académique* in the Academy's annual report, AASIP 1780 Pars Posterior, *Histoire de l'Académie*, pp. 109–110, printed 1783.

J.C.-E. Stén, *A Comet of the Enlightenment*, Vita Mathematica 17,
DOI 10.1007/978-3-319-00618-5_9,

9.1 Prelude to the Journey

Ever since medieval times, *peregrinatio academica* or academic travelling[2] has been an essential element in the training of a scientist and scholar. For a young scientist, personal meetings with the great masters may be more important than reading of their books. However, before the mid-eighteenth century, scientists rarely had a chance to go abroad without some kind of sponsorship. The customary *grand tours* on foot could take many years, since travelling by sea and by stagecoach was only affordable for royalty and the upper strata of society in the main.

In the mid-eighteenth century socio-economic conditions began to improve and the mindset to change: a foreign trip for educational purposes was no longer unusual and could be adorned with various cultural virtues [7]. Privileged artists and poets absorbed inspiration from foreign countries, such as Goethe from Italy, as described in his romanticised *Italienische Reise*, in 1786–1788. Lexell's curiosity had undoubtedly been stimulated by the European journeys of his compatriots, particularly the astronomers Anders Celsius and Fredrik Mallet as well as the natural philosophers Johan Welin [151] and Bengt Ferner (Ferrner) [97], the orientalist J. J. Björnståhl[3] and more recently by his colleague and friend from Åbo, the philologist Henrik Gabriel Porthan, who visited Göttingen in 1779 [60]. In 1777, Johann III Bernoulli had met Lexell on a personal visit to Saint Petersburg [12]. With the reputation Lexell already had made in the learned world, it seemed only natural that he should also travel.

The earliest indication of Lexell's desire to travel is found in a letter to his colleague in Åbo, Professor Anders Planman, dated 12 September 1774.[4] Complaining about his wearing dispute with the Viennese astronomer Maximilian Hell on the determination of the solar parallax (cf. Sect. 5.1) and of the sarcastically written anonymous review article in the *Allgemeine deutsche Bibliothek* (cf. Sect. 5.4), Lexell continues:

---

[2]The term may refer to studies abroad, often in a famous institution, or a form of "brain drain" in general. Academic mobility was preceded, in the thirteenth century, by travel to the monastic *studia generalia* of Paris, Oxford, Cologne and Bologna.

[3]Jakob Jonas Björnståhl (1731–1779) was a travel journalist and Professor of Oriental Languages and Greek at the University of Lund, Sweden. From 1767 onward Björnståhl was constantly on the move in Europe and Turkey. His encounters were collected and edited posthumously by Royal Librarian C. C. Gjörwell: *Resa til Frankrike, Italien, Sweitz, Tyskland, Holland, Ängland, Turkiet och Grekeland* (1780–1784). In the fourth part (publ. 1782), p. 101, Björnståhl gives an account of his meeting with Christian Mayer in Mannheim in February 1774. On that occasion, Mayer had spoken about his agreeable journey through Sweden in 1769, praising both Wargentin and Lexell. He had also showed Björnståhl many letters written by them in Latin, which were filled with calculations.

[4]Apparently, Lexell did not withhold his plans from his closest academic colleagues: there are mentions of the journey which Lexell was planning to undertake in the correspondence between J. A. Euler and his uncle (by marriage), the Secretary of the Prussian Academy of Sciences Samuel Formey.

> Thus, I would rather prefer not to be obliged to publish anything or to stay in contact with scholars anymore; If I only were wealthy enough to get on with my life, I would prefer to work in peace, and if something would come out worthy of bringing into light, those who survive me could take care of its publication.
> For a long time I have wanted to travel, and that desire has now come to my mind stronger than ever. If a suitable offer presents itself for me to travel with a Russian *seigneur*, I will quit Petersburg. I would prefer to know that my position stayed open for me until I return, but if that is not possible, I would not hesitate to resign. This may sound paradoxical, but it is my earnest intention.

These are strong words by a fresh academician. Clearly, Lexell was upset by the unjust criticism of his work, and the thing he wanted the least was to be ridiculed as Euler's "calculator". Although he felt that it was his responsibility to defend his honour and integrity against unfair insinuations, he soon found out that the opponents were not playing a fair game. His naivety wore off with time, but not his sensitivity and fear of being misunderstood, which seems to be typical of many a great mathematician (Newton is perhaps the best example).

Several years of hard work evidently took their toll. The pace of work at the prestigious Petersburg Academy of Sciences was demanding, with at least two conferences held every week. The academicians were expected to produce as many outstanding papers as possible each year, and additionally they were obliged to occupy themselves with various practical tasks, such as making astronomical, meteorological and hydrological observations, as well as attending to different scientific commissions for the imperial court. The situation at the Academy worsened with the introduction of Sergey Domashnev as its director, whose strict leadership and high-handed attitude annoyed the academicians, who at that time were accustomed to a certain degree of autonomy and academic freedom in their activities.

Only once between 1768 and 1780 was Lexell granted a holiday long enough to make a trip to his home town. This happened in January–February 1773, when he, according to a letter to Wargentin, stayed in Åbo for 14 days. Later in the same spring, Lexell suffered from serious health problems; according to his correspondence with Wargentin (letter dated 12 May 1773), probably an infection or gangrene (in the original Swedish: *röt feber*). He also indicated that he had lost some weight.

In 1775, Lexell had been appointed Professor of Mathematics at the Academy of Åbo in his absence. The expectations of his return to occupy his chair were growing year by year, a fact that Lexell knew, being the conscientious man he was. Already in 1777, on the occasion of the royal visit in Saint Petersburg (see Sect. 7.4), the Swedish Minister and Chancellor of the Royal Academy of Åbo, Ulrik Scheffer, had uttered to Lexell the words: "We hope to have you back next year", alluding to his appointment to the Chair of Mathematics in Åbo. Lexell, baffled by the unexpected remark, could not produce a quick reply. Feeling concerned about the incident, he explained his point of view to Wargentin (letter dated 14 July 1777; Bergianska brevs. Vol. 17, pp. 260ff.). The leave of absence from his chair in Åbo had already been extended twice, but when his leave was coming to an end for the third time, he started talking more urgently about his wish to travel to Wargentin in a letter dated 20 August 1779:

> Until now I have not decided about my move to Åbo. Nevertheless, I have started to complete some work which I have committed myself to finish off, and which I cannot leave uncompleted without disgrace, if I were to leave. I would very much want to travel through Germany, France and England for a year before returning home, but I feel reluctant to ask for permission.

A few months later, in a letter to Wargentin dated 29 November 1779, Lexell says he has reason to fear that a further extension of his leave of absence from the Royal Academy will not be granted anymore, and that he therefore has reached the decision to quit Saint Petersburg and settle down in Åbo. On the other hand, he also felt exceedingly worn out, and so, complaining about his fragile health,[5] he writes to Wargentin:

> Now that my decision is taken to go home to my Fatherland, I hope that the grace will not be denied me to stay for some time in a milder climate, the more so as the constitution of my weakened body demands it.

In a letter to his colleague and friend in Åbo dated 6 January 1780, Professor Pehr Adrian Gadd, Lexell mentions in a resigned tone (KVAC, Berg. brevs. Vol. 19, pp. 511–517):

> I cannot say yet if I will go to Åbo the next spring, but it is most likely that it will not happen, as I am going to apply for His Majesty's permission to travel in Germany and France, in particular to visit the baths at Spa or Aachen for the sake of my health. I hope that this permission will be granted, but if it is not, it does not matter whether I am going to be buried in Petersburg or in Åbo.

On 14 January 1780, he writes to Wargentin regarding the application to travel which he has submitted to the King:

> I long to hear what news Count [Ulrik] Scheffer may have concerning my permission to travel, which in view of my health will be highly necessary.

In March 1780, still impatiently waiting for a letter of permission to journey from the King of Sweden, Lexell writes about his decision to resign:

> I have thus resigned from my position here in Petersburg to go to Åbo. The future will tell if I have acted wisely or unwisely, but I fear that I will be very bored in Åbo; not to mention the prize I have to pay for the move. If only Mr Domashnev had allowed me to travel while maintaining my salary here, I would never have chosen to go to Åbo.

In the ensuing letter of 24 April 1780 Lexell talks of the developments in a somewhat distressed manner:

> When so much mail from Sweden has been delayed here, I can easily imagine that the letter containing my application for His Majesty's permission to travel has only recently arrived. I admit in my letter the reasons why my application was written so late, but afterwards I find that I have nonetheless acted imprudently in not sending the application earlier, as I was

[5]There are no indications as to what illness he might have suffered from.

> already determined to travel. It might not be a big deal, but what makes it unpleasant for me is that I will arrive so late at Spa, where I intend to profit from the healthy waters.
> I cannot remember exactly if I mentioned in my last letter the proposition made by Mr Domashnev after his arrival, namely, if I agreed to stay in the Academy for a short time up to 10 years, I could be sure of my position.

Apparently, Lexell's decision to quit Saint Petersburg came as something of a surprise to the Director of the Academy Sergey Domashnev, although his commitment to the Chair of Mathematics in Åbo was by no means a secret. Lexell being one of the most respected members of the Academy, the news of his resignation was not a welcome one, and thus the director proposed another solution for him, as Lexell writes to Wargentin on 19 May 1780:

> Our Lord Director [Domashnev], who has recently departed to Moscow to inspect his estates, finally turned out more generous than I had expected. He gave me 200 Roubles for the journey and a gold medal as a souvenir, and, moreover, he offered me a rise of 200 Roubles in salary when I returned.

Although the offer was evidently a tempting one, Lexell stood by his decision to return to Åbo. Meanwhile, he received a letter from Åbo, which seems to have added the finishing touch, as he details in a letter to Wargentin dated 22 May 1780:

> [W]ith the latest mail I received a letter which totally deprives me of the willpower to go to Åbo, which was not strong even before. For about 13 to 14 years ago the Cashier of the city of Åbo, Mr Boerman,[6] left a deficient balance, and since according to Swedish law the councillors are responsible for the balances of the Cashiers, the County Administrative Court of Åbo at the request of the Audit office now require the respective councillors or their heirs to pay for the deficit of the balance, which is 5500 *Riksdalers*. Unfortunately, my Father was a councillor back then, hence the unpleasant duty falls upon me to compensate for the deficit, how much I do not know yet, but be it much or little, it is quite unpleasant for me. [. . .]
> The news arrived strangely enough just as I was about to travel, and I am inclined to think that it was a hint by divine providence to refrain from doing so. In any case, it is certain that if I were obliged to pay back, there would be nothing left, since I have spent so much on the journey and my move to Åbo. It might seem unwise of me to spend up to 600 or 700 Roubles or more of my own savings only on the journey, and, in addition, 300 to 400 Roubles for buying me a new wardrobe and for furnishing my home in Åbo. If Mr Domashnev were here, he might perhaps grant me the permission while keeping my salary here; but he has already departed and before he returns, the best season is already over. As it is, I am inclined to submit my resignation from the chair in Åbo and stay here, and to let divine providence care for the rest.

Having carefully considered his chances, Lexell sent his letter of resignation to the King and started to prepare the route of his journey, notifying his relatives in Åbo and asking them to arrange for bills of exchange to be sent via Wargentin to Johann Bernoulli in Berlin. After the return of his chief from Moscow, he writes again to Wargentin on 30 June 1780:

> After the return of Director Domashnev last Wednesday, I have visited him with a proposal of a permission to travel while maintaining my salary, having first applied for resignation in

[6]Gustaf Johan Boerman (–1784).

> Sweden. He was entirely satisfied with it and accepted my proposal. Now there remains nothing that can hold me back, except that my companion[7] has not yet obtained an acceptance of his resignation. [...]
> In fact, my plan was first to go via Berlin, but now in view of the season being so advanced already I would rather go by sea to Lübeck and thence directly to Spa.

However, in a letter dated 14 July 1780, Lexell announces his new commission:

> In order to give me more reasons to travel, Mr Domashnev has commissioned me to report to the Academy on the new findings in astronomy, geography and physics etc. I am quite pleased with it, as I may in this way perhaps be more useful to the Academy than if I were to stay the year in Petersburg.

Indeed, on 22 June 1780 [141, Vol. III, p. 476] a message from the director was read out at the conference, announcing that

> [...] His Excellency the Director has granted to Professor Lexell the right to travel for the improvement of his health for one year, counting from the day of his departure; he has been ordered to visit the most celebrated astronomical observatories which are not situated too far away from his route, in Germany, France and England, in order to acquire their plans and drawing in detail, to learn about the new instruments in mathematics and physics which he finds in the places he visits, chiefly in London. He shall obtain information on the geographical and hydrographical maps and on plans and drawings recently published etc., and finally, he shall observe anything which he deems to be of importance and to instruct the Academy by correspondence either with His Excellency the Director or with the Secretary, especially during his stay in the main cities. The Secretary will accordingly submit to him a formal instruction, signed by His Excellency, which will serve as a *plein-pouvoir* in his commissions.[8]

Although not a secret, Lexell's commission was evidently a kind of a scientific espionage or information-gathering expedition. As no one at the Petersburg Academy of

[7]Lexell's incidental companion, at least during the voyage to Germany, seems to have been Lieutenant Ulrik Karl Stierneld (1751–1816), a Finnish-Swedish nobleman. He was born and educated in Åbo, later he studied privately in Uppsala [88] and made a rapid military career. Having killed his opponent in a duel (duels were prohibited in Sweden) he fled to Russia, where he served as a Cornet in a Russian dragoon regiment until 1780. Thereupon he served in the French *Régiment Royal Hesse-Darmstadt*. In 1784, Stierneld was allowed to return to Sweden.

[8]Le Directeur vient d'accorder à Monsieur le Prof. Lexell la permission de voyager pour rafermir sa santé, pendant une année en comptant depuis le jour de son départ: qu'il l'a engagé de visiter les observatoires astronomiques les plus célèbres qui ne sont pas trop éloignés de sa route, en Allemagne, en France et en Angleterre, d'en acquerir les plans et descriptions en détail, de prendre des informations sur les nouveaux instrumens de mathématiques et de physique qu'il aura occasion de voir dans les lieux par où il passe, principalement à Londres; de se procurer des notices sur les cartes géographiques et hydrographiques, sur les plans et dessins nouvellement publiés etc. Enfin d'observer tout ce qui pourroit lui paroître de quelque importance; et d'en instruire l'Académie en correspondant soit avec Son Excellence Monsieur le Directeur, soit avec le Secrétaire des Conférences, surtout pendant le séjour qu'il fera dans les villes principales. Le Secrétaire lui expédiera en conséquence une instruction formelle, qui signée par Son Excellence lui servira de plein-pouvoir dans les commissions dont il est chargé.

Sciences had made such a long visit to the European scientific centres for decades, there was an increasing need to keep up with the rapid developments in the sciences elsewhere in Europe. Lexell felt it his responsibility to act as the eyes and ears of his academic colleagues and to convey to them useful information. Also Lexell himself, who had to abandon his original intention to enjoy a health treatment in Spa, seemed quite relieved and happy with his new commission, not least with the financial support of the Academy. He saw the commission as a personal favour by Sergey Domashnev, with whom he tried his utmost to stay on good terms (but as it later turned out, in vain).

At last, in a letter to Wargentin on 6 August (26 July o.s.) 1780, he announces with some satisfaction:

> Now I have the honour to report some progress as regards my journey; it is that today, I will embark if the wind blows favourably. If everything goes well I may be in Stettin in 9 to 10 days. But as the season is already this much advanced, I will not be arriving in Paris before mid-October. From Berlin I plan to journey via Göttingen, Cassel, Frankfurt, Mannheim and Strasburg.

The Journey was about to begin.

9.2 Berlin: Work, Theatre and Sightseeing

Lexell's Journey lasted more than a year, from August 1780 to October 1781. His itinerary was the following: Saint Petersburg → Swinemünde → Stettin → Berlin^{P,S} → Potsdam^{P,S} → Leipzig → Göttingen^{P} → Kassel → Mannheim^{P} → Strasbourg^{P,S} → Paris^{P,S} (Versailles) → Ostende → London^{P,S} (Oxford) → Ostende → the Hague^{P} → Brussels^{B} → Amsterdam → Hanover → Hamburg^{P} → Kiel → Copenhagen → Stockholm → Uppsala → Åbo (Turku)^{P,S} → Saint Petersburg. We have indicated with the superscripts P, S, B, the places where Lexell wrote his reports to Saint Petersburg (addressed to either J. A. Euler or S. Domashnev), Stockholm (Wargentin) and Berlin (J. Bernoulli), respectively (Fig. 9.1).

Lexell most probably also wrote letters to his friends and family in Åbo, as Porthan in his correspondence mentions the latest news from Lexell, but the original letters have not been discovered. In the following, we give an account of Lexell's journey as described in his letters to Wargentin, occasionally augmented with details from his more ample correspondence with Saint Petersburg.

In the first letter Lexell reports on his successful journey by sea to Swinemünde (today Swinoujscie in Poland), which had taken 8 days. During his journey, "in order not to lose touch with the work", he pursued his astronomical calculations of the orbit of the comet seen in 1763,[9] which he had started in Saint Petersburg. In his letter to Saint Petersburg, which was dated in Berlin on 26 August 1780 (АРАН Фонд 1, опись 3, дело № 65, л. 71–74об.), he communicates the elements of the

[9]Comet "C/1763 S1".

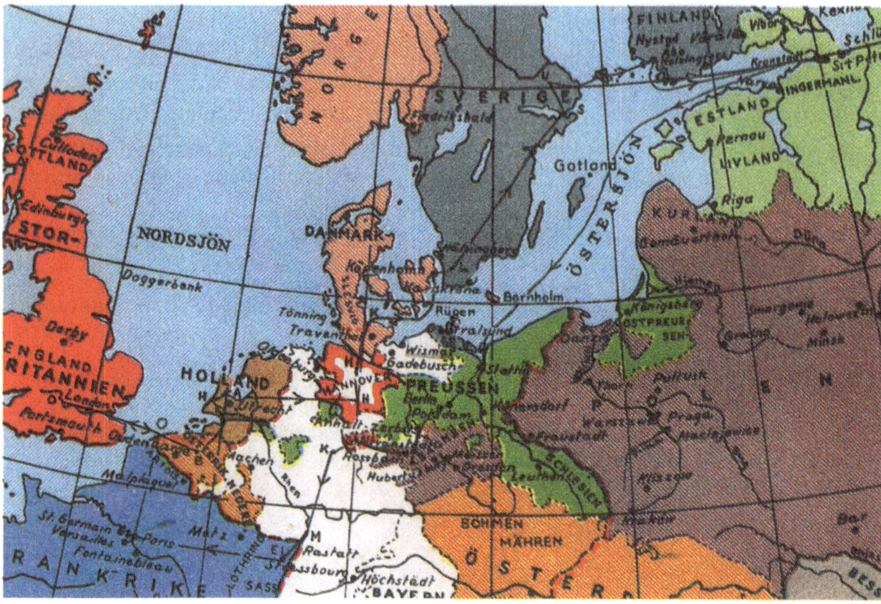

Fig. 9.1 Map indicating the approximate route and the main stops on Lexell's European journey in 1780–1781. Abbreviations: L = Leipzig, G = Göttingen, K = Kassel, M = Mannheim, O = Ostende, B = Brussels, H = the Hague, A = Amsterdam, H = Hanover and Hamburg, K = Kiel, L = Lund, S = Stockholm, U = Uppsala

comet he had calculated, mentioning that it was perhaps the first time anybody had made astronomical calculations in a tavern in Swinemünde, an example of Lexell's dry humour.

From Swinemünde Lexell continued via Stettin (today Szczecin, Poland) to Berlin, the capital of Prussia, where he stayed for nearly a month. Compared to the old cities of London and Paris, which Lexell would visit during his journey, Berlin was a small but rapidly growing capital with about 100,000 inhabitants, i.e. of approximately the same size as Saint Petersburg.

> Mr Bernoulli handed me the letter from you, Herr Secretary, about eight days ago. The letter from Count Scheffer, which was forwarded to me from Saint Petersburg, says that my resignation [from Åbo] will be granted as soon as His Majesty the King has returned to Stockholm from his journey. I am completely satisfied with this, because an extension of my permission which was proposed by Baron Nolcken would only have caused much unhappiness among those who think they have been treated unjustly because of my absence. My journey to Swinemünde by sea went both happily and well, but advanced subsequently more slowly from Swinemünde up here [to Berlin], a journey which took me more than eight days. I arrived here in Berlin on the 18th of August, so that I have been here for almost three weeks and I am now ready to leave this place, which up to now has pleased me much and would undoubtedly be even more agreeable, as soon as one came to know it better. The acquaintances I have made here are principally among the learned men, in particular the mathematicians of the Academy of Sciences, namely Mr La Grange,

Bernoulli, Castillon, Schultze and Bode.[10] The first one mentioned is a very courteous man, whose mere physiognomy suggests an incomparable genius. Among the other members of the Academy of Sciences I have made acquaintance with are Mr Merian[11] and Formey, who have received me very amiably.

As I have been commissioned to visit observatories and to make notes on astronomical subjects, I have made a careful inspection of the present observatory, which as regards both its construction and supply of instruments is nothing out of the ordinary. A mural quadrant of radius 5 feet, a passage instrument of 3 ft, a Dollond tube with a threefold objective lens [=compound lens with three glass layers], one telescope and four timepieces constitute the most exquisite instruments here, the rest being on the whole unserviceable. As to the mural quadrant, there is a facility to suspend it both towards the north and the south. From the *Ephemerides* of Berlin you, Herr Secretary, may have noticed that Mr Silberschlag[12] has invented a new astronomical instrument he calls the *Uranometer*, which he has kindly agreed to show me. However ingenious this instrument may be, I have difficulty in believing that it can be very accurate, not to mention comparable to a quadrant of radius 60 ft, as Mr Silberschlag himself claims. At the slightest movement of the observer, or the gentlest touch of the micrometer, the tube starts to tremble, which cannot but make the observations unreliable. The only advantage of the instrument is that it can easily be applied both towards the north and the south without deranging the instrument and also that observations which require finding the azimuth may be properly and accurately executed. On the occasion of my visit to Mr Silberschlag to see his *Uranometer*, he also showed me his models of the old throwing machines i.e. *Ballistae catapultae*, and also a model to explain the Deluge. Mr Silberschlag is without a doubt a clever man, but he is moreover, unless I am mistaken, a bit of a charlatan. He told me for example that, as according to the Book of Moses, it rained for 40 days, the water stayed for 110 days and an additional 120 days would pass for it to flow away; so he had also discovered that, having poured water into his model for 40″, the water had stayed there for 100″ and flowed away during the next 120″. I find it difficult to believe that such a precise correspondence could take place, unless it occurred by sheer chance.

Mr Bernoulli has asked me to send the enclosed advertisement to Professor Planman for distribution in Åbo,[13] but as I do not have time to write to Planman this mailing day, I request you, Herr Secretary, to send it to Mr Planman on a suitable occasion.

I have the honour to be with a profound respect,

Honourable Herr Secretary,

Your

humble servant

Lexell

Berlin, 5 September 1780

I intend to leave [Berlin] in the beginning of the next week, before which I am going to write once more. Now I am waiting to hear if I will have the good fortune to be presented to the King, and as soon as that is over I will move on.

---

[10] Johann Karl Schulze (1749–1790) and Johann Elert Bode (1747–1826) were mathematicians and astronomers in Berlin.

[11] Johann Bernhard Merian (1723–1807), Swiss philosopher, member of the Prussian Academy of Sciences.

[12] Johann Esaias Silberschlag (1721–1791), theologian, natural scientist and inventor, a member of the Prussian Academy of Sciences.

[13] A prospectus for Bernoulli's Journal from his travels [12].

In his letter to Saint Petersburg Lexell enters into more detail, knowing that many of the persons he encountered were friends of the Eulers, and also because Johann Albrecht Euler was more at home in Berlin, where he had grown up and lived for the most of his life, than in his native town of Saint Petersburg [37]. Being like Johann Albrecht Euler a devotee of the theatre, Lexell also conveys a vivid description of the play given in the *Comédie Allemande* and its actors (Berlin, 26 August 1780; АРАН Фонд 1, опись 3, дело № 65, л. 71–74об.):

> ...we went to the *Comédie Allemande* [German Comedy]. There was a performance of a play which was a translation of the English *School for Scandal*.[14] It was like most English plays usually are, filled with a lot of action but without a proper plot. The women who were playing in this comedy¸ namely Miss Döbbelin,[15] Mrs Böheim and Mrs Mecour, performed very well, but a certain Miss Seifert was only mediocre. Among the men who were playing, Mr Brucker and Mr Langerhans[16] seemed rather talented. A certain Mr Unzelman is a big fellow with a handsome face; he also played the role of a debauched person very credibly; but he was not able to express emotions, since when he was meant to move to tears, he made people laugh instead, just as I have noticed him do since in another play. The other comedians who acted in this play performed very badly.[17]

Other plays Lexell mentions to have seen were *Les Hollandois* and *Les six plats*.[18] During his stay in Berlin the weather was stifling, and Lexell stayed inside during daytime. He enjoyed walking in the park (*Tiergarten*), admiring the views of Berlin—in today's language, sightseeing. He was particularly impressed by the broad streets, the like of which he had never seen before, such as the *Potsdamer Strasse* and *Unter den Linden*. Having a taste for fine arts, he visited the Academy of Arts and the art museums of the town and dined with the director of the Royal Porcelain Manufactory Johann Georg Grieninger, who was a friend of the Eulers. At an assembly of the Academy of Sciences to which he was invited he noted several other visitors present, namely the Polish politician and author, Count Józef

---

[14]The play was written by the Irish dramatist and playwright Richard Brinsley Sheridan (1751–1816).

[15]Karoline Maximiliane Döbbelin (1758–1829), Maria Anna Böheim (1759–1824) and Susanne Mecour (1738–1784) were actresses [164].

[16]Karl David Langerhans (1748–1810), actor and director, and Karl Wilhelm Ferdinand Unzelmann (1753–1832), actor, singer and dancer [164].

[17]...nous allions à la Comédie Allemande. On jouoit une pièce traduite de l'Anglois, nommée *l'École de la Médisance*. Elle est comme les pièces Angloises pour la pluspart sont, remplie de beaucoup d'action mais sans aucun plan. Les femmes qui jouoient dans cette comédie, sçavoir Mlle Döbbelin, Mme Böheim et Mme Mecour faisoient très bien, mais une certaine Mlle Seifert ne jouoit que médiocrement. Parmi les hommes qui jouoient, Mrs Brücker et Langerhans paroissent avoir assez de talent. Un certain Mr Unzelman est un grand garçon et d'une assez belle figure, il jouoit aussi le rôle d'un débauché dans cette pièce avec beaucoup de vérité; mais pour exprimer des sentiments il n'en est pas capable, en sorte que lorsque il doit toucher aux larmes, il fera plustôt rire, comme je l'ai remarqué depuis dans une autre pièce. Les autres comédiens qui jouoient dans cette pièce, s'acquittèrent fort mal.

[18]Play written by Gustav Friedrich Wilhelm Grossmann (1746–1796).

Maksymilian Ossoliński (1748–1826), the Italian diplomat and favourite of the King, the Marquis Girolamo Lucchesini (1751–1825), and the natural scientist and explorer Johann Reinhold Forster (1729–1798). In the evenings, he dined mostly with the families of Bernoulli and Formey.

On a visit to the young astronomer Johann Elert Bode, Lexell had the opportunity to meet Christine Kirch (1696–1782), "la Doyenne de tous les Astronomes" as he called her in his letter to J. A. Euler (АРАН Фонд 1, опись 3, дело № 65, л. 83). She had been trained in astronomy by her parents, the astronomers Gottfried and Maria Kirch, first as an assistant in observations, later in calendar calculations. Lexell testified that despite her considerable age she had been calculating only two years earlier (at the age of 82 years), but now her eyesight was too feeble, although she could still see with her right eye. She had fond memories of the Eulers, especially Leonhard who had been the de facto director of the Berlin Academy of Sciences. Kirch's duties at the Almanac office had been taken over by Bode, who was married to one of her grandnieces.

9.3 King Frederick Denies an Audience

The first adverse experience on Lexell's tour occurred in Potsdam, some 25 km from the centre of Berlin, where he, on the suggestion of his colleagues in Berlin, had ventured to ask for an audience with King Frederick of Prussia in the magnificent rococo palace of *Sanssouci*. However, the King flatly refused him an audience. Lexell took this as a shameful setback and regretted his "foolish" attempt, as he explains to Wargentin:

> I was planning to write to you, Herr Secretary, before my departure from Berlin yesterday, but for several reasons I had to postpone it, so that I now have the honour to write to you from Potsdam. My purpose in coming here was to be presented to the King, wherefore I had first corresponded with Mr de Catt,[19] who is the reader (*lecteur*) of the King, and later on his recommendation asked for this favour in a letter to the King himself. Despite the efforts made by Mrs de Catt and Lucchesini to obtain this favour for me, my request was denied in a reply signed by the King, although in the most gracious terms possible. Nevertheless, it is quite unpleasant for me that this happened, and I would have preferred a thousand times not to have attempted this at all instead of being denied. Many may judge that I have asked for more than I have reason to, and when I think about how incomplete my knowledge is in every respect, it now really seems to me that I have committed an inexcusable folly in endeavouring to obtain a favour which was denied to me. Though in my defence I can say that first of all it cannot be regarded as imprudent that I had wanted to see the King, and also I may be excused for having thought the most decent and not to mention the surest way would be to be presented to him. Later on I was encouraged when I heard that such a presentation would be of no difficulty and that the usual etiquette requires a written request to the King.

[19]Henri Alexandre de Catt (1725–1795), Frederick's confidant until 1780.

The most unpleasant thing for me is that my attempt to be presented is so largely known in Berlin; wherefore I fear that many will see me as a self-opinionated fool. I presume, nevertheless, that whoever knows my way of thinking is convinced that my greatest ambition is not to pursue something I am unworthy of and that nothing can be more unpleasant for me than to be made a fool. In other respects, I am quite content with Berlin. I have been shown so much courtesy and friendship here, that I am perfectly satisfied in this regard. I have been especially delighted to make the acquaintance of M^{rs} La Grange, Schulze and Tempelhof, all of whom are quite skilled mathematicians, especially the first one.

I found the Berlin Observatory to be in a state which one easily might imagine for a place where no great expenses are made for astronomy. A 5 ft mural quadrant and a Dollond tube with a threefold objective constitute the most exquisite instruments there. Presently both Mr Bernoulli and Schulze are making observations, at any rate as long as Mr Schulze can stay in the rooms Mr Bernoulli has prepared for him. Nevertheless, these gentlemen must sometimes put up with inconveniences, such as, for instance, during my stay in Berlin, when General Prittwitz's[20] carriage was parked right in front of the Observatory, so that the door could not be opened. The Observatory is situated in the same building as the royal stables as well as the stables of the gendarmes. I have seen the *Uranometer* of the Councillor of the Consistory Mr Silberschlag and heard him demonstrate its benefits, but I am still not convinced that it possesses the requisite steadiness and security. It seems to me that both the hollow cylinder, which constitutes the axis of the instrument, and the assembly of the instrument in the same, cannot be sufficiently steady and trustworthy; at least it is certain that the slightest movement in Mr Silberschlag's small observatory will make the tube tremble considerably, which, it appears to me, will cause a change in the micrometer.

I mentioned in my last letter Mr Bernoulli's advertisements of the account of his journey, but I forgot to include them in the letter; now I have an opportunity to ask you again, Herr Secretary, to send them either to Professor Planman or to Professor Porthan; though I can easily imagine that no market can be expected for them in Åbo.

During my stay in Berlin, Mr Bernoulli persuaded me to have my silhouette made by a master, who does it as his occupation, having already done about 20000 silhouettes. I find my silhouette [see Fig. 9.2] quite well done and the others think so, too. Thus, I send one example of it enclosed and ask that you, Herr Secretary, if you consider it worthwhile, would let it be copied for yourself as a souvenir, and thereupon send it to my brother-in-law in Åbo, to whom I take the liberty of enclosing a letter.

My time constraints oblige me to stop here with an assertion of the constant affection by which I have the honour to be,
Honourable Herr Secretary,

Your

humble servant
Lexell

Potsdam, 14 September 1780

[20]Joachim Bernhard von Prittwitz (1726–1793), Prussian military commander.

Fig. 9.2 Silhouette of Anders Johan Lexell cut by an engraver in Berlin in 1780. The signature "Haf" suggests that the artist was probably Johann Lorenz Haf (1737–1802) [173] (Johann III Bernoulli Briefe, UBB Basel Ms L I a 703 f° 106. Published with permission)

9.4 Leipzig, Göttingen, Mannheim and Strasbourg

While continuing his journey across the German states, Lexell managed to overcome his disappointment in Potsdam. He focused his attention on the landscape passing by the window of his stagecoach, and with his rudimentary knowledge of natural history he tried to make himself useful. In his next letter to Wargentin sent from Strasbourg he describes the varying terrain and makes some surprising observations of the geology of these lands.

> My last letter to you, Herr Secretary, was, if I remember correctly, sent from Potsdam, just before my departure. Now that I am obliged to stay in Strasbourg a few days longer than I had intended, until I can arrange lodgings for myself in Paris, I can use this time to write a brief account of my journey from Potsdam to Strasbourg, which went through Wittenberg, Leipzig, Göttingen, Cassel, Frankfurt and Mannheim.

I am not versed enough in natural history to make sufficiently accurate remarks on the change of the soil which is seen in the numerous German provinces I have traversed. Nevertheless, I wish to communicate what an observant traveller cannot pass without taking notice. In large parts of Pomerania and Mark Brandenburg the soil consists of mere sand all the way up to Saxony, where a mixture of clay commences. All the heights and hills in this land consist of sand, and I found it quite strange to find at our arrival in Swinemünde large amounts of the finest sand.[21] In my opinion this completely refutes the hypothesis according to which the basic substance of mountains is sand, because then there would be no rocks and stones in those countries where there is sand.[22] Saxony is quite a wonderful and blessed country, where the soil consists mostly of clay and earth, especially delightful are the surroundings of Leipzig. In this country there are only a few hills, but the most magnificent fields, which often have a size of three or four German miles. As is well-known, the surroundings of Hanover are quite hilly. As my journey did not go past the Harz mine works, I am unable to say what kind of rock they are, but the rocks I had the occasion to see at Schwartzfeld and Osterode were mostly limestone and shale. In Hesse the land remains mountainous, but now the rocks change to sandstone except for the copious volcanoes that exist all the way to Frankfurt. There is a volcano next to Münden, a town between Cassel and Göttingen, from which stone has been excavated to improve the main road, stone which no doubt must be lava. Also the famous building in Cassel called the *Winter Kasten* is for the most part built out of lava stone, which is why it cannot withstand the changing weather for a long time. The rocky ground upon which this building has been erected is without doubt an extinguished volcano as it consists almost exclusively of lava. As is well-known, the Palatinate is considered to be the most pleasant part of Germany, and as far as I can judge, considering the late season, this is not without good reason, so happily is this land suited for farming and viticulture. Along the river Rhine there is a ridge of mountains [the Vosges] that traverses Alsace probably down to Basel, where the more considerable Swiss mountains begin.

About his stay in Leipzig Lexell reports:

In Leipzig I stayed a few days, and I spent them quite well, partly because I quite liked this town, and partly because I found a rather good-mannered and serviceable gentleman there, *Magister* Hindenburg,[23] who arranged for me an opportunity to see everything significant in Leipzig; since it is a commercial city the sites cannot be abundant in the sciences. However, the Senate Library[24] is quite extraordinary and surpasses the one that belongs to the University; moreover, it is set up in a quite beautiful hall, and also the building itself

[21]The sandy beach of Swinoujscie (Swinemünde) is 5 km long.

[22]Here, Lexell in effect went against Linnaeus, who believed that stones grow, though not as living creatures, but by sticking together of earthy particles. Thus, sandstone is created from grains of sand lumped together, granite from marl, and limestone from clay. Linnaeus also thought that bedrock is formed from hardened gravel [101]. It is possible that Lexell's analysis was inspired by the book on geology published a short time previously by his colleague academician Pallas, entitled *Observations sur la formation des montagnes* (St. Petersburg, 1777), a copy of which he owned [47].

[23]Carl Friedrich Hindenburg (1741–1808) worked especially on combinatorial mathematics by developing a "polynomial theorem" as a generalisation of the ordinary binomial theorem. He became in 1781 Professor of Philosophy in Leipzig and edited, in part together with Johann III Bernoulli, several mathematical journals [55].

[24]In the Swedish original: *Råds-Bibliotheket*. The building of the *Ratsbibliothek* with its baroque façade, generally appreciated as the most exquisite in Germany, was destroyed during the Second World War.

> is the finest in Leipzig. A merchant named Winckler[25] owns a large and beautiful collection of paintings. Around this city there are quite neat and beautiful gardens, of which the most exquisite one is the property of a merchant named Richter.[26]

It is interesting to compare this account with Lexell's report to Saint Petersburg written in Göttingen on 27 September 1780 (АРАН Фонд 1, опись 3, дело № 65, л. 101–103 — see Appendix section "Letters" for the original French quotation of the letter), which he embroidered with some details:

> Going from Berlin to Göttingen, I preferred the route passing through Saxony to the one which goes via Magdeburg and Halberstadt, both to see Saxony, which is quite a beautiful country, and also to make the acquaintance of learned men in Leipzig and especially of a certain Mr Hindenburg, who is a very skilled mathematician. Thus, having left Potsdam on 15 September, as I had the honour of noting in my last letter, I arrived in Leipzig on 17 September, and having stayed there until the 19th, I could not write from Gottingen until 22 September. In Leipzig there is almost no astronomical activity at all, and there are even no astronomical instruments, except for the remainder of those which used to belong to Mr Heinsius,[27] and which are now in the possession of Mr Borz,[28] the Professor of Mathematics. The two main libraries are those of Leipzig University and of the Magistrate, the first of which does not appear to be very extensive and contains for the most part books on theology; the other is much more beautiful and contains forty thousand books, the largest collection being of books on history. The Library is also housed in a very beautiful house belonging to the city. The surroundings of Leipzig, both natural and cultivated, are beautiful, the city being surrounded by several beautiful gardens, the finest of which belongs to a merchant named Richter. Another trader named Winckler has a fine collection of nine hundred paintings, which even includes some of the best Italian, Flemish and German masters.

After Leipzig, Lexell arrived in Göttingen, which at the time was a popular destination for students from Sweden [21, 60]. The university, which had only been founded in 1737 [65], was already renowned for its open atmosphere, where philosophical ideas could be debated freely. Göttingen belonged to the Electorate of Hanover, whose Duke during that period was also the King of England. Only a few decades later, in the nineteenth century, the town developed into the celebrated "Capital of Mathematics" owing to the fact that inter alios Gauss, Riemann, Dirichlet, Hilbert and Klein lived and worked there. However, at the time, the town seemed not to have been quite to Lexell's taste, partly because of the style of the houses built of wood, partly because of the bitter disagreements and feuds between the academicians of the town.

Abraham Gotthelf Kästner, to whom, as a mathematician, Lexell was principally attached, had a reputation for missing no opportunity to run down his colleagues

---

[25]Gottfried Winckler (1731–1795), merchant and art collector in Leipzig.

[26]Georg Wilhelm Richter (1735–1800), wine trader, land owner, and coffeehouse owner in Leipzig.

[27]Gottfried Heinsius (1709–1769), Professor of Mathematics in Leipzig and member of the Petersburg Academy of Sciences, where he had been an astronomer in 1736–1743.

[28]Georg Heinrich Borz (1714–1799), Heinsius's successor in Leipzig as mathematician and astronomer.

acidly, especially his enemy, the philosopher, theologian and orientalist Johann David Michaelis (1717–1791) [21]. This view is confirmed in Lexell's account. Among other academic celebrities in Göttingen, Lexell met the experimental physicist and philosopher Georg Christoph Lichtenberg (1742–1799), the historian August Ludwig Schlözer, who had been instrumental in his employment in Saint Petersburg, as well as his compatriot, the botanist Johan Andreas Murray (1740–1791).[29]

In order to assess the personalities he met as objectively and accurately as possible, Lexell consulted Johann Caspar Lavater's (1741–1801) famous book on physiognomy in four volumes, *Physiognomische Fragmente* (1775–1778), which in 1777 had been acquired by the Petersburg Academy of Sciences. Lavater was a reformed clergyman and philosopher from Zürich, who maintained that the appearance and expression of the face mirrors the condition of the soul and that the movement of the body expresses the human character. Lavater was not the first (or the last) one to advance such a pseudo-scientific theory, and his book was considered dubious even in his times [11]. Also Lexell was soon to discover that Lavater's analysis rarely corresponded to the actual character of the persons he met. In fact, towards the end of his sojourn in Paris he concludes (letter to J. A. Euler dated 24 March 1781): "Having seen the faces of ten talented mathematicians one finds such marked differences in their appearances, eyes, noses, mouths etc., that it would be impossible to conclude that they occupy themselves with the same science".[30]

The letter from Göttingen to Wargentin continues:

> Göttingen is a rather unpleasant place and the town itself is constructed in a flat manner, without a single massive house, but all of them built out of timber frame. Thus, for my part, I would not be willing to choose this place for my stay, the more so as there is not the least bit of harmony and agreement between the local learned men, and it rarely happens that two or three of them meet each other. My acquaintances here were with Kästner and Lichtenberg as mathematicians, with Murray as a compatriot, and with Schlözer and Beckmann[31] since they have been in Russia. Kästner has shown me much friendship and politeness. Many regard him as an honest and well-disposed man, although his sarcastic nature[32] often brings him into conflict with his colleagues, as has recently happened with Beckmann. Even without having read Lavater's *Physiognomy*, it is not difficult to judge that Kästner has a strong penchant for sarcasm, so extraordinary is his appearance when one sees him. Lichtenberg is quite cheerful and polite, no matter how little the constitution of his body shows it, because he is crooked. He has told me a lot about the instruments of the English observatories, which has pretty much excited my curiosity. Our compatriot Professor Murray has shown me much courtesy and I for my part have the best reasons to

[29]Two short letters by Lexell, written during his stay in Göttingen in 1780, are preserved among Murray's correspondence at the National Archives (Riksarkivet) in Stockholm (Murrayska släktarkivet, Vol. 67).

[30]АРАН Фонд 1, опись 3, дело № 65, л. 191–194об.

[31]Johann Beckmann (1739–1811), Professor of Economy in Göttingen, historian of science and technology, and coiner of the term "technology" [55]. He had studied in Sweden and had also taught at the *Gymnasium* of Saint Petersburg.

[32]In the original Swedish: "satyriskt lynne".

> be satisfied with him, but otherwise the rumour in Göttingen has it that he has almost no friends and that he is rather greedy, both of which qualities he shares with the majority of his colleagues. Their self-interest has gone so far that even the famous Michaelis offers his courses for subscription and even haggles with the students to make them choose more of his courses. The Observatory in Göttingen is not in an impressive state; nevertheless they have some fine instruments there, but as long as Kästner has many other things to attend to, he probably will not take part in any observations except as a distraction.

From Kassel, Lexell reports:

> In Cassel, I became acquainted with Matsko[33] as a mathematician, Professor Stegmann[34] who is a known and able mechanic and Professor Forster the younger,[35] a rather polite and courteous man, who has received me with much friendship. Mr Matsko supervises the Observatory, which is pretty much decayed and also not very rich in instruments, in particular those which are needed for determining the time. Now the Margrave has ordered a new observatory to be built in the tower where in bygone days Margrave Wilhelm[36] used to observe, situated alongside the costly building the Margrave wishes to have built, to establish there his so-called museum, containing the library and its collection of antiquities, artefacts, astronomical and physical instruments, natural curiosities etc. However, this tower may have the inconvenience of not being trustworthy enough, as a major road runs next to it, on which all the carriages pass on the way to the Margrave's out-of-town palace called *Weissenstein*, situated 2 hours or 1 $\frac{1}{2}$ Swedish quarter miles from Cassel. This palace is known for the extraordinary building erected on the mountain and for the fountain, which is driven by water running down from a spring located on top of the mountain, rising up to 160 feet.

In his letter to Saint Petersburg, dated Göttingen, 27 September 1780 (АРАН Фонд 1, опись 3, дело № 65, л. 101–103 — for the French original, see Appendix section "Letters"), Lexell reports in more detail his observations about the scientific equipment of the Göttingen Observatory:

> The Observatory of Göttingen is built on a tower which was once part of the fortifications of the city. It is 18 to 22 feet in diameter, and its height is only 10 ft at most. The observatory is very narrow since the stairs leading to the Observatory are in the middle. The horizon is on all sides lined with mountains, thus no observations can be made for elevations less than 3° above the horizon. The instrument collection is very nice and contains: 1. A quadrant with a radius of 6 ft by Bird,[37] which is placed to face the south, but the wall to which it

---

[33]Johann Mathias Matsko (1717–1796), mathematician and astronomer.

[34]Johann Gottlieb Stegmann (1725–1795), mathematician, physicist and inventor.

[35]Johann Georg Adam Forster (1754–1794), German naturalist, "travel journalist" and revolutionary, son of Johann Reinhold Forster, whom Lexell had recently encountered in Berlin. In his letter to Saint Petersburg from Mannheim dated 12 October 1780 (АРАН Фонд 1, опись 3, дело № 65, л. 97–100об.), Lexell characterises Forster as an able natural scientist and recommends him for some of the Russian expeditions: "En cas qu'on voudroit faire quelque expédition pour l'histoire naturelle dans la Russie et qu'on y vouloit employer quelque étranger, je suis persuadé que Mr Forster accepteroit cette proposition très volontièrement, tant il aime à voyager."

[36]Wilhelm IV (1532–1592), the Landgrave of Hesse-Kassel, was a keen astronomer.

[37]John Bird (1709–1776), maker of scientific instruments. He specialised in so-called divided instruments, the scales of which were divided manually by bisection [139].

is attached has moved slightly so that the instrument deviates somewhat from the vertical plane. Opposite this wall there is another wall to align the instrument to the north. 2. An English quadrant by Sisson[38] with a radius of 2 ft [3] A quadrant with a radius of 2.5 ft, made by Campe[39] in Göttingen in the French style, but as to the telescope, the construction is more like those used for wall quadrants. 3. Three achromatic refractors, of which the main one was made by Dollond in London: This telescope is only five or six foot long, but it has a very nice magnification and is far superior to the other two, which were made in Göttingen by Baumann.[40] 4. Five pendulum clocks, among which there is one with a compound pendulum, made in London by Shelton[41]; the others were made in Göttingen. Shelton's clock also has a counter by the same craftsman. 5. A clock showing the half seconds, and also the quarters and eighths of a second. This instrument is a gift from the Queen of England. 6. A clock which displays the thirds of a second by Campe.

Lexell also mentions that Leibniz's mechanical calculator or arithmetic machine for computing multiplications was conserved at the Observatory.

On his way to France, Lexell next arrived in Mannheim, where he met an old acquaintance, Father Christian Mayer. The reunion seems to have been warm and cordial, as Lexell describes Mayer's observatory and instruments in great detail and admiration (Fig. 9.3):

I had heard Professor Mayer's Observatory in Mannheim been much praised, but I have to admit that I found it much more perfect than I had anticipated, and I am convinced that anyone who knows astronomy must give Father Mayer the credit of having set up a building as practical and well-organised for astronomical observations as the situation possibly can permit. The Observatory has three floors, each one of about 34 to 36 ft. In the lowest room, which is the amplest, is the entrance and above that the living quarters of Father Mayer's Adjunct, who is a certain Father König. After that, on the second floor, which is somewhat more inward than the lower, is the Observatory, where the mural quadrant is positioned; on the third floor, which is less wide than the second, there are two levels; on the lower one are Father Mayer's living quarters and in the one on top he has set up a sector towards zenith, of which he has not hitherto made much use, as he had had some doubts about its verifications. On the top of the roof there is a small dome with a moving cover, where he has set up the quadrant he had brought with him to Russia. Of remarkable instruments, there are no more than 1° the large 8-foot quadrant with an achromatic tube, 2° the mentioned sector, whose movement is effected by a rather strange and maybe somewhat untrustworthy construction, and 3° a wonderful timepiece by Arnold,[42] whose pendulum has a rather curious structure. The attached figure [see Fig. 9.3] may cast some light on the matter. AG, ab, cd, ef, and BH are steel rods, attached to the crossbars of brass AB, CD and EF, of which the two first ones are attached to each of the steel rods, but EF only to the three middle ones, and it can

[38]Jonathan (1690–1760) and Jeremiah Sisson (1736–1788), father and son, were makers of scientific instruments [139].

[39]Franz Lebrecht Campe (1712–1785), construction developer, instrument maker and Senator of Göttingen.

[40]Johann Christian Baumann (1711–1782), maker of scientific instruments in Göttingen.

[41]John Shelton (1712–1777), English clockmaker, famous for his so-called regulator clocks used for the timing the observations of the transits of Venus.

[42]John Arnold (1736–1799), inventor and maker of chronometers.

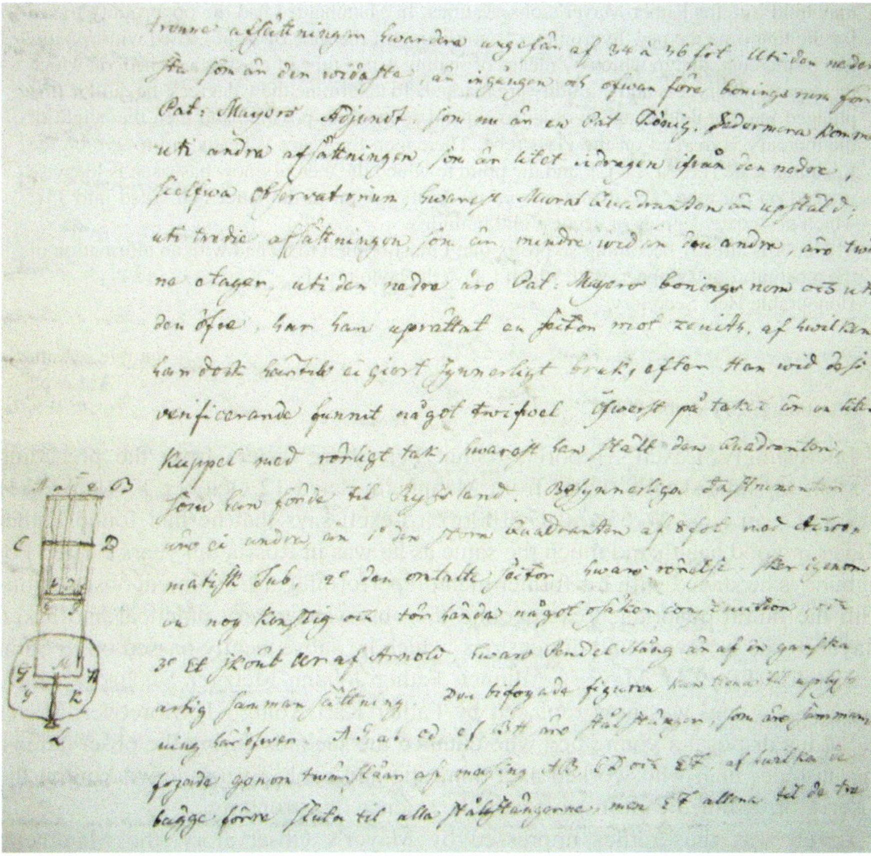

Fig. 9.3 Drawing of the construction of Christian Mayer's pendulum clock in the margin of a letter to Wargentin (KVAC, Royal Swedish Academy of Sciences, Stockholm. Photo: the author. Published with permission)

be pushed between the rods AG, BH, which go through the ball and are attached to each other with a brass plate GH, the ball is, as it were, cut in two pieces, which have a section at the centre, and these two parts are joined together with a plate of brass IK, through which as well as through GH there is a screw ML, which enters into the ball through a knob at L. Mayer claims that this timepiece has not differed more than 3 to 4 seconds a year from the *motus fixarum* (motion of the fixed stars).

Ever since the time Mayer set up his mural quadrant he has made a great number of observations with it and he still rarely misses an opportunity when the sky is clear to observe the passage of the fixed stars across the meridian. Nevertheless, it seems to me that a kind of superabundance has taken place with such observations, since in my view observations are useful only as long as there is a purpose to achieve with them and as long as the observations themselves are perfectly fit for this purpose. Thus, it is strange that one publishes passages of the fixed stars without having first applied the necessary conversion or made them applicable to astronomy. One can rest assured that hardly anyone would undertake the calculation of the observations Mr Maskelyne has published, and the same may hold true for Father Mayer's observations. In Mannheim I had the opportunity to see for the first time the new hygrometer,[43] consisting of the tube of a quill filled with *spiritus vini* or mercury, and to which by means of sealing wax a tube of glass is attached, on which the degrees of dryness and humidity are marked. To determine these degrees the quill is first plunged into ice cold water in order to expand as much as possible and then the height of the mercury is marked on the glass tube. Thereupon the water is heated up to 25° on the Réaumur scale [i.e. 31.25°C], and the point to which the mercury now has risen is likewise marked on the tube. The distance from the former point to the latter is divided into 75°, which designate degrees of dryness and humidity.

My letter is already becoming so prolix that I must bring it to an end with an affirmation of the most profound respect, with which I have the honour to be,

Honourable Herr Secretary,

Your

humble servant

A: Lexell

Strasbourg, 20 October 1780

Interestingly, Lexell's report to Saint Petersburg differs from the preceding account in some details (letter from Mannheim dated 12 October 1780; АРАН Фонд 1, оп., д. № 65, л. 97–100об.). Lexell says that he had found Father Mayer in good health and much the same as he was in Russia ten years earlier. For eight days he stayed with his former mentor, performing astronomical observations with the mural quadrant. Lexell found the observatory very practical and asked Mayer to see its layout and dimensions, which he subsequently passed on to Saint Petersburg (Fig. 9.4). Mayer's Adjunct, Father Johann Metzger SJ, had recently passed away and was now replaced by Father Karl König SJ. Moreover, Mayer had at his disposal a young boy who counted the seconds during the observations, and an older man who worked as a domestic and cook. Everyone working at the observatory seemed well instructed in the science of astronomy.

Lexell was thus rather impressed by Mayer's Observatory, the Mannheim *Sternwarte*, as well as by its collection of physical instruments. The collection

---

[43]In a letter written in Strasbourg to the Petersburg Academy of Sciences (фол. 104–105об.), and in another letter addressed to the Director Sergey Domashnev (which has not been recovered), Lexell ascribes the instrument to the French physician Gabriel Hubert (Noël) Retz and the Genevan geologist and meteorologist Jean André Deluc: "Dans ma dernière lettre à M^{r} le Directeur, j'ai parlé d'un nouveau hygromètre de Mr de Retz, imaginé après celui de M^{r} [Jean André] de Luc. Cet hygromètre consiste d'un tuyau de plume au quel on applique un tuyau de verre très fin et délié. On remplit le tuyau de plume de mercure et on l'enfonce dans de l'eau refroidie jusqu'au point de la congélation; en sorte que par ce moyen le tuyau de plume puisse se dilater autant qu'il est possible, et le mercure étant descendu dans le tuyau de verre, on marque le point où il est descendu. Ensuite on enfonce l'hygromètre dans de l'eau tiède à la température de 25° selon Réaumur, et le mercure ayant monté dans l'hygromètre, on marque le point, jusqu'au quel il est monté; après on divise l'intervalle entre les deux points tellement marqués en 75° et on continue les divisions encore plus loin, et voilà le hygromètre arrangé. Une remarque très curieuse de l'Abbé Hemmer, qui a la direction du cabinet des instrumens physiques à Manheim, c'est que l'hygromètre montant 60°, on peut toujours être sûr que l'air est électrique, le conducteur donnant des étincelles, quoique le ciel est très pur et serein."

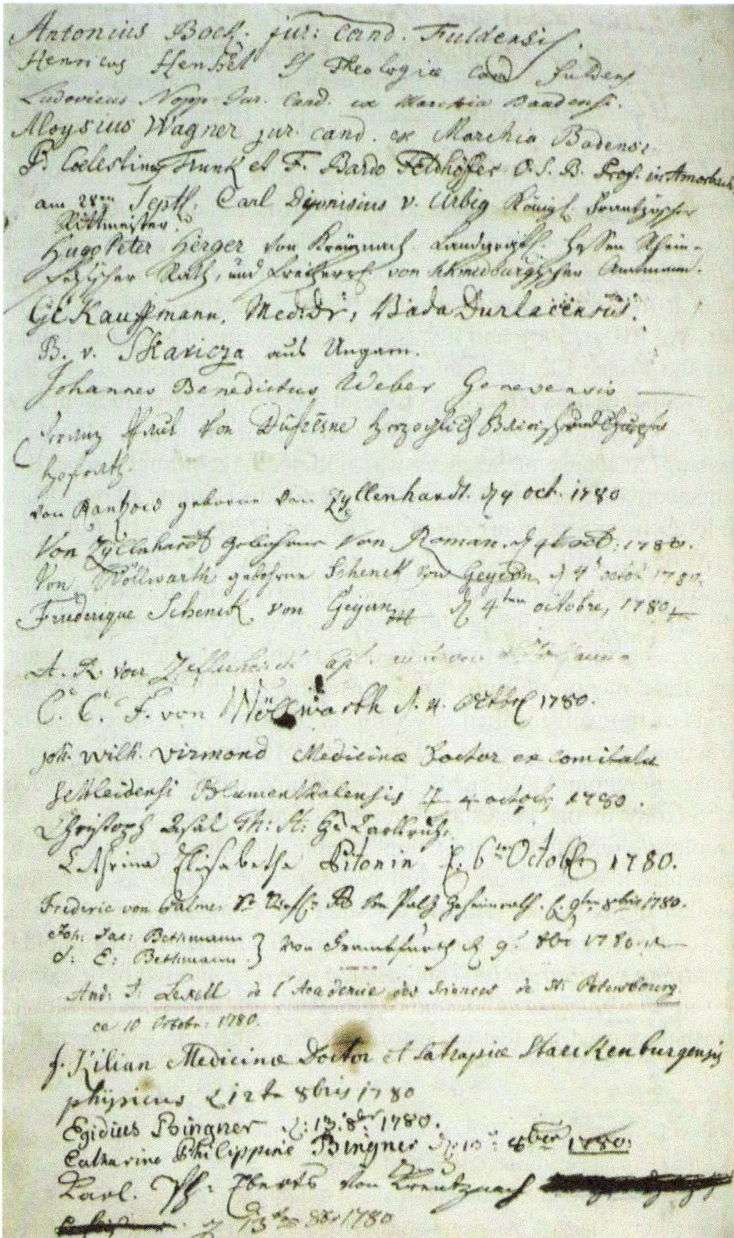

Fig. 9.4 Lexell's signature (*underlined*) in the *Besucherbuch* of the Mannheim Observatory. Most of the visitors were actually not scientists. For instance, two years earlier, on 16 November 1778, the guestbook contains an entry by Wolfgang Amadeus Mozart (1756–1791), calling himself *Maître de Chapelle*. At the time, Mozart had returned from Paris and was searching employment in the courts of Europe (Landessternwarte Heidelberg. Published with permission)

was supervised by Father Hemmer,[44] who told Lexell that his chief, the Prince Elector,[45] had ordered the construction of meteorological instruments, barometers, thermometers and hygrometers for distribution all over Europe in order to obtain observations made with similar instruments [5]. Hemmer asked Lexell if the Petersburg Academy of Sciences would also be willing to accept some of these instruments and to entrust them to persons capable of making the observations. Lexell wrote a letter to J. A. Euler, asking him to discuss the matter at the academic conference, and mentioning that he had already written about the proposition to Director Domashnev (in a letter which has not been found). The matter was deliberated at the conference of the Petersburg Academy held on 17 November 1780 [141, Vol. III, p. 500], and it was decided to write to Mayer to learn on what conditions the Prince Elector wanted to donate the instruments to the Petersburg Academy of Sciences. As it was, the proposal was well received in Saint Petersburg, and a close cooperation between the Meteorological Society in Mannheim[46] and the Petersburg Academy of Sciences was initiated. Accordingly, several boxes of instruments were delivered to Saint Petersburg for distribution all over Russia.

In the letter from Strasbourg dated 21 October 1780 (АРАН Фонд 1, опись 3, дело № 65, л. 104–105об.), Lexell wrote that he was already impatient to get to Paris and that he was forced to spend some extra days in Strasbourg while waiting for the diligence to Paris. Being already concerned about his fragile health, he did not want to try out any alternative and more adventurous means of reaching Paris.

Having thus plenty of time for sightseeing, Lexell includes in his letters descriptions of the well-fortified city with its old Gothic architecture. The town's main attractions were, according to Lexell, the Münster Cathedral, which is of an extraordinary height and also richly decorated, and the famous mausoleum of the Count of Saxony in the Protestant church of Saint Thomas.[47] Even if he knew that the sculpture representing the Count stepping into his grave was much praised and admired as a work of art, Lexell had some objections as to its execution, as he detailed in his letter.[48] He twice visited the *Comédie françoise*, where they played the comedy *Les amans généreux*—a French adaptation of Gotthold Ephraim Lessing's *Minna von Barnhelm*—but seemed not very amused by it. Among other curiosities in Strasbourg Lexell found a young boy of 22 years, born without hands, who nevertheless used his feet to perform all that others do with their hands. The

---

[44] Johann Jakob Hemmer (1733–1790), court chaplain and physicist [120].

[45] Elector of the German Palatinate, Prince Karl Theodor (1724–1799).

[46] The *Societas Meteorologica Palatina* was founded in 1780, see [120, p. 328ff.].

[47] The mausoleum of Marshal Maurice de Saxe (1696–1750) is a work by the French sculptor Jean-Baptiste Pigalle (1714–1785).

[48] On sçait que le Comte y est représenté descendant au tombeau et que la France en voulant empêcher, implore la mort, qui lève le couvercle du cercueil. Dans cette représentation l'artiste a commis deux fautes, 1° en plaçant le cercueil trop près des marches par lesquelles le Comte descend, 2° En tournant le couvercle vers le côté, duquel le Comte vient, ce qui est très mal conçu, au moins lorsque on ouvre une porte pour quelqu'un, on ne place pas celui, qui doit entrer du côté, vers lequel la porte s'ouvre.

boy was very good at drawing, he said, and added that he already had sent a specimen of the boy's drawing to the director (the letter and the drawing have not been discovered). This boy was said to be quite well educated and he spoke several languages with ease — not unusual for the region of Alsace, for that matter.

The learned men of Strasbourg with whom Lexell made acquaintance were the mathematicians Johann Jeremias Brackenhoffer (1723–1784) and Johann Ludwig Alexander Herrenschneider (1760–1843), the physicist Jacob Ludwig Schurer (1734–1792), the naturalist Johann Hermann (1738–1800) as well as the pharmacist and chemist Jakob Reinbold Spielman (1722–1783). Continuing his meticulous observations of the persons he met, Lexell mentions that the latter, in his speech and general behaviour, resembled the philosopher Johann Bernhard Merian in Berlin.

9.5 Paris: "... an incomparable capacity to take hold of ideas"

Lexell arrived in Paris in the end of October 1780. His first known letter from there was written to the Petersburg Academy and was dated 10 November 1780 (АРАН Фонд 1, опись 3, дело № 65, л. 107–114об.), followed by a letter to Stockholm almost 3 weeks later, on 29 November. On that date, many of the scientists of the *Académie Royale des Sciences* had only just returned from their country manors. As a corresponding member of the Academy, Lexell was allowed to participate in the assemblies, which commenced on 15 November. In a letter to J. A. Euler dated 7 January 1781, which has been edited and published in [15, 106, 108],[49] Lexell gave a vivid description of the academicians and the course of the conferences, which were held in the antechamber of the royal residence in the Louvre. Even the order of seating of the academicians, as well as the works of art and the statues in the conference hall were described by Lexell in some detail, not least because one of the paintings represented his King, Gustav III of Sweden, by the famous portrait painter Alexander Roslin.

Among the first distinguished scientists Lexell encountered in Paris was Jean d'Alembert. Knowing that the Eulers had met d'Alembert in Berlin almost 20 years earlier, he begins (Paris, 10 November 1780. Фонд 1, опись 3, дело № 65, л. 107–114об.):

> Since you already know Mr d'Alembert, it is hardly necessary to say that nothing in his outward appearance suggests a genius and I am sure that all of Lavater's rules fail in his case. Otherwise, I have good reason to praise his civility towards me. He received me very graciously, and as I was visiting him one day when there was a meeting at his home, he invited me to come back as often I wanted and what is more, he even came to visit me. He asked most sincerely after you and your Father and gave an assurance of both his esteem for,

[49]Lexell's often cited letter (10 November 1780. Фонд 1, опись 3, дело № 65, л. 26–34об.) was published in toto in [108] and, annotated, in [15].

and attachment to him. So I am reassured, if ever there have been any differences of opinion between these two great mathematicians, that age, which changes our thoughts on many things, has been able to extinguish all the feelings of jealousy which Mr d'Alembert may once have harboured towards your Father. Since Mr d'Alembert has lost his good friend Mme Geoffrin[50] and a further lady[51] whom he mourned so sincerely, he arranges meetings at his home three times a week, which are attended by some members of the Academy and a number of other intellectuals, among whom there are also some ladies.[52]

Other acquaintances included the Marquis de Condorcet, the Abbé Bossut, Cousin[53] and Laplace. Their physiognomies were described as follows:

> The Marquis de Condorcet is tall and handsome. He and the Abbé Bossut are extremely attractive in appearance, but I like that of the Abbé Bossut the best, since his exterior indicates a really agreeable fellow. Monsieur Cousin is tall and even too big to look like a good mathematician, that is, he is too portly. Monsieur de la Place has reddish hair, he is slim and nothing in his physiognomy suggests a mathematical genius.[54]

Lexell having been in Paris only for a short time, his observations were indeed very superficial. However, on examining closer the characters of these scholarly personalities during his stay, Lexell seems to have been attracted most to that of Condorcet. One reason may have been the modest demeanour the two men had in common [106].

---

[50]Marie Thérèse Rodet Geoffrin (1699–1777), hostess of the most celebrated *salon littéraire* of the Enlightenment in Paris [61]. She was a friend of d'Alembert's mother, Claudine-Alexandrine Guérin de Tencin (1682–1749), known for her famous salon of the time.

[51]Most probably alluding to Jeanne Julie Éléonore de Lespinasse (1732–1776), hostess of a famous literary salon in Paris and an intimate friend of d'Alembert [61]. In fact, the death of Mme Lespinasse and, soon afterwards, of Mme Geoffrin, were severe setbacks for d'Alembert, who apparently never regained full working capacity, especially in mathematics.

[52]Vous connoissez M^r d'Alembert, ainsi il n'est pas nécessaire de dire, que sa figure n'a rien qui marque un grand génie et je suis assuré, que toutes les règles de Lavater se trouveront en défaut avec M^r d'Alembert. Au reste je n'ai que les plus fortes raisons de me louer de la civilité et du gracieux accueil de M^r d'Alembert envers moi. Il m'a reçu très gracieusement, et comme j'y venois un jour, lorsque il y avoit assemblée chez lui, il m'a prié de revenir si souvent que j'en avois loisir et ce qu'il y [a] encore plus, il est même venu me rendre la visite. Il s'est informé avec le plus sincère intérêt de Vous et de Monsieur Votre Père et il a témoigné autant d'estime que d'attachement pour lui. Aussi suis-je bien persuadé, que s'il y a eu quelque différend entre ces deux grands mathématiciens, l'âge, qui change nos sentiments sur plusieurs objets, a bien été capable de détruire tous les mouvemens de jalousie, qui s'auroient pû trouver chez M^r d'Alembert vis-à-vis de Mr Votre Père. Depuis que M^r d'Alembert a perdu ses bonnes amies Madame Geoffrin et encore cette autre Dame, qu'il a pleuré si sincèrement, il donne lui-même des assemblées chez soi, trois fois la semaine, où quelques membres de l'Académie et plusieurs autres personnes d'esprit viennent, parmi lesquelles il y a aussi des Dames.

[53]Jacques-Antoine-Joseph Cousin (1739–1800), mathematician and physicist.

[54]Le Marquis de Condorcet est grand et bien fait, c'est lui et l'Abbé Bossut qui ont l'extérieur le plus avantageux, mais celui de l'Abbé Bossut me plaît encore mieux; c'est que son extérieur marque un caractère de bonhomme. Mr Cousin est grand et même trop grand, pour avoir la mine d'être bon mathématicien; c'est à dire il a trop d'embonpoint. Mr de la Place a des cheveux rougeâtres, est un peu maigre et n'a rien dans sa physionomie, qui exprimeroit le génie d'un mathématicien.

Lexell's favourable reception in Paris is also echoed in his account to Wargentin, which we give here translated in its entirety:

> My previous letter to you, Herr Secretary, was from Strasbourg, if I remember correctly. After my arrival in Paris I have not wanted to write before I had the honour to report something of interest.
> In a town that sets the pace for the whole of Europe there are undoubtedly a lot of peculiarities for the inexperienced to see, but at a certain age these peculiarities start to lose their value; yet there are still a great deal left in the arts and sciences, which for a foreigner are well worth taking notice of.
> It is well-known that the French nation must be praised for receiving their visitors most politely, and when it comes to me I must confess that I have been accepted in a way which my merits do not even remotely justify. In particular Mr d'Alembert has shown me the most wonderful favour of allowing me to join the assemblies he arranges in his home three times a week, namely on Monday, Wednesday, and Saturday evenings, and where I have the opportunity to meet and to become known by people that are here the most distinguished for their ingenuity and talent. Likewise I have been quite well received by President Saron,[55] at whose house I have once had a meal, when he showed me his beautiful astronomical tube along with a tripod he has invented himself, by which the tube can be put in any position whatever without any difficulty. He also possesses another peculiar instrument, which is the dividing engine belonging to Ramsden,[56] which has a rather simple but nonetheless clever construction. The other members of the Academy that I have visited are Cassini father and son,[57] le Monnier,[58] de la Lande, le Roy,[59] the Marquis de Condorcet, Pingré, du Séjour,[60] l'Abbé Bossut,[61] Jeaurat,[62] le Gentil,[63] Bailly,[64] Bezout,[65] Vandermonde,[66] Cousin, Messier

[55]Jean-Baptiste-Gaspard Bochart de Saron (1730–1794), representative of the Parliament of Paris, astronomer, honorary member of the Academy of Sciences and its President. He became a victim of the reign of terror and was guillotined.

[56]Jesse Ramsden (1735–1800), English optical and mathematical instrument maker and inventor. His dividing engine is a mechanical device to mark graduations on scientific instruments [139].

[57]César-François Cassini de Thury, called Cassini III (1714–1784), and his son Jean-Dominique Cassini (Cassini IV) (1748–1845). The Cassini astronomer dynasty was founded by Giovanni Domenico Cassini (1625–1712) and his son Jacques Cassini (Cassini II) (1677–1756), who was a known supporter of the theory of an elongated form of the Earth.

[58]Pierre-Charles Le Monnier (1715–1799), astronomer, participated in Maupertuis' geodetic expedition to Swedish Lapland in 1736–1737.

[59]Jean-Baptiste Le Roy (1720–1800), physicist who supported the single-fluid theory of electricity of Franklin, with whom he also corresponded.

[60]Achille Pierre Dionis du Séjour (1734–1794), mathematician and astronomer.

[61]Charles Bossut (1730–1814), mathematician and priest, student and associate of d'Alembert.

[62]Edme-Sébastien Jeaurat (1724–1803), astronomer and geographer.

[63] Guillaume-Joseph-Hyacinthe-Jean-Baptiste Le Gentil de la Galazière (1725–1792), astronomer, assistant of Cassini II.

[64]Jean-Sylvain Bailly (1736–1793), astronomer, member of the *Académie française* and author of *Histoire de l'astronomie ancienne* (1775) and *Histoire de l'astronomie moderne* (1779–1782). After the revolution, he was engaged politically as the acting mayor of Paris. Later, during the reign of terror, he was guillotined on the *Champ de Mars*.

[65]Étienne Bézout (1730–1783), mathematician, known for his work in number theory and algebraic geometry.

[66]Alexandre-Théophile Vandermonde (1735–1796), mathematician, music theorist and chemist, known in particular for his work in linear algebra and the theory of determinants.

and de la Place. As far as the sciences are concerned my closest acquaintances have been with de la Place and Messier. There is no doubt that de la Place is next to Euler and de la Grange the greatest mathematician of our time, and it is quite extraordinary that he is so little appreciated. He is still young and has started early. Mr Messier is a rather polite and obliging man, who does not make more noise about himself than he deserves, thus he is in this respect the antithesis of de la Lande, who at every opportunity pretends to be advancing something new, although his researches for the most part are quite unreliable. Before my arrival here I was perhaps slightly disappointed with de la Lande, on account of a rather serious letter I had written to him with regard to his doubts about the comet of the year 1770. All the same it is not my business to harbour ill will against any man, and so my heart was totally seduced by him in our first conversation, on which occasion I also had the opportunity to say to him that I believed I had a reason for the distrust I had been showing him. Since then he has visited me and promised me anything for friendship and willingness to assist me, and I for my part would be most happy to believe all this, if his countenance would not contradict the thoughts his mouth interprets to me. He has what I would call a rather unpleasant countenance, but when he talks to me I can almost read in his eyes his desire to hurt me, if he only had the opportunity and authority to do so. Otherwise, he is almost universally hated by the other Academicians and that is mainly because of his charlatanism. Only recently, at the first assembly of the Academy, he read a dissertation on the inclination of the ecliptic and its change, which he asserted to be 35″ in a century, from which he concluded that the mass of Venus must be less than half of the mass of the Earth. It does not appear but very peculiar that one could infer a change for a period of a hundred years from 20 years of observations, and since many of the mathematicians and astronomers present here thought otherwise than Mr de la Lande and were above all not pleased with the praise he gives himself for having established such an important element, it caused a lot of arguments in the Academy. Mr Cassini the Elder read an article on the occasion of Mr de la Lande's memoir, in which he wanted to demonstrate on the one hand that Mr de la Lande has been using the discoveries of others without properly acknowledging them, on the other hand, as to the presented conclusions in dispute concerning the inclination of the ecliptic, that so far no certain conclusions could or should be drawn. I hope this will bring this conflict to an end, if it does not come up again today. Nevertheless, on the occasion of these conflicts, Mr de la Lande made a rather tedious performance, having almost nobody to take his side. Another charlatan in the Academy is Abbé Rochon,[67] who at the opening ceremony of the Academy proposed a new kind of instrument to measure the elevation of the Sun especially at solstices, having encountered I do not know how many different difficulties and uncertainties with the usual instruments. This instrument of his is a mirror placed along the meridian line to reflect the Sun's image. Having as yet no clear picture of how this mirror is to be positioned, I cannot pronounce anything certain in this regard, but what I can say with some confidence is that he who claims before an audience of several learned men that such an instrument can be as precise as a quadrant of 4000 ft, must be downright impudent. It is difficult to believe, that in a community as enlightened as the Paris Academy of Sciences, such absurdities could take place.
During the short time I have been here it has appeared to me that the gentlemen of the Academy of Sciences are not quite unanimous, that controversies and disagreements arise almost continuously, and that somebody's fondness of one person makes him condemn another of a different opinion. As for myself, I presume that I would not be able to act against my moral conviction by not recognising the true merits of a person I was dissatisfied

[67]Alexis-Marie de Rochon (1741–1817), ecclesiastic, astronomer, explorer and geographer, inventor of optical instruments. He was the curator of the royal instrument cabinet in the *Muette* palace, which no longer exists. He built prismatic binoculars using a double-refracting crystal he had brought from Madagascar.

> with. I think the reason why intrigues are much more frequent in the Paris Academy of Sciences than in our Academy in Petersburg is that the Paris Academy of Sciences is incomparably more liberal than ours. When, at our Academy, all things, at least the most important things, depend on one person only, who is the head of the Academy, its President or Director, there is no point in intriguing, because nothing could be gained by it. In a more liberal society it is certain that parties, intrigues and plots are bound to arise. Certainly there have been plots at the Petersburg Academy of Sciences also in times when the Director was less concerned with the matters of the Academy than he is today.

Thus, while reflecting upon the remarkable differences between the cultures of the Academies of Paris and Saint Petersburg, Lexell takes the opportunity to justify his stand in the current controversy between the director and the academicians in Saint Petersburg about the developments of which Johann Albrecht Euler had undoubtedly kept him updated. The letter to Wargentin continues:

> One of my chief concerns since my arrival here has been to pay visits to the local observatories, i.e. the Royal Observatory, de la Lande's observatory at the *Collège Royal*, the observatory of Mr le Monnier, and that of Mr Dagelet[68] at the Military School. That of Mr Messier I have not yet seen. I went to see Mr Pingré, but he told me that there was no point in visiting his observatory. No observations are made anymore at the Luxembourg [l'Observatoire de Paris] or at the *Collège Mazarin*, nor are there any instruments left. The Royal Observatory is a rather large and splendid building, but less than practical for making observations; hence, a smaller building next to the large one has been used for making observations, but since this building has turned out not to be solid enough and practical, it has been decided that a couple of new rooms shall be built with a solid enough foundation and sufficient practicality to be able to set up a mural quadrant and a passage instrument. These rooms are presently under construction and in the meantime the instruments have been taken into the main building, where they are now standing crisscross. In other respects the supply of this observatory contains mostly quadrants, smaller and larger ones, and a few fairly usable tubes. Mr le Monnier has in his observatory two mural quadrants, one portable of 5 ft radius, which he has described in the *Déscription des Arts et Métiers*[69] (article *Quart de Cercle*) and another one of 8 ft radius, which he, in spite of his attempts, has not been able to make portable. For me and many others it may seem odd that somebody should try to make a mural quadrant portable; but Mr le Monnier seems to be entirely convinced that this is the best way. For somebody who does not know astronomy it might also seem futile to attach such a quadrant to a wall, when it should be moveable, as long as one really can endow it with both a more agile and more secure motion than if it was to carry a great lump. Mr le Monnier is a stubborn old man with a very high opinion of himself and his observations, which he claims to have been used by Mayer, de la Caille and by who knows which astronomer, but otherwise rather unclear in his way of expressing himself. I for my part have all the reason to be satisfied with his behaviour, but yet this does not prevent me from pursuing the truth in my judgement, and so I find that I have to excuse some of Mr le Monnier's weaknesses on account of his age. In de la Lande's observatory there are two sectors in constant use, one 6 ft long and the other $4\frac{1}{2}$ ft; additionally there is a passage instrument, which is small and unreliable. De la Caille has used the larger one for his observations at the Cape. The tripod of this instrument is too fragile. Mr de la Lande

---

[68]Joseph Le Paute Dagelet (1751–1788?), astronomer, student of Lalande [38], was appointed astronomer in Count La Pérouse's ill-fated expedition around the world. He perished in a shipwreck in the Pacific Ocean.

[69]Series of technical descriptions (printed in folio), published by the *Académie Royale des Sciences*, of instruments and manufacturing processes of the time.

makes no observations himself, instead he uses his students for this, hence it can be easily imagined how these observations will look. It was noon when I was in his observatory. The passage of the Sun across the meridian was to be observed; without first making sure that the sector was in a vertical plane (at least the ball was hanging far away from the plane of the instrument), it was aimed at a mark in the meridian and subsequently to the Sun, which touched the vertical wire at the stroke of noon.
Mr Dagelet has a beautiful 8 ft mural quadrant, with which he has made and still makes a lot of observations, but as to their reliability I am not sure, although I want to believe the best of them, the more so as Mr Dagelet seems to have adequate knowledge and to be capable of analysing his observations with a critical eye. He does not want to hasten the publication of his observations; this is quite sensible, the more so as his observatory will have to go inside at least for some time. Talking of astronomy, I must not forget to mention the elements I have found for the comet[70] Mr Messier has been observing [...]
I am grateful for the notification of Mr Prosperin's calculations of the comet of the year 1779,[71] but I am rather doubtful about them, because one cannot determine the *tempus periodicum* of a comet, the observations of which can be fitted with parabolic elements within 2′ of accuracy, as Mr Messier has shown me. As long as it is possible to commit an error of 2 minutes in these observations, Mr Prosperin's calculation does not seem to prove what he has attempted. A crucial fact is that one can find older observations for this comet than those of Mr Messier, namely by Mr Schulze in Berlin, which begin eight days before Mr Messier's. If these observations, when combined with Mr Messier's, turn out to give other elements than those found by Professor Prosperin, that certainly would cast a doubt on his claim about the *tempus periodicum* of this comet. I do not want to discuss Mr Bode's observations, they will not do; but Mr Schulze is a capable man and has observed with great accuracy. Generally speaking, I think that when the *tempus periodicum* is as long (such as 1000 years) one would have to be more certain about the observations than one can reasonably hope for.
As for the favourable reception I have been given amongst the persons with whom I have made acquaintance here I am quite satisfied with Paris, but in other respects as a traveller coming from Petersburg I find many things unusual, if not plainly uncomfortable. First of all there is this way of heating, which is rather uncomfortable and insufficient, since either they use only the stove, which does not give much heat, or else they use small ovens which give much smoke and where the fire must be constantly tended, if one wants to keep warm. Moreover, as a foreigner, being unable to always afford a carriage, I find it rather difficult to walk in the faeces and dirt which lie on the streets and being all the time splashed over by passing wagons, as well as running the risk of being run over on their narrow streets, where two carriages can barely go past each other. Although I have not often been able to hire a carriage, it has cost me more than 120 Livres already. These inconveniences will undoubtedly be forgotten through habit and a longer stay, but there is one particular example of moral laxity which I have noticed with discontentment, which is the too broad-minded consideration of religion, which commonly goes as far as denying any kind of deity. I was surprised to hear Mr de la Lande conjecture that the Empress does not believe in God, considering that she had received Mr Diderot so gracefully. Thus, anyone here is regarded as a fool unless he renounces all the prejudices about God and the afterlife that we have been brought up with since we were children. One may easily imagine the moral philosophy of a nation which embraces thoughts of this kind. With this in mind it is always dangerous for a benevolent father to allow his son to visit this perilous place as regards religion and ethics. Even the noblest heart in the world can hardly escape being seduced here, unless he has the moral strength to ignore all the harsh and ludicrous

---

[70]Comet "C/1780 U2", discovered by Messier.

[71]Comet "C/1779 A1", discovered by Johann Elert Bode.

remarks made here on religion and ethics. The reason for this free-thinking in these matters derives without a doubt from the Catholic religion, where so much is incomprehensible to the common sense, that even their priests are hypocrites, who do not believe in the things they are supposed to preach, who use the religion to cover up their darkest deeds, and who against their instruction appropriate themselves large and profitable mansions and through their ignorance, indolence and weakness have made themselves both despicable and hated. The most regrettable fact is that the most enlightened part of the nation consists of free-thinkers, if not downright atheists, and that the other part, having still confidence in religion, probably approaches the other extreme of a foolish innocence, religious hatred and persecution.

I have not as yet had the chance to study the temperament of the French nation to be able to pronounce anything definite in this respect; nevertheless I think I am not mistaken if I say that the majority of them have a rather happy and joyful temperament, they can be as happy as children over nothing, such that both men and women carry on long discussions on the most trifling subjects such as food, garments and so on, that they often talk more than they think, that they do not worry much about the future, and that they are zealous in their endeavours, but do not have the patience to persevere for a long time. The temperament of a nation often reveals itself through simple and petty things, thus I believe that no example does better reflect the frivolity of the Frenchmen than a recent incident at the so-called Italian Comedy. During the intermission between the two plays which were performed, somebody sitting in the stalls had noticed a feather floating around and found it so funny that he announced his discovery to the rest of the audience, who took part in his delight to the extent that not only did they burst into laughter but some even started to applaud. Everyone can judge for himself whether such a discovery in the stalls occupied by Englishmen would ever have caused a similar reaction. To make the English public laugh, a much more certain effect may be achieved by throwing some punches with the fist. Which nation is more in the right is not my concern.

I am now waiting for the news on my resignation from you, Herr Secretary, and a reply to this letter, which I might expect here in Paris, as I plan to stay here at least until the month of February. I have asked my brother-in-law to send 100 *Riksdalers* to you, Herr Secretary, with which you would be so kind as to arrange for me a bill of exchange to London, sent at earliest in February and addressed to the Swedish Minister, so that I can have it when I arrive in London.

The fact that I am not writing as often as I would like to is due to the more official correspondence I have been instructed to maintain with Petersburg. I sincerely hope that my letters might contain something my friends would appreciate knowing, but as yet I do not know if I have achieved my purpose. I think this long epistle will have to do for this time, and I hope that I will have the pleasure of writing to you at least a few times before my departure.

I remain with the most profound respect,
Honourable Herr Secretary,

Your

humble servant
A: Lexell

Paris, 29 November 1780

During Lexell's journey, Crown Prince Friedrich Wilhelm of Prussia, the nephew of Frederick II, was sent on a diplomatic mission to Saint Petersburg. One of the Prince's objectives was to visit the Imperial Academy of Sciences, where he had a *tête-à-tête* meeting with Euler and his family. The event was interpreted as an act of reconciliation between Frederick and Euler, since relations between them had broken down in 1766 with known consequences. The Prince honoured the

Academy by attending its conference held on 30 (19) September 1780. However, the conference session took a dramatic turn [141, Vol. III, pp. 491–492]: During the lecture given by Nikolaus Fuss the Prince began to feel faint and was taken to a window for fresh air. Nevertheless, the Prince eventually collapsed, but quickly regained consciousness, whereupon the conference was continued.

In a letter to Lexell dated 10 October in Saint Petersburg (which, like most of the letters addressed to Lexell, has not been preserved in the archives), J. A. Euler gave a detailed account of the encounter between his father and the Prince. Lexell passed on the letter to the Marquis de Condorcet, who immediately wrote an article about the event for the *Journal de Paris*.[72] Lexell's letter, given in its entirety in the Appendix, expresses a particular concern about the attitude of the French mathematicians to Euler's health and working capacity (20 November 1780; АРАН Фонд 1, опись 3, дело № 65, л. 131–134).

In the letter to J. A. Euler, Lexell writes that, although he already had mentioned the respect with which the local mathematicians speak of Euler, a decisive proof of this was given in the above-mentioned issue of the *Journal de Paris*, containing the details of the meeting between the Prince of Prussia and Euler (as described by J. A. Euler to Lexell). Lexell believed that the article had been written by the Marquis de Condorcet, and although not entirely accurate, he hoped it proved that every mathematician in Paris appreciated Euler's great talents and merits. Lexell also told that he wished that the Marquis had not been so precise about Euler's hearing, because he had not implied that Euler was deaf, only that he was hard of hearing. As to the number of papers Euler was turning out for the Saint Petersburg proceedings, Lexell had given only a rough estimate; nor did he say exactly how long it took for Euler to write a paper, but in general it took about a week. All the inaccuracies apart, Lexell thought that the article proved that the French mathematicians universally recognised Euler's merit. He was also certain that Condorcet wrote the article in concert with d'Alembert, because on the same day as the journal appeared, the following happened: On that day there had been an assembly of the Academy of Sciences, and as it was also one of those days of the week when d'Alembert arranged meetings at his home, Lexell went there to find out what d'Alembert had to say about the article. As he entered the house, d'Alembert asked him immediately if he had read the article and named its author, which Lexell had already guessed. And as there were some at the meeting who did not know of the article, d'Alembert read it out aloud and charged Lexell to give his compliments to the Eulers. Lexell further asserted that every mathematician he met that day and who had read the article was very pleased with it and the interpretation of the Marquis de Condorcet.

Seen as an envoy, more or less, of the Petersburg Academy of Sciences and especially the Eulers, it is easy to imagine how content Lexell must have been when

[72]Daily literary magazine appearing in Paris since 1777. The issue in question containing the article concerning Euler is preserved in St. Petersburg АРАН Фонд 1, опись 3, дело № 65, лист. 137–138об.

he learned how sincerely the French mathematicians revered Euler and his talent. Obviously, Lexell's courteous reception in Paris was partly due to his association with Euler, as his intimate friend and colleague. On the other hand, it is possible that Lexell's excessive distrustfulness also reflects Euler's prejudices vis-à-vis the French.

The letter continues with Lexell's description of the opening ceremonies of the *Collège Royal*[73] and the *Académie Royale des Sciences*. At the *Collège Royal*, Le Monnier presented a memoir on the different densities of air (a talk which Lexell regarded as very confused), Vauvilliers[74] read out an extract from Thucydides, containing the funeral oration by Pericles in honour of the warriors of the first Peloponnesian war, and J. J. de Lalande presented a part of his treatise on the ebb and flow of the sea, a talk which according to Lexell contained some dubious claims. The assembly was closed by Abbé Delille[75] reciting a beautiful poem. Lexell proceeds by describing the opening ceremony of the *Académie des Sciences*. The assembly opened with a eulogy on Adjunct Bucquet[76] pronounced by the Marquis de Condorcet, whereupon J. J. de Lalande read out a memoir, in which he claimed that the obliquity of the ecliptic diminishes no more than $35''$ in a century and that the duration of a year is at present 365D. 6 h. $48'.48''$, and that it has become $4''$ shorter in 2000 years. From this fact Lalande claimed to deduce that the mass of Venus cannot be but half that of the Earth. To this, Lexell adds that many a Parisian thinks that Lalande ought not to meddle in this matter and that he lacks the patience and discernment to form an opinion on it. According to Lexell, Lalande seems to be not much liked by his fellow academicians because of his charlatanism (Lexell's expression for somebody who is either pretentious or not very reliable scientifically). Next, Charles Messier delivered a report of the most recently discovered comet. The Abbé Rochon continued the assembly by describing a new instrument for measuring elevations of the Sun; it is a levelling instrument placed in the meridian plane and reflecting the rays of the Sun. To Lexell it seemed that the Abbé had been too hasty in publicising this instrument without due tests and preferring it to larger meridian instruments. Next, Vandermonde read a memoir about music, and finally, Cornette[77] presented results of his and Lassone's[78] experiments relating to the production of phosphorus out of bone. Lexell added that

[73]Nowadays *Collège de France*.

[74]Jean-François Vauvilliers (1737–1801), poet and Professor of Greek at the *Collège Royal*.

[75]Jacques Delille (1738–1813), poet and member of the *Académie française*.

[76]Jean-Baptiste-Marie Bucquet (1746–1780), Professor of Chemistry in Paris. Bucquet's *éloge* by the Marquis de Condorcet appeared in the *Histoire de l'Académie pour l'année 1780*, p. 60.

[77]Claude-Melchior Cornette (1744–1794), chemist, student of Lassone (see note below).

[78]Joseph-Marie-François de Lassone (1717–1788), physician and chemist, *Premier médecin* of Louis XVI and Marie Antoinette.

also Lavoisier[79] was said to have given two talks at the Academy about phosphorus. "Chemistry is at the moment the most fashionable of all the sciences", he concludes.

Lexell continues the letter by describing his visit to the Royal Library, where the porter showed him around in the collection of sculptures of celebrated personalities made by Jean-Antoine Houdon.[80] Among the works of art, Lexell mentions a statue of Voltaire sitting in an armchair,[81] Diana the Huntress [*Diane chasseresse*], as well as busts of many royalties, d'Alembert, Rousseau,[82] Franklin,[83] the Duke de Praslin,[84] Turgot, the *Gardien des Sceaux*,[85] le Noir,[86] Chevalier Gluck,[87] Palissot,[88] Paul Jones,[89] J. J. de Lalande and Vitinghöf.[90] Finally Lexell mentions that he has been preoccupied with calculations of the orbit of the comet which Messier has observed since October 1780, that is C/1780 U2 [82].

All letters written by Lexell from Paris testify that he did not hesitate to take part in all the literary events to which he was invited. He was received at the soirées in d'Alembert's home and at the salon of the Lavoisiers. In a letter to J. A. Euler he writes (Paris, 7 January 1781. АРАН Фонд 1, опись 3, дело № 65, л. 26–34об.):

> Monsieur Lavoisier is a young man with a pleasant face; he is a very able and industrious chemist. He has a beautiful wife,[91] who also loves literature and who presides over the

---

[79]Antoine Laurent de Lavoisier (1743–1794), natural scientist and chemist, "Father of modern chemistry". In 1777, he recognised phosphorus as an element. He also realised that burning means that the substance in question combines with oxygen. The maxim attributed to him; *Rien ne se perd, rien ne se crée, tout se transforme* expresses the principle of conservation of mass. Lavoisier was executed during the reign of terror.

[80]Jean-Antoine Houdon (1741–1828), prominent French sculptor of late Baroque and Neoclassicism.

[81]Catherine II of Russia, who for a long time had been fond of Voltaire's writings, had recently purchased Voltaire's library to Saint Petersburg. She commissioned the statue for the Library in 1780.

[82]Jean-Jacques Rousseau (1712–1778), born in Geneva, was a leading philosophical author and critic of the Enlightenment.

[83]Benjamin Franklin (1706–1790), scientist and inventor, politician and American ambassador in France.

[84]Marquis César Gabriel de Choiseul (1712–1785), French military officer, diplomat and statesman.

[85]Probably alluding to Armand Thomas Hue Marquis de Miromesnil (1723–1796), Minister of Justice in Louis XVI's government.

[86]Jean-Charles-Pierre Lenoir (1732–1807), head of the police, *Lieutenant-Général de Police de Paris*.

[87]Christoph Willibald von Gluck (1714–1787), opera composer.

[88]Charles Palissot de Montenoy (1730–1814), French dramatic author.

[89]John Paul Jones (1747–1792) was a Scottish sailor and sea captain in the American Navy during the American War of Independence.

[90]Baron Otto Hermann von Vietinghoff (1722–1792) was a German-Baltic statesman and Director of the Medical College of St. Petersburg.

[91]Lavoisier's wife Marie-Anne Pierrette Paulze (1758–1836) was also a chemist.

> assemblies of the academicians, when they meet at their home after the Academy to have some tea. I have attended these assemblies several times.[92]

Even the honorary members and directors of the Academy, the enlightened noblemen showing interest in the learned and their discoveries, received Lexell's approbation, perhaps even admiration:

> There are only two honorary members by whom I have the honour to be known, namely the Duke d'Ayen[93] and President Saron. The Duke is a very gracious and well-mannered *seigneur*. The President also possesses the same excellent qualities, but he is at the same time so gentle, so considerate and his manners are so obliging, that everyone who appreciates honesty will become attached to him. There are a few Academicians who dine with him every Sunday and I go there quite regularly, not for the dinner, but for the good company. As to these two *seigneurs* I have noticed, and I believe that it will hold true generally, that French noblemen who apply themselves to science are much less pretentious that the learned men themselves, and that when it comes to paying their respects to the learned, they are not at all mean-spirited; on the contrary, they consider themselves honoured by the interest learned men take in them.[94]

On the other hand, the "literary characters" Lexell encountered were described with unsparing scrutiny of physiognomy (Paris, 15 January 1781; АРАН Фонд 1, опись 3, дело № 65, л. 156–160об.).

> Father Boscovic[95] is a strong, robust old man, who is in good shape for his age. I do not think he is the most sublime of geniuses, but here he is less appreciated than he deserves. Still, he has obtained, I know not how, profits amounting to an annual income of 8000 Livres. His conversation is not very engaging, since he talks of his own ideas to an excess, and he is very stubborn and confused at the same time.[96]

---

[92]Mr Lavoisier est un jeune homme d'une figure agréable, très habile chymiste et laborieux. Il a une belle femme, qui en même temps aime la littérature et qui préside aux assemblées des Académiciens, lorsqu'ils vont chez lui après l'Académie prendre du thé. J'ai été plusieurs fois à ces assemblées.

[93]Jean Louis Paul François d'Ayen, Duc de Noailles (1739–1829), military commander and chemist, President of the *Académie Royale des Sciences*.

[94]Parmi des honoraires je n'ai l'honneur d'être connu que de deux, sçavoir le Duc d'Ayen et le Président Saron. Le Duc est un seigneur très gracieux et poli. Le Président ne possede seulement ces bonnes qualités, mais il est en même temps si doux, si prévenant et ses manières sont si obligeantes, que chacun, qui aime les sentiments honnêtes, doit lui être attaché. Il y a toujours deux ou trois des Académiciens, qui dînent chez lui tous les dimanches et j'y vais assez régulièrement, non pour le dîner, mais pour la bonne compagnie. Par rapport à ces deux seigneurs j'ai fait la remarque et je crois qu'elle se trouvera vraie généralement, que les seigneurs François qui s'appliquent au[x] sciences ont beaucoup moins de prétention, que les sçavans eux mêmes, et que, faisant quelques civilités aux sçavans, ils n'en tiennent pas compte; au contraire, ils se croient honorés de ce que les sçavans viennent les voir.

[95]Jesuit Father Rudjer Josip Boscovic (1711–1787), polymath, physicist and astronomer, was born in Dubrovnik, but lived in many places in Europe, especially Italy and France. He proposed *inter alia* the first mathematical model for the atom. In 1780, he served the French Navy as a specialist in optical instruments.

[96]L'Abbé Boscovich est un vieillard fort et vigoureux, qui pour son âge se porte très bien. Je ne le crois pas d'être un des plus sublimes génies, mais aussi est il ici moins estimé, qu'il ne mérite, c'est qu'il a obtenu ici, je ne sçais comment, des bénéfices, qui lui valent 8000 Livres de revenu, par an.

About Diderot Lexell reports that, because of bad health, his physiognomy has changed somewhat since his visit to Russia. He received Lexell rather courteously and said nothing about atheism, to Lexell's satisfaction.[97] The account continues:

> Monsieur Marmontel, tall, with a handsome head, but rather lifeless eyes. An extreme egoist, but at the same time flattering those he wishes to make friends with.[98]
> The Abbé Raynal[99] is a small man, passionate and full of life. I have visited him at one of his very elegant lunches frequented by so many Princes, Dukes, Duchesses, etc. that my head started to spin. His book is said to be a mere compilation, borrowing text from all kinds of authors, word for word; astronomy, physics, chemistry and so on. Consequently, writing such a book is not a great merit. Nevertheless, I would like to have written it, but maybe I would at least be honest enough to say exactly what I have borrowed from others.

Lexell describes Turgot as tall and somewhat resembling Prince Grigory Orlov (known as the lover of Catherine II), and well versed in science, literature and several languages. Nevertheless, Lexell regrets that he had a different opinion on many things, mainly on the merit of Shakespeare, whom Lexell seemed to rate as highly as did the English.[100]

Sa conversation n'est pas trop agréable, puisque il Vous parle toujours, même jusqu'à dégoûter, de ses propres idées, et qu'il est très entêté et confus en même temps.

[97] Mr Diderot, sa figure Vous est assez connue, pour que j'en parle. Elle a pourtant un peu changée, depuis que Vous l'avez vû, car il s'est porté mal après son voyage. J'ai été le voir. Il m'a reçu très poliment et pas un mot de l'athéisme, dont j'ai été très content.

[98] Jean-François Marmontel (1723–1799), historian, writer and one of the encyclopedists, and member of the *Académie française*. Lexell describes him as follows: "grand et assez belle tête, mais point de vivacité dans les yeux. Égoïste au dernier degré, mais en même temps flatteur de ceux, dont il veut ménager l'amitié."

[99] Guillaume-Thomas François Raynal (1713–1796), political author who following his criticism of French absolutism in his work *Histoire philosophique et politique des établissements et du commerce des Européens dans les deux Indes* was forced into exile in May 1781. Lexell's description bears witness to a particularly active social life only a few months before the exile: "[Il est] un petit homme plein de feu et de vivacité. J'ai été chez lui à un de ses déjeuners très élégants, où il se trouvèrent, tant de Princes, Ducs, Duchesses, etc. etc., que la tête commençoit à me tourner. On dit que son livre n'est qu'une compilation, qu'il a emprunté de toutes sortes de gens des morceaux entiers, qu'il y a uns de mot en mot, Astronomie, Physique, Chemie etc.; on conclut que le mérite n'est pas grand d'avoir fait un tel livre. Cependant je voudrois bien l'avoir fait, mais peut-être serois-je assez sincère, pour dire ce que j'ai emprunté des autres."

[100] "J'ai été la semaine passé chez Mr Turgot avec le Marquis de Condorcet. Il est grand et d'une belle figure. Il ressemble un peu au Prince Orlov. Il est très instruit dans les sciences et dans la littérature. Il sçait aussi plusieurs langues. Il m'a prié de Vous saluer, aussi bien que Mr Votre Père. Je ne sçais s'il étoit trop content de moi, puisque je n'ai pas été de son avis sur plusieurs choses et principalement sur le mérite de Shakespeare, que j'estime presque autant, qu'un Anglois". Later in the same spring (on 24 March 1781), Lexell writes regarding Turgot's death on 18 March (АРАН Фонд 1, оп. 3, д. № 65, л. 191–194об.): "On vient de faire une grande perte par la mort de Mr Turgot, qui est universellement regretté, mais principalement par ces amis, comme un homme de très rares talents, des grandes mérites et du caractère le plus noble et le plus aimable. Il y a déjà quelques semaines qu'il a eu un accès de goutte, qui étoit sur le point de le suffoquer, mais étant ensuite revenu, on espéroit qu'il se rétabliroit; cependant il a à la fin succombé à son mal. Le Marquis de Condorcet est très affecté de la mort de son ami et pour moi je n'ai pû m'empêcher de la regretter plus, que aucun de mes connoissances ici, quoique je n'avois eu l'honneur de le voir

The Baron Holbach[101] looks like an *honnête homme* [cultivated and learned]. He does not pretend to be witty, but he really is, and everything he says he puts so gently, that it is a delight to hear him speak. If I am not mistaken, he must be the most tender and affectionate husband and father.

Sometimes Lexell spiced up his descriptions with humorous anecdotes (Paris, 12 February 1781; АРАН Фонд 1, опись 3, дело № 65, л. 166–171), such as this one from the Marquis de Condorcet:

Since my last letter, Father Boscovic arranged for me the honour of being acquainted with Cardinal de Luynes,[102] who is an honorary member of the Academy of Sciences. He is a very respectable old man judging by his face and by the gentleness of his character, but he appears to suffer from slight infirmities of old age. At the last election of foreign members to the Academy of Sciences, the candidates were Mr Pringle and Mr Hunter.[103] Cardinal de Luynes, who was attending this election and wished to vote for Hunter, asked Marshal de Richelieu,[104] who was seated near him, how to write in English the name Hunter. To this, the Marshal answered: P, r, i, n, g, l, e. This anecdote I learned from the Marquis de Condorcet, who as Secretary collected the ballot papers.[105]

que deux fois. Il est très rare de trouver dans ce païs des gens qui ont une réputation universelle de honnêteté, mais Mr Turgot l'avoit. Même ceux qui n'approuvoient pas sa manière à penser, disent que c'étoit un homme, qui avoit les meilleurs intentions du monde, un fond d'honnêteté rare ou plustôt unique, des mœurs très pures et vertueuses. Il a donc emporté des regrets universels et même à la cour on en a parlé avec beaucoup de sensibilité. Comme il étoit beaucoup ami de Mr Votre Père, j'éspère qu'il ne manquera pas de prendre parte à cette perte, qui est ici très vivement sentie de tous les honnêtes gens."

[101]Paul-Henri Thiry, Baron d'Holbach (1723–1789), philosophical author and participant in the *Encyclopédie*, renowned proponent of materialism and atheism. His main work was *Le Système de la Nature* (1770). With this in mind, it is remarkable how respectfully Lexell describes him: "[Il] a la figure d'un honnête homme. Il n'affecte pas d'avoir de l'esprit, mais il en a beaucoup, et il met tant de bonhomme dans tout ce qu'il dit, qu'on est charmé de l'entendre. Si je ne me trompe, il doit être un époux et père très tendre et affectueux."

[102]Cardinal Paul d'Albert de Luynes (1703–1788), astronomer and member of the *Académie française*. In his youth, when Bishop of Bayeux, he had commissioned his secretary, astronomer Abbé Réginald Outhier, to participate in Maupertuis' Lapland expedition [135].

[103]Royal physician, Baronet John Pringle was elected a foreign member on 26 February 1778, replacing the recently deceased Linnaeus. Royal physician William Hunter (1718–1783) was elected a foreign member on 3 February 1782 [72].

[104]Louis François Armand de Vignerot du Plessis, Duc de Richelieu (1696–1788), military commander who rose to the rank of *Maréchal de France*. He was a great-great nephew of the famous cardinal and statesman Richelieu and an associate and intimate friend of the Marquise Du Châtelet.

[105]Depuis ma dernière lettre, l'Abbé Boscovich m'a procuré l'honneur de faire connoissance avec le Cardinal de Luynes, qui est des Honoraires de l'Académie des Sciences. C'est un vieillard très respectable par sa figure et par la douceur de son caractère, d'ailleurs il paroît qu'il se ressent un peu des infirmités de la vieillesse. À la dernière élection d'un membre étranger de l'Académie des Sciences, il a été question de Mrs Pringle et Hunter. Le Cardinal de Luynes assistant à cette élection et voulant donner sa voix à Hunter, demandoit au Maréchal de Richelieu, qui étoit assis près de lui, comment on écrivoit en Anglois Hunter, à quoi le Maréchal répondoit: P, r, i, n, g, l, e. C'est un anecdote que je tiens du Marquis de Condorcet, qui comme Secrétaire rassembloit les voix.

or this one, featuring the alleged arrogance and fiery temperament of the young Pierre Simon de Laplace (Paris, 7 January 1781)

> Monsieur de la Place is in my opinion the one [here] who has the greatest talents in mathematics, at least for the part which concerns calculus. He has achieved some beautiful and sublime things. And he seems to know this well himself. He has some knowledge of other sciences as well, but it seems to me that he puts it to wrong use, because he wants to have the final say on everything at the Academy. Moreover, he is very opinionated and stubborn. His acid and sometimes offensive temper is perhaps due to his having had more than his fair share of bad luck. Rumour has it that, because of his indiscretion, the Spanish Ambassador, who is a very violent man, was about to throw Mr de la Place down the stairs; because he had begun to malign the Spaniards, who were in his view idiots and ignoramuses.[106]

In the same letter Lexell recognises another mathematical talent, Gaspard Monge,[107] in the following equivocal words:

> Monsieur Monge is a very capable man, he is a newly elected member of the Academy. He has written a memoir on bodies, the surfaces of which may be developed in the plane,[108] and another on partial differentials,[109] which are highly regarded. His appearance is not charming, he is very dark, he has a frowning expression, and his upper lip is curved. Besides, he seems to have a high opinion of himself, because the first time I met him he told me that it is he alone who deals with geometry in this country.[110]

When Lexell's sojourn in Paris was coming to an end, he summed up his experience of the French people in his letter to Wargentin dated 30 March 1781.

> It has been a while since I last had the honour of writing to you, Herr Secretary. Due to my rather frequent correspondence with Petersburg as well as the abundance of amusements in Paris, as they are called, I have long enough come to neglect this precious responsibility

[106]Mr de la Place est certainement à mon avis celui, qui a les plus grands talents pour les mathématiques, au moins pour la partie, qui est purement du ressort du calcul. Il a fait des choses belles et sublimes. Aussi paroît-il le sçavoir lui-même. Il a des connoissances encore dans les autres sciences, mais il me semble, qu'il en fait un abus, car à l'Académie il veut décider du tout. D'ailleurs il est très opiniâtre et entêté. Son humeur aigre et quelque fois répugnant[e] vient peut-être aussi de ce qu'il est très mal partagé de la fortune. On prétend qu'à cause de ses imprudences M^{r} l'Ambassadeur d'Espagne, qui est un homme très violent, a été sur le point de faire jeter Mr de la Place par son escalier; c'est qu'il s'est mis à parler mal des Espagnols, en les traitant comme des sots et des ignorants.

[107]Gaspard Monge (1746–1818) became *Adjoint Géomètre* of the Academy on 14 January 1780 [72]. He worked especially on differential geometry and laid the foundation for descriptive geometry. He was also one of the founders of the *École Polytechnique* [55].

[108]"Mémoire sur les propriétés de plusieurs genres de surfaces courbes", *Mémoires de Mathématique et de Physique*, Tome IX, 1780, pp. 382–440.

[109]"Mémoire sur les fonctions arbitraires continues ou discontinues", ibid., pp. 345–381.

[110]Mr Monge est un homme très habile, il est nouvellement agregé à l'Académie. Il a donné un mémoire sur les corps, dont les surfaces peuvent être développées dans un plan, et un autre sur les différences partielles, qu'on estiment beaucoup. Sa figure n'est pas prévenante, il est très noir, les sourcils froncés et la lèvre superieure recourbée. D'ailleurs il s'estime assez soi-même, car la première fois que je l'ai vû il m'a dit, que c'étoit lui seul qui traitoit la géomètrie dans ce païs-ci.

of mine. I am now ready to leave this place after a stay of five months, which has been quite an interesting time for me, principally since I have met and become known to most of the learned men at least within the Academy of Sciences. In spite of all the faults the French nation might have, one feels obliged to admit that there is no nation which is as agreeable to deal with, and as free and natural, and I am equally convinced that nowhere can a lover of the sciences expect an appreciation as high as in France. Even if the rule is monarchical, less difference is made here between people according to their descent, titles and reputation in the political circles, than in many other monarchies. On the contrary I have noticed that men of high descent and rank value it as an honour to interact with the learned, and that they are often more obliging than the learned themselves, who for the most part are unbearable egoists. The details I gave in my previous letter may be in the need of some correction, as I can infer from your response, Herr Secretary. First, as to Mr Rochon, I may have pictured him in my last letter as a charlatan and there is no doubt that, considering the paper he read at the Academy, my conclusion was right. Nevertheless, I must confess that he is a mechanical genius; the three instruments he has invented are quite clever. The first one gives a means to measure the refrangibility of light in different substances using two crystal prisms, which consist of the two halves of a small crystal cylinder cut along the diagonal BC. The second one based on the same idea is for measuring small angles. He claims to be able to measure the diameter of Jupiter to within $\frac{1}{10}$ of a second, which I would like to see for myself before I believe it. The third one is a machine to print letters, quite clever, but notwithstanding all the improvements he has applied to it, it remains only a curiosity. Thus, when I give him some justice with regard to these inventions, yet I believe that what he has claimed about the instrument in measuring the elevation of the Sun is much exaggerated. De la Place is certainly the best mathematician in Paris, at least when it comes to calculus. I have given your compliments to Mr le Monnier. I would gladly believe that the old man can make good observations, but he has made quite a few really bad ones, too. The worst thing with him is that he meddles with things beyond his capacity, and that he finds faults in the doings of more talented men, such as Mayer and de la Caille. He has recently published a volume of *Opuscules* containing varying subjects, physics and astronomy, intertwined.[111]
I have likewise accomplished the commissions to Mr Cassini the younger and to Mr du Hamel.[112] Du Hamel is old and forgetful, but is still going strong considering his age — he is over 80 years — working rather assiduously and rarely misses the Academy. He has recently published his description of fishes.
Before Christmas there was a rather fiery dispute in the Academy between Mr Tillet[113] and Sage[114] regarding the question whether nitric acid dissolves gold or not. It would be too much to report all that has come to pass in this issue, which was more about persons than science, but what I want to say is that those who were against Mr Sage maintain 1° that nitric acid does not really cause a dissolution, but instead the gold which is exposed for a long time to the action of this solvent is so to speak rubbed off, which is later recovered in the form of thin foils 2° that this dissolution, if it may be called so, is of no importance in the usual metal separation applied in the mints, which was the principal issue in the research, of which Mr Sage had made a lot of fuss. As this matter concerns our Swedish chemists, too, especially Professor Bergman, I have considered it best to append the report written

[111]Referring probably to Le Monnier's *Mémoires concernant diverses questions d'Astronomie et de Physique*. Paris, 1781.

[112]Henri-Louis Duhamel du Monceau (1700–1782), botanist and anatomist who collaborated with Count Buffon. His work *Traité général des pêches* was published between 1769 and 1782. He had also written a treatise on naval architecture.

[113]Mathieu Tillet (1714–1791), botanist, director and treasurer of the *Académie*.

[114]Balthazar-Georges Sage (1740–1824), chemist and mineralogist, opponent of Lavoisier's chemical theories, contributed to the creation of *École des Mines*.

by the commission nominated by the Academy to investigate the matter. Chemistry is the most fashionable science here today, so much in fact that the public lectures given by Mr Sage are attended by a few hundred persons, of which around 30 or 40 are women. He is not without merits, but not very profound.
The mathematical classes consist generally of quite capable men, but unfortunately they are far from unanimous; in particular, there are two parties, in one of which there are Mr d'Alembert, Bossut and Condorcet, and in the other Mr Bezout, Chevalier Borda[115] and de la Place; apart from all the merits of Mr d'Alembert one cannot but blame him for not always being reasonable and for being too prone to find faults in other people. In view of the frivolity of the French nation it may seem a bit odd that the French more than any other nation cultivate the mathematical sciences; the reason I find to be that this nation has an incomparable capacity to take hold of ideas, and that they can learn things with much less effort than the others.
This time I must not be very prolix as I am going to leave this place tomorrow and I still have several letters to write. I will remain in London till the end of July and next I will continue to Gothenburg by sea, thence to Stockholm by land, and further to Petersburg.
I remain with the most profound respect,
Honourable Herr Secretary,

Your

humble servant
A: Lexell

Paris, 30 March 1781

In his last long letter from Paris to Saint Petersburg (24 March 1781; АРАН Фонд 1, оп. 3, д. № 65, л. 191–194об.) Lexell mentions a *déjeuner* which Abbé Raynal the past week had offered in honour of Princess Catherine Dashkova, a *confidante* of Catherine II, who at the time was visiting Paris.[116] From the wording of the letter it seems as if Lexell was not invited. However, he knew that the guests who were presented to the Princess included Lalande, Le Roy, the Lavoisiers, Chevalier Bory, du Séjour, Bailly, Bézout, Desmarest,[117] Laplace, Vandermonde and Monge. Moreover, the meeting was attended by Benjamin Franklin, the Duke de La Rochefoucauld,[118] Count de Verdun[119] and Lord Morton, the son of the President of the Royal Society of London. Lexell informed his colleagues in Saint Petersburg that the Princess planned to travel to Italy the following week by going through Switzerland. Eventually, she entrusted to Lexell some letters which were to be delivered to England, from where she had arrived in Paris.

---

[115]Chevalier Jean-Charles de Borda (1733–1799), captain of the Navy, physicist and mathematician.

[116]Princess Ekaterina Romanovna Vorontsova Dashkova (1743–1810) was an intimate friend of the Empress. Owing to her interest in literature and science, she was appointed Director of the Petersburg Academy of Sciences in 1783 [20,31], see Chap. 10.

[117]Nicolas Desmarest (1725–1815), geologist and mineralogist.

[118]François Alexandre Frédéric de La Rochefoucauld (1747–1827).

[119]Possibly Jean-René Antoine de Verdun (1741–1805), explorer and naval officer.

9.6 London: "... wherever I go, there is a comet waiting for me"

Lexell's arrival in London was delayed until the beginning of May 1781 because of the hostilities between France and England, which forced him to take a ship from Ostende in the Southern (Austrian) Netherlands, today's Belgium. Before his journey, Lexell had been studying the English language, but found it difficult. However, he seems to have enjoyed the English lifestyle and London in particular: its architecture and relatively developed infrastructure seem to have made an impression on him.

When Lexell arrived in London, the Royal Society had recently moved from their premises on Crane Court to the newly built palace of *Somerset House* (situated between today's Waterloo Bridge and the Strand). In spite of not being a fellow of the Royal Society, he was warmly welcomed to their meeting in the capacity of a *stranger*. Lexell does not explicitly describe the nature of the meetings he attended, as he did in Paris, but as he indicates in his letters, the mathematicians were strikingly few. As we know, the Royal Society had evolved during the eighteenth century into a forum for experimental physics and chemistry, quite different from the Paris Academy of Sciences [64, 69]. London was at that time also famous for its makers of high-quality scientific instruments, such as pendulum clocks and astronomical tubes. As a member of both the Stockholm and the Petersburg Academy of Sciences, Lexell had been commissioned to present orders for both Academies of telescopes from the optical company of Peter Dollond — son of the inventor of the achromatic telescope John Dollond — and precision chronometers from the equally famous clockmaker John Arnold (1736–1799) [139].

In London, Lexell's attention was likewise caught by the discovery of a new moving celestial object, which at first was believed to be a comet. However, as Lexell was the first to discover, it resembled more a planet, as it was moving in a nearly circular orbit.

This is his first letter to Wargentin:

> I have been in London for more than six weeks now without having written to you, Herr Secretary, of which I am quite ashamed, and I have to reproach myself for my neglecting to escape being reprehended by you, Herr Secretary.
> The astronomical tube that has been ordered for the Observatory of Uppsala is now ready, but has not been built according to the instructions, because it has a length of 10 ft and a double lens; although it has the requested magnification, my opinion is that the commission has not been adequately executed. The reason is that nowadays Mr Dollond, in want of flint glass, does not manufacture astronomical tubes with three lens glasses, at least not with an aperture of 40 lines [ca. 84 mm], which he did about six or seven years ago. But on the other hand Mr Maskelyne had been able to commission a shorter achromatic tube of length $3\frac{1}{2}$ ft, which also magnifies 120 times. I have ordered such a tube from Mr Dollond, equipped with a micrometer lens, for which I must pay 47 Guineas. The tube for the Uppsala Observatory costs only 25 Pounds. Mr Lindgren has received the order to pay for it, but it can be delivered no earlier than in a month.
> I find it peculiar that wherever I go, there is a comet waiting for me. It so happened that a few weeks before my arrival in London, a musician named Herschel in Bath had discovered a

small star in [the constellation of] *Gemini*, which he noticed had a movement of its own — at that time of about one minute [of an arc] a day — and which subsequently has been observed by Mr Maskelyne and in Paris by Mr Messier (after notification by Mr Maskelyne). Its position was on 17 March: longitude[120] 2s.24°.29′.46″, latitude north 11′.48.4″ and on 11 May: longitude 2s.26°.24′.42″ and latitude 11′.35.6″. There are no visible signs of the orbit, coma or tail, of this comet, as it is comparable to a star of the fifth magnitude, very clear and bright. This circumstance, as well as its slow movement following the ecliptic makes it rather extraordinary. I have been occupied with calculations [of its orbit], but so far I have not achieved a great deal, except that it is quite far away from the Sun, certainly 10 times further away than the Earth. A parabolic orbit, whose parameter is twice the radius of the ecliptic, agrees tolerably with the observations and corresponds to a distance from the Sun of 14 to 15 times the Earth's distance. But a circular orbit of 18.928 times the Earth's mean distance from the Sun fits even better. I believe, however, that the only thing which can be said about the comet from the observations is that it is quite distant from the Sun and that it has described a rather small angle around the Sun while being visible. If this extraordinary star was visible after its conjunction with the Sun as expected, then it could also be decided if it moves in a more or less eccentric orbit.

I have been quite well received here by those who profess the mathematical sciences, Maskelyne especially has shown me much courtesy and I have visited him out in Greenwich four to five times. The Greenwich Observatory is rather well organised and possesses two mural quadrants, one transit instrument, one zenith sector, two equatorial sectors, which are quite useful and highly necessary, when observations are made outside the meridian. Since the axis of this instrument is parallel with the axis of the Earth, it follows that the aiming line of the tube will always remain within a plane that traverses this axis, or which is the same thing, the tube always moves in the hour circle. I have already seen many such instruments and one particularly complete is to be found at the King's Observatory in Richmond,[121] which also possesses many other fine instruments, such as a zenith sector, a transit instrument and a quadrant, the arc of which is 135°, but which I for my part do not like particularly. My other acquaintances include the astronomer Mr Aubert,[122] who is quite a polite gentleman and owns a beautiful observatory, Cavendish[123] who is quite versed in mathematics, Dalrymple[124] is a good geographer, Russel[125] is an astronomer, Admiral Campbell[126] who observes much and is very much Mr Hornsby's[127] friend, Doctor Shepherd,[128] Professor at Cambridge, but who stays here most of the time,

[120]Lexell was still employing the ecliptic coordinate system, in which longitudes are expressed by the number of zodiac signs (counting from the vernal equinox) plus the number of degrees within the following sign.

[121]King George III was a keen amateur astronomer. His private observatory was in Kew Gardens, Richmond, London.

[122]Alexander Aubert (1730–1805), merchant and amateur astronomer.

[123]Henry Cavendish (1731–1810), experimental physicist and chemist who anticipated several theoretical concepts and laws concerning electricity, measured the density of the Earth and discovered hydrogen as one of the constituents of water.

[124]Alexander Dalrymple (1737–1808), Scottish geographer and hydrographer.

[125]Possibly John Russell (1745–1806), painter and amateur astronomer.

[126]John Campbell (1720–1790), sea captain, naval officer who studied astronomical methods to determine longitude at sea.

[127]Thomas Hornsby (1733–1810), Savilian Professor of Astronomy in Oxford.

[128]Anthony Shepherd (ca. 1721–1796), astronomer.

and Chevalier Schuckburg.[129] My other acquaintances are Sir Banks,[130] Doctor Solander,[131] who have shown me much courtesy, Mr Planta[132] the Secretary of the Society, Mr George Loid,[133] Kirwan[134] a good chemist, likewise Doctor Crawford,[135] Doctor Priestley,[136] Doctor Price,[137] Baron Bennett,[138] Doctor Blagden,[139] Doctor Gray[140] and others I cannot remember now. I am in all respects quite satisfied here, and I would be even more so, could I better speak English. It has never been easy for me to learn languages, and I speak and write satisfactorily all those I can. But then I find that for me, at my age, it is more difficult to learn a new language than if I were ten years younger. Nevertheless, it is an advantage for me to understand most of what is spoken and to be able to help myself.

There is no place as expensive as London, and thus it is not wise to stay here any longer than necessary except for those who have a lot of money. I would very much like to stay here a bit longer, but it seems that I have to return back north in the end of the next month or the beginning of July. I plan to make the journey either to Gothenburg, or Holland, or Hamburg, but in any case travel across Sweden.

I have ordered a small quadrant[141] of one ft radius for the Academy in Åbo, and therefore I will now write to my brother-in-law asking him to send to you 100 *Riksdalers*, with which you, Herr Secretary, should be so kind as to order a bill of exchange to be addressed to Doctor Solander or alternatively to instruct Mr Arfwidson and Tottie to give Mr Grill and Lindgren[142] the commission to pay Mr Ramsden[143] when the instrument is finished, which may take place in a couple of years, if Mr Ramsden is as much engaged with me, as he usually is.

---

[129]George Augustus William Shuckburgh-Evelyn (1751–1804), experimental physicist.

[130]Joseph Banks (1743–1820), botanist who participated in Captain Cook's first voyage to the South Pacific.

[131]Daniel Solander (1733–1782), Swedish-British botanist and disciple of Linnaeus, who likewise participated in Captain Cook's voyages. His assistant on the voyage was Herman Diedrich Spöring (ca. 1733–1771), a skilled instrument maker born in Åbo. Spöring perished on Cook's first voyage and was buried in the Indian Ocean.

[132]Joseph Planta (1744–1827), Principal Librarian of the British Museum and one of the secretaries of the Royal Society.

[133]Probably George Lloyd (1707/08–1783), a merchant and manufacturer from Yorkshire who was elected to the Royal Society in 1737 for being "well skilled in mathematical knowledge and Natural Philosophy".

[134]Richard Kirwan (1733–1812), Irish chemist and geologist.

[135]Adair Crawford (1748–1795), chemist and thermodynamist.

[136]Joseph Priestley (1733–1804), chemist and experimental physicist.

[137]Richard Price (1723–1791), political and moral philosopher.

[138]Richard Henry Alexander Bennett (ca. 1743–1814), Member of Parliament.

[139]Sir Charles Blagden (1748–1820), physicist and physician.

[140]Possibly Charles Gray (1696–1782), Member of Parliament, lawyer and antiquary.

[141]This quadrant is currently on view in the Old Observatory of the University of Helsinki (see Figure 9.5).

[142]Swedish trading companies lead by Carl Kristoffer Arfvedsson, Anders Tottie and Claes Grill [129]. Claes Grill was also acting Swedish consul in London.

[143]Jesse Ramsden (1735–1800), maker of scientific instruments.

Fig. 9.5 This quadrant of one foot radius ordered by Lexell in London is among the oldest scientific instruments at the Observatory Museum of the University of Helsinki (Photo: the author)

I would still like to write some more, but this letter goes through the care of Baron Nolcken, who takes more or less one Shilling for it. Thus, I will stop here with an affirmation of the invariable friendship with which I have the honour to be,
Honourable Herr Secretary,

Your

humble servant
Lexell

London, 25 May (5 June) 1781
Professor Linné[144] has been here for fourteen days now and he has been received very courteously, especially as Mr Solander is taking care of him.

In his letters to Saint Petersburg, Lexell continues his analyses by employing the precepts of physiognomy (London, 4 May 1781; АРАН Фонд 1, опись 3, дело № 65, л. 195–197об.):

—Sir Joseph Banks is very easy-going and without any pretensions, like most Englishmen; he is taller than average, and there is something very serious in his appearance, although he seems to be of a good and gentle character. There is a portrait of him by Reynolds,[145] which has been engraved, but which does not bear much resemblance. In this painting he seems sad, which in reality he is not.[146]
—Doctor Solander has a round face and a Swedish physiognomy, but manifesting a lot of candour, and I also believe that the Doctor has a very good character, at least as much as

[144]Carl von Linné the younger (1741–1783), Swedish naturalist and successor to his father, Carl Linnaeus, as Professor of Medicine [98]. He collaborated with SOLANDER on his main work *Supplementum plantarum* (1781).

[145]Joshua Reynolds (1723–1792), influential English painter of portraits, President and founder of the Royal Academy of Arts.

[146]Sir Joseph Banks est comme sont presque tous les Anglois, sans aucune façon et même sans aucune prétention, il est d'une taille plus que médiocre et a quelque chose de très sérieux dans sa physionomie, cependant il paroît être d'un bon et doux caractère. Il y a un portrait de lui par Reynolds qui a été gravé, mais qui n'est pas très ressemblant. Dans ce portrait il paroît plus triste, qu'il n'est pas.

I have made his acquaintance until now. He appreciates good company and is also very outgoing.[147]
—Maskelyne is not tall, but he has quite a pleasant physiognomy. He has received me very well and I am quite happy with him.[148]
—Mr Cavendish is said to possess a lot of knowledge, but he has great difficulty expressing himself.[149]
—Mr Kirwan is one of the best chemists in London; there is nothing remarkable in his physiognomy.[150]
—Dollond's physiognomy is more reminiscent of a large bourgeois gentleman, and he is not very clever. As to John Arnold, the physiognomic rules are quite in error; one would take him for a brewer, but he certainly is quite ingenious and a clear thinker.[151]
—Mr Crawford, who published a year ago the *Treatise on animal heat*, is a very gentle and very modest young man. He performed some experiments on the heat of common air and dephlogisticated air, which I attended, but with which I was not quite satisfied; I will talk about that another time in more detail.[152]

In fact, before his journey in 1780, Lexell had acquired Crawford's book *Experiments and observations on animal heat* (1779), which he had found interesting to such a degree that he thought its basic principles, mathematically formulated, could serve as a foundation for a rigorous theory of thermodynamics. He even wrote a brief essay on the basic laws of thermodynamics in his letters to Wargentin, Pehr Adrian Gadd and Johann III Bernoulli, but these abstracts were never published. However, after having met Crawford personally and attended his demonstrations in London, Lexell seemed to be less convinced of the solidity of the empirical basis for thermodynamics than he was before.

In the subsequent letter to Wargentin, dated 24 July 1781, Lexell delivers a detailed account of his commissions for his compatriots in Uppsala and Stockholm:

In a letter sent one month ago or so in the envelope to Baron Nolcken, I wrote about the developments of the commissioned achromatic tube for Uppsala, but as some mail was lost at that time, it may well be that my letter has been destroyed, and so I have to write

[147]Doct. Solander a un visage rond et une physionomie Suédoise, mais qui marque beaucoup de candeur, aussi crois je que le Docteur a un très bon caractère tant que j'ai eu occasion de le connoître jusqu'à présent. Il aime la bonne société et il est aussi très répandu.

[148]Maskelyne n'est pas grand, mais il a une physionomie assez favorable. Il m'a reçu très bien et j'en suis assez content.

[149]Mr Cavendish possède à ce qu'on dit beaucoup des connoissances, mais il a une grande difficulté de s'exprimer.

[150]Mr Kirwan est un des meilleurs chemistes à Londres, sa physionomie n'a rien de remarquable.

[151]La physionomie de [Peter] Dollond est plustôt d'un gros bourgeois, aussi dit qu'il a fort peu de génie. Chez [John] Arnold les règles physionomiques se trouvent tout à fait en défaut, on le prendroit d'après son extérieur pour un brasseur, il est cependant certainement un homme très ingénieux et même un homme qui raisonne fort bien.

[152]"Mr Crawford qui publia il y a un an, le *Traité sur la chaleur animale*, un jeune homme d'un caractère très doux et très modeste. Il a fait quelques expériences sur la chaleur de l'air commun et de l'air déphlogistiqué auxquelles j'ai assisté mais dont je n'ai pas été tout à fait satisfait; j'en parlerai une autre fois plus en détail." (According to the phlogiston theory of heat, "dephlogisticated air" — a precursor of what was later identified as the element oxygen — was apt to combine with phlogiston and thus able to support combustion longer than "common air").

some more about it. Mr Maskelyne had commissioned the tube already before my arrival here, but not in accordance with the instruction, since instead of an achromatic lens with a threefold glass, Mr Maskelyne had ordered an achromatic tube 10 ft in length, which may well magnify 120 times, but in other respects it does not meet the requirements. The reason why Mr Dollond had decided to make the tube in this way is probably that he does not nowadays make any tubes with three-fold objective glasses with an aperture of more than 30 lines. In order not to give Mr Maskelyne a denial, I was obliged to accept the commission as it was submitted, and so I have paid for the tube 18 Guineas, or 18 Pounds and 18 Shillings, and asked Mr Dollond to send it to the Consul Grill, and so it has been done. I have bought Walker's[153] *Doctrine of the sphere* to Professor Mallet, it costs 12 Shillings. Of the *Nautical Almanac* I took one for the years 1781–86, which costs 1 Pound 1 Shilling or 3 Shillings 6 Pence apiece. I am aware that the Longitude Commission has decided to send to you, Herr Secretary, these Ephemerides as well as other recent publications, which is why I have not wanted to buy more than one of these. Having withdrawn from Consul Grill the 25 Pounds sent by you, Herr Secretary, I am now 4 Pounds and 9 Shillings in debt, which I will settle in Stockholm in due course.

However, more than anything else, the new "comet" in the sky, which in fact turned out to be the planet Uranus, captured Lexell's attention:

I have made quite a lot of calculations relating to the comet which was discovered in the month of March, and probably more than was needed, as I for some reason was inclined to think that the comet could not have a parabolic orbit of such a large distance from the Sun at perihelion as it really seems to have, because I have found that a parabolic orbit with a distance at perihelion of 16 times the distance of the Earth from the Sun matches fairly well the observations, as can be seen in the following table (see Fig. 9.6):

Dist. perih.	8	10	12
Posit. perih.	$6^s.5^\circ.34'.8''$	$5^s.24^\circ.43'.15''$	$5^s.14^\circ.5'.55'$
Time of perih.	3101D.7468	3244D.4691	
Time Greenw.	calc.		
1 April $9^h.58'$	$2^s.24^\circ.47'.8''$	$47'.36''$	$48'.5''$
18 April $8^h.0'$	$2^s.25^\circ.20'.0''$	$20'.43''$	$21'.20''$
22 May $9^h.6'$	$2^s.26^\circ.59'.15''$	$59'.23''$	$59'.37''$

Dist. perih.	14	16		
Posit. perih.	$5^s.3^\circ.16'.34''$	$4^s.21^\circ.43'.53''$		
Time of perih.	3200D.5067	2960D.6933		
Time Greenw.	calc.		observ.	
1 April	$48'.5''$	$48'.27''$	$48'.50''$	$49'.30''$
18 April	$21'.20''$	$21'.50''$	$22'.20''$	$22'.14''$
22 May	$59'.37''$	$59'.47''$	$59'.53''$	$59'.50''$

One should note that these elements have been calculated so that they always satisfy the observations for 17 March at $10^h.40'$ and 28 May at $9^h.8'$, when the longitude of the comet was, respectively, $2^s.24^\circ.29'.46''$ and $2^s.27^\circ.19'.59''$. I have also found that a circle of radius 18.928 times the distance between the Earth and the Sun matches the observations

[153] George Walker (ca. 1734–1807), English ecclesiastic and mathematician.

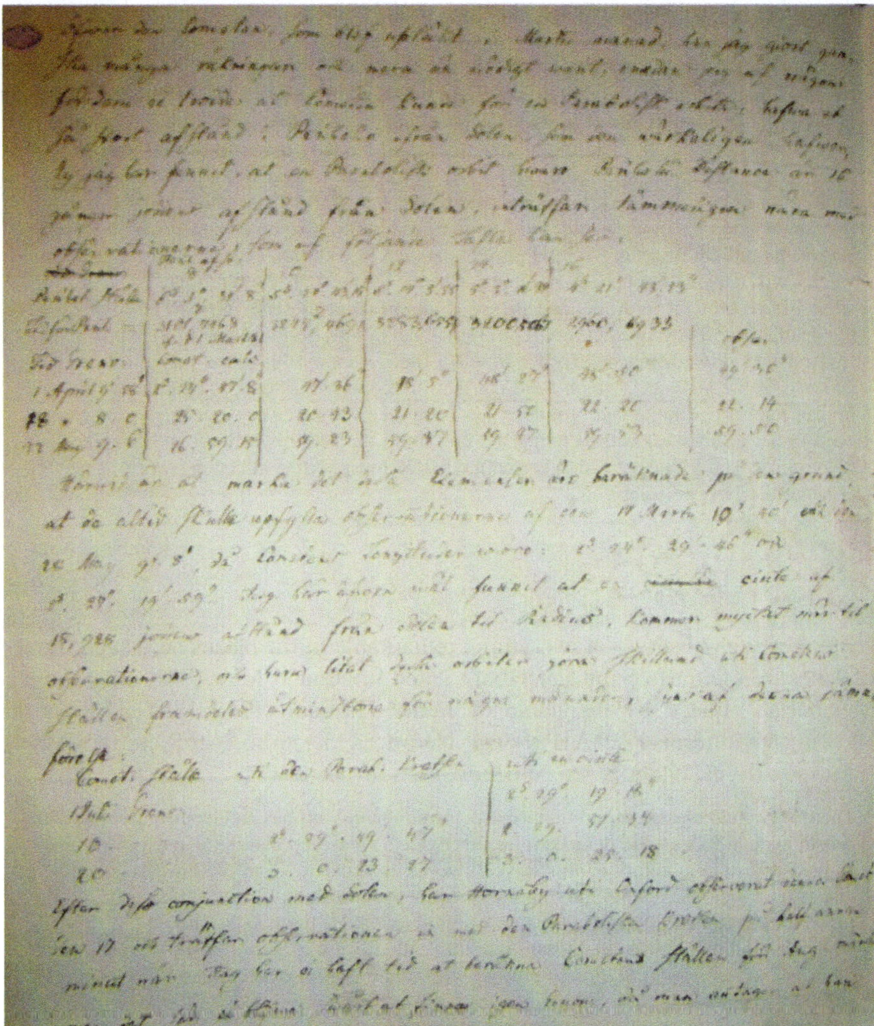

Fig. 9.6 Page of the letter from London to Wargentin containing the newly-compiled data of the motion of the new planet (KVAC, Royal Swedish Academy of Sciences, Stockholm. Published with permission)

very well, and the following comparison below shows how small the differences in the positions will be between these two orbits:

Position of comet	in parabolic orbit	in circular orbit
1 July Greenw:	$2^{s}.29^{\circ}.19'.18''$	
10	$2^{s}.29^{\circ}.49'.47''$	$2^{s}.29^{\circ}.51'.34''$
20	$3^{s}.0^{\circ}.23'.27''$	$3^{s}.0^{\circ}.25'.18''$

Following its conjunction with the Sun Mr Hornsby in Oxford observed this comet on 17 July, and this observation agrees with the parabolic orbit to an accuracy of within one and a half minutes. I have not had the time to calculate the positions for the month of August, but it will not be difficult to find it by assuming that it advances approximately one degree in 20 days. The other comet Mr Méchain[154] has observed in Paris has not been found here as it moves very quickly and the observations were already rather old when they arrived here, so that it has not been possible for Mr Maskelyne to find it. I am now too short of time to be able to tell you everything noteworthy here for an astronomer, which is more than anywhere else in the world, but I hope to be able to give a verbal account of it when I have the delight of seeing you, Herr Secretary, in Stockholm, because at the end of this week I will depart from here to Hamburg, whence I go to Lübeck to forward to Petersburg what I have brought with me, and thence if an occasion presents itself to go to some of the southern ports of Sweden, I will go by sea, but otherwise I will go with the post either to Stralsund or Copenhagen. To begin with I did not enjoy this place very much mainly because of the language, which I did not learn to cope with, but now I can already make myself understood quite well. Linné[155] has been received here very graciously, and in fact he is rather celebrated, as the English are nowadays the nation which the most appreciates the sciences and Linné the Elder. I remain with the most perfect affection,
Honourable Herr Secretary,

Your

humble servant
A: Lexell

London, 24 July 1781

Being a keen observer of everything remarkable in the places he visited, Lexell describes in a letter to Saint Petersburg dated London, 4 May 1781, a flamboyant "Mesmerian" quack by the name of James Graham (1745–1794) [48] in the following vivid terms (АРАН Фонд 1, опись 3, дело № 65, л. 195–197об — for the original citation, see Appendix section "Letters"):

There is a great charlatan here named Doctor Graham, who has an electric machine with enormous conductors. He demonstrates this twice a week and makes the most extravagant claims, and quite contrary to modesty. However, in order to spare the blushes of women who attend, he extinguishes the candles when he starts making his pronouncements on the influence of electricity in connection with reproduction. He has bought a house in Pall Mall, where he claims to have established what he called *celestial beds*, by the door of which there is the inscription *Sacred to Hymen* with many other extravagances such as statues and related emblems. However, he has been obliged to remove some nude statues from the front of the house, but it would have been prudent to require him to erase the inscription too, which judging by his advertisement in the newspapers is very risqué and in dubious taste. It is one of the most unusual amusements to hear this charlatan perform. His clothing fits the rest, when I saw him he was wearing a velvet coat with very fine lace and hair with enormous long curls. His carriage is also decorated and entirely covered in symbolic figures.

On the whole, Lexell seems to have appreciated the English, except for their excessive drinking. Wondering if the success of the English was due to the climate, the healthy upbringing of children regardless of gender, or the nutritious food, he

[154]Pierre François André Méchain (1744–1804), astronomer who began his career as Lalande's assistant [38].

[155]Carl von Linné the younger.

writes (London, 15 June 1781. АРАН Фонд 1, опись 3, дело № 65, л. 208–210об.):

> I cannot say if it is due to the climate or their upbringing that English children have a very different character from that of other countries; maybe the two causes unite in forming this spirit of boldness, courage, love of freedom and independence, which one finds in young children here. I was surprised to find children of three or four years climbing ladders as boldly as boys of fifteen years in other counties. Not even the gender makes any difference because the girls as much as the boys run around in the streets, and even children of the better homes are taken out for a walk several times a day, which is good and useful for their health. The demeanour of the English is generally very similar to that of other peoples of the north, that is, they seem very calm and unaffected and they do not have the sophistication of the French, but it would be unfair to say that they are not as helpful as the French; on the contrary, I am persuaded that they consider it an honour to show every possible courtesy to strangers, and when one has gained their confidence, they are extremely frank and unreserved. I would almost venture to say that there is no nation as educated as the English, because there are people in all kinds of circumstances who have some knowledge of several sciences; but as to real scientists, that is, those who are able to make discoveries, their number is very limited here, and mainly in mathematics, where there are no more than two or three with some reputation. And I am surprised they are not all that stupid, because their main source of nutrition is meat, which is of a better quality than in any country of the world. Observing that the English eat a lot of meat, drink strong wine such as Port, breathe an extremely thick air filled with smoke of coal, it is not surprising that they become melancholic, sad and drowsy.

However, Lexell continues, the English also have some peculiar habits:

> Apart from the Dutch, there is no nation so keen on cleanliness as the English, mainly when it comes to clothes, and if I am not mistaken, it is so for a good reason, since due to the coal smoke, one has to change the linen every day. Women here are almost always dressed in white during summer, and as their clothes are shining white, it makes a very nice impression. I quite like the way of life of the English, except that they stay too long at the table and drink strong wines; because even if you do not have to drink with others when you feel indisposed, it is difficult not to drink toasts with the other guests, if you do not want to look peculiar. Other unique customs of the English are the following. Upon entering a house, you leave your hat in the antechamber, and if you are at the table with women, it is regarded courteous of a man to ask if any of the ladies would like to have a drink with him. After the dessert the women leave the table and then the bottles begin to be passed round. One of the strangest habits of the English is to let their fingernails grow almost as long as the Chinese.

Summarising Lexell's enjoyable visit in England, his almost eulogistic description of Oxford is worth quoting:

> Although I have not seen much of this country, all that I have seen is beautiful and excellent, and it seems to me that every place from London to Oxford is just a cultivated garden. The situation of Oxford is charming and I have never seen a university town so beautiful. Leipzig comes the closest, but I prefer Oxford. The Oxford Observatory is the most magnificent temple that has ever been devoted to the service of Venus Urania,[156] both in view of elegance and utility. Besides the Observatory there are also several other beautiful buildings

[156] *Venus Urania* (a celestial nymph) represents the science of Astronomy.

in Oxford, such as Christ Church, in which the front of the library[157] is certainly a splendid piece of architecture, the Radcliffe Library,[158] which has a beautiful dome and All Souls College, which is also exquisite. Mr Hornsby has shown me much civility during my short stay at Oxford, and he showed me the Observatory and the construction of its instruments, with which I was very pleased. There is, among other instruments, an achromatic telescope by Dollond with an apparatus of six eyepieces, which magnify from 50 to 300 times, increasing in steps of 50. I observed Saturn with the eyepiece which magnifies 200 times, and I have to admit that I have never seen it so distinctly although it was at an elevation of only 2°.

9.7 Last Visit to Sweden and Finland

In the next letter, which was written in Åbo, Lexell talks about his arduous journey across Sweden, where he was received by his King at his palace and paid a visit to his colleagues in Uppsala and Wargentin in Stockholm (Fig. 9.7). He distinctly observes that the exact sciences in his home country are in a less than creditable state. Before returning to Saint Petersburg he pays a visit to Åbo, where he meets his family and friends for the last time.

For many mail days I have been planning to write to you, Herr Secretary, but every time I have been distracted so that it has been a great effort to find a free moment for this purpose. I owe you, Herr Secretary, my deepest gratitude for all the friendship you have shown me both before and during my stay in Stockholm and I wish I could somehow find an occasion to really show my appreciation.

Having left Stockholm I came to stay in Uppsala longer than I had foreseen, namely four days, with the consequence that I missed the mail [post-boat] over the Sea of Åland, which caused me some additional expenses and some delay, so that I arrived here at Åbo no less than fourteen days after my departure from Stockholm. Of the mathematicians in Uppsala I found Mr Melanderhielm to be rather unchanged, cheerful in his temper, but lazy and feeble, Mr Mallet fat as ever, and very much happier in spirit than the last time I saw him, Mr Prosperin rather content, nevertheless it is clearly visible that he must be badly affected by hypochondria at certain times, Mr Duraeus has gone down totally and lost his memory. Bergman[159] was then said to be recovering, but if I had to hazard a guess whether he still be alive next spring or not, I would reply in the negative, although I wholeheartedly wish him the longest life possible. It would cause irreparable damage to science, and for our country the loss would be even greater, as there is nobody to replace him. It is regrettable that he himself has paved the way for this fate by undue assiduity and incessant work.

Here in Åbo I have been delighted to meet my family and friends in good health and for almost a week now I have attended so many parties that it is starting to feel tedious, and I already wish to have some peace. I do not know what they will think of me in Petersburg when I am so late, in particular I am a bit worried about things I was commissioned to send from England, which have not arrived there yet; however, I am not to be blamed for this, but the Russian Consul, who had accepted the task of delivering these things. I could pay the 30 *Riksdalers* I owe the Royal Academy of Sciences from here, but since I have already

[157]The Christ Church Library was completed in 1772.

[158]The *Radcliffe Camera* housed the Oxford scientific library.

[159]Torbern Bergman died only 3 years later, 48 years old [98].

Fig. 9.7 Wargentin's Observatory and home on the Brunkeberg ridge in Stockholm, which was on the outskirts of town in the eighteenth century. The original building was designed by the architect Carl Hårleman [139] (The wooden tower was built later). Nowadays, the Observatory houses a science museum maintained by the Royal Swedish Academy of Sciences (Photo: the author, 2011)

been commissioned to send some things from Petersburg to Stockholm, I will choose this way to compensate the Academy, which I will write about when I arrive in Petersburg.

Although His Excellency Baron Scheffer has settled the dispute between Professors Lindquist[160] and Kreander[161] to the advantage of the former, yet the latter being a good advocate has raised an issue in view of the statement of the Chancellor on the present occasion, which was not entirely appropriate. As there is no one who can be more obliged to take up Mr Lindquist's cause, I have promised him to write to His Excellency and also I ask you, Herr Secretary, to talk to Baron Scheffer about it, as well as to the Royal Secretary Mr Fredenheim,[162] who is believed here [in Åbo] to be on Kreander's side. If Herr Secretary could be so kind as to say to the Royal Secretary Fredenheim that whatever he can do in favour of Mr Lindquist shall be regarded as a favour to me personally, and that nothing can be given as a greater proof of his friendship than the fact that he takes interest in Mr Lindquist, but in any case not in Mr Kreander.

The future prospects of the Academy here now seem somewhat happier after the nomination of some young and capable professors, such as Porthan, Calonius,[163] Lindquist and Hellenius.[164]

The comet discovered in May cannot be the same as the comet of 1770, I think; as to the more recent one, more observations are needed. I am with undying respect,

Honourable Herr Secretary,

Your

humble servant

A. J. Lexell

20 November 1781

[160] Johan Henrik Lindquist, Lexell's locum in the Chair of Mathematics.

[161] In 1780, Salomon Kreander (1755–1792) was appointed Pehr Kalm's successor in the Chair of Economy in Åbo against the preference of the Consistory, which had favoured Professor Pehr Adrian Gadd to the chair.

[162] Carl Fredrik Fredenheim (1748–1803), son of Archbishop Mennander.

[163] Matthias Calonius (1737–1817), lawyer, professor at the Faculty of Law in Åbo, future member of the Supreme Court in Stockholm [81, 158].

[164] Carl Niclas Hellenius (1745–1820), extraordinary Professor of Natural History and Economy in Åbo [81, 158].

This polished letter may be compared with the more outspoken correspondence with Johann Albrecht Euler in Saint Petersburg (dated Åbo, 8 November 1781; АРАН Фонд 1, опись 3, дело № 66, л. 161–162об — for the original, see the Appendix section "Letters"):

> My journey to Stockholm through Sweden during an extremely rainy period was quite tiring and painful, but I was amply compensated by the most friendly reception from many of my compatriots. I was presented to the King at Drottningholm, but I could see very clearly that the sciences in general and mathematics in particular are not very much appreciated, either at court, or by the most distinguished people in Sweden. Mr Wargentin has become old, his hearing is weak. Your old friend Mr Wilcke[165] is also unwell, he was indisposed because of the gout. He always thinks of you and your Father with gratitude. From Stockholm I went on a visit to Uppsala, where I found all my friends in good health, except Mr Bergman, who after a severe illness was still extremely weak and whose life seemed not to be quite out of danger. It would be a considerable loss for the science of chemistry if he died, as he is certainly one of the most distinguished chemists. Mr Melanderhielm is still the same, extremely lazy and so fond of taking it easy that he moves from his chair to the door only with great effort. Mr Mallet, Professor of Geometry, has put on some weight and seems much better since he married. Mr Prosperin is a real *honnête homme*, but an extreme hypochondriac. The trip from Uppsala here was also quite unpleasant, but I had the great pleasure of finding my sister [Magdalena Catharina Pipping] and her family in good health. I was deeply touched by seeing my sister surrounded by so many beautiful children; she has ten of them, and I wish more than anything in the world to be rich enough to guarantee a fortune for all my nieces. I beg you, dear colleague, to apologize to the Academy if I have to stay here a fortnight longer. It is quite uncertain when I may have the pleasure of seeing my sister again, and thus it will not be found strange if yielding to my nature, I can not tear myself away so soon from the arms of my family.

From the views he expressed on the decline of the exact sciences in Sweden during the reign of Gustav III, it indeed appears that Lexell had no regrets as to his decision to remain in Saint Petersburg, which was the final destination on his journey.

---

[165]Johan Carl Wilcke (1732–1796) was a Swedish experimental physicist of German origin. In 1755 he arrived in Berlin to study with his mentor Franz Aepinus. When Aepinus subsequently moved to Saint Petersburg, Wilcke pursued his career as lecturer in Stockholm and Uppsala. He succeeded Wargentin as Secretary of the Royal Academy of Sciences in 1784. Wilcke is best known for determining the latent heat of different substances, for the discovery of the *electrophorus* before Volta, and for anticipating the concept of electric polarisation [55].

Chapter 10
Return to an Academy in Crisis

Although Lexell had to forsake the therapy he had planned to enjoy at some of the health resorts of Europe, he returned from his academic journey seemingly in good health and in good spirits. Despite the physical fatigue he must have felt from the inconveniences of travelling, the personal meetings and experiences during the journey had strengthened his self-confidence. He was eager to return to work. However, by the time Lexell returned, the Academy was virtually paralysed by the conflict between Director Sergey Domashnev and the staff of the Academy of Sciences. Salaries had not been paid for months, the Academy's publications were way behind schedule and the sales revenues were falling. The new crisis was triggered by the director, who, without mentioning it to anyone, had taken possession of a package of Swedish minerals donated to the Academy by Baron Scheffer in 1780. The scandal grew into the proportions of a diplomatic conflict when the Swedish envoy in Saint Petersburg Baron Nolcken began to inquire about the gift, the existence of which no academician seemed to be aware of [141, Vol. III, pp. 577–579, 4 March 1782].

On 14 December 1781, having arrived from his journey four days earlier, Lexell returned to his place as an academician at the conference. A few weeks later, he writes to Wargentin:

> Having stayed for three weeks in Åbo and two weeks on my way here [to Petersburg], I arrived about three weeks ago. I should have announced my arrival much earlier, but I have been partly occupied by many visits and renewals of my acquaintances and partly prevented by an indisposition, which I have now overcome.
> My arrival coincides with a time when the Academy here is in a considerable turmoil, because the so-called Academic Commission is in an open conflict with the Director of the Academy, Mr Domashnev, a quarrel that in many an opinion will end in such a way that Mr Domashnev will be forced to resign from the leadership. Even if I have no part in these quarrels it is nevertheless very unpleasant for me inasmuch as the affair will not be finished soon, since it may well be that I will lose my fee of 200 Roubles, which I have left at the Academy.

J.C.-E. Stén, *A Comet of the Enlightenment*, Vita Mathematica 17,
DOI 10.1007/978-3-319-00618-5_10,

I have presented to the Academy Professor Schenmark's[1] table of prime numbers, which was received with great pleasure, and the Academy would probably have taken the decision to print it had not Adjunct Fuss said that Mr Bernoulli had announced from Berlin that Professor Hindenburg of Leipzig already has a table of primes up to two million ready for printing.[2] If this turns out to be the case, Professor Schenmark's table would be less useful to print, which is why the Academy will not take on this additional expense. Nevertheless, I have been charged to thank Professor Schenmark on behalf of the Academy and to ask whether the Academy can offer him a favour in return, such as if he would like to receive some of the Academy's printed works. If I remember your words correctly, Herr Secretary, there were some costs for having this table copied. As soon as I know the sum, I will arrange payment for this work.

I am eager to hear the latest developments of Mr Lindquist's argument on the salary against Mr Kreander, who is now said to be in Stockholm. Once more I would like to recommend Mr Lindquist to you, Herr Secretary, and plead you to assure His Excellency Baron Scheffer that everything the Baron does for the benefit of Mr Lindquist will be considered a favour for me personally.

I have the honour to hold you in the most perfect respect,
Honourable Herr Secretary,

Your

humble servant
A: Lexell

Petersburg, 30 December 1781

If you, Herr Secretary, could be so kind as to communicate to me some observations of the remarkable comet, two observations for each month, it would give me great pleasure.

In April 1782, the dispute with the Director took a turn to the worse. After Domashnev had removed the academician Semyon Kotelnikov from the supervision of the Natural Cabinet without a warning, an extraordinary meeting without the Director was convened on 20 (9) April in Euler's house to discuss possible countermeasures. At the following ordinary conference held on 28 October 1782, the three academicians Kotelnikov, Rumovski and Lexell having formally protested against the publication of four new prize competitions in the name of the Academy, insisted that their protest should be inserted in the minutes of the meeting as a response to the proposal of the Director made at an earlier conference, at which they had not been present. Lexell also demanded that all the academicians and adjunct members should be obliged to sign the minutes of the conferences where they had been present, or to state in writing their reasons for not doing so [141, Vol. III, p. 631].

At the conference held on 29 (18) November 1782 [141, p. 635], the printing of the calendar for the following year was deliberated, although the Director had expressly forbidden the Academy to take action in the matter on its own. Then,

[1]Nils Schenmark (1720–1788) was Professor of Mathematics in Lund. On his way from Copenhagen to Stockholm, Lexell visited the university of Lund and met Schenmark, who donated a table of prime numbers to the Petersburg Academy of Sciences, apparently in the hope of having the work printed.

[2]Hindenburg published his method, a refined version of Eratosthenes' sieve, in his *Beschreibung einer ganz neuen Art, nach einem bekannten Gesetze fortgehende Zahlen durch Abzählen oder Abmessen bequem und sicher zu finden*, Leipzig 1776; his table was never printed.

Fig. 10.1 Sergey Gerasimovich Domashnev, Director of the Imperial Academy of Sciences 1774–1782. Engraving after contemporary portrait (Stählin [154])

unexpectedly, the Director entered the conference hall, asking the Secretary Johann Albrecht Euler what was being discussed (Fig. 10.1):

—The calendar, the Secretary replied.
The Director reproached the Secretary for taking up the matter for discussion contrary to his orders. Then Lexell began to speak, asking in what way this interfered with the conference. The Director — not used to spontaneous interjections in the style of the Paris Academy of Sciences — ordered him to be quiet and not to interrupt him.
—I have not interrupted and I am not going to be quiet, Lexell replied. Then, Domashnev, now even more determined, shouted:
—*T a i s e z - v o u s*! (be quiet!).

After this unexpected treatment, Lexell, humiliated, stood up and left the conference hall.

Before leaving the hall himself, the Director terminated the scene by dictating to the Councillor Protasov some words in Russian. Thereupon, the conference ordered the Secretary to insert the aforesaid procedure of the Director in the present minutes and to ask Lexell to return. Lexell came back on this invitation, and having resumed his seat in the conference hall, he demanded the Secretary to ask everybody present: "Did he (Lexell) interrupt the Director?" Everyone except Protasov and Zuyev[3] answered no: the former saying that "His Excellency had hardly finished speaking, when Mr Lexell began to speak" and the latter saying that "the Director had addressed his words to the Secretary and that it was up to him to reply and not to Lexell".

Although conference sessions were arranged regularly, the Academy was paralysed by the tension, as nothing could be printed and no salaries were paid. Finally, a few months after the incident, an extraordinary assembly was convened on Saturday 8 February (28 January o.s.) 1783 [141, Vol. III, p. 647], where

> [T]he Secretary opened and read out an *ucase* [указ, decree] of Her Majesty the Empress to the Academy of Sciences, signed by the First Secretary [of the Ruling Senate] Lashkevich on 27 January 1783, containing the nomination of a new Director of the Imperial Academy of Sciences, Her Excellency Madam Princess Catherine Romanovna Dashkova, *Dame d'Honneur* and *Chevalier* of the Order of Saint Catherine.

The beginning of the Princess's tenure was promising: According to her autobiography [31], her introduction at the Academy was a memorable event. She had personally persuaded Leonhard Euler, who for almost a decade had avoided the conferences and all the disputes under Domashnev's leadership, to introduce her to the academicians. After her initial presentation, the academicians took their seats. As Minister von Stählin habitually seated himself next to the Director, she pronounced to Euler: "Please be seated anywhere, and the chair you choose will naturally be the seat of honour". This manoeuvre pleased everyone present, except of course the complacent von Stählin.[4]

The reforms initiated by the new Director started immediately. First, a Latin translation of the title of the Director was to be decided. This was followed by deliberations concerning a proposal for the uniform for the academicians.[5] On 21 (10) February 1783, the academicians finally received their medals commemorating the inauguration of the equestrian statue of Peter the Great in August 1782 (see Sect. 7.1). The grand medal was bestowed upon Leonhard Euler, the other academicians received a medium sized medal. The Director also announced that she, on an imperial order, had allowed some books left by Domashnev to be burnt, as they had been found to contain "indecencies". At the same time, Princess Dashkova was engaged in the preparations for the establishment of a new important institution for the study of Russian language and literature, the Russian Academy, of which she became the first President [31] (Fig. 10.2).

---

[3]Vasily Fedorovich Zuev (Зуев; 1754–1794), Pallas's Adjunct of Natural History [119].

[4]The elderly Jakob von Stählin (born in 1709) was nevertheless one of the Academy's most frequently participating members.

[5]The following was decided as a uniform of an academician: "Cloth of deep violet, lined with a sequined material. The front and the cape of light blue velvet, the jacket and the breeches of sequined woolen cloth, with gilded buttons made of brass."

Fig. 10.2 Director of the Imperial Academy of Sciences of Saint Petersburg 1783–1794, Princess Catherine Dashkova, née Vorontsova, after a contemporary engraving. The Dashkov and Vorontsov families were among the most influential in imperial Russian politics (Daschkoff [31])

A public assembly of the Imperial Academy of Sciences was organised on Saturday, 22 (11) March 1783, beginning at noon by the adjudication of the prize in botany. This was followed by lectures given by the academicians Ivan Lepyokhin and Lexell, the first one in Russian and the second in French. Lexell's lecture was entitled *Recherches sur la nouvelle planète, découverte par Mr. Herschel et nommée Georgium Sidus*. The Academy decided this talk would be printed and distributed separately [Lexell 82], as it was obviously regarded as a major achievement.

At the ordinary conference of the Academy held on 24 (13) March 1783, Lexell presented on behalf of his Finnish student Martin Platzman[6] a mathematical study entitled *Problemata nonnulla de lineis curvis* as a specimen of his skills for the Academy's consideration [141, p. 660].

At the conference session held on 18 (7) April 1783 [141, p. 666], Lexell read an extract from a letter of Joseph Banks to Director Dashkova. According to the letter, Herschel had proposed at a meeting of the Royal Society of London that our planetary system probably is in motion towards the constellation of *Hercules*

---

[6]Martin Platzman (1760–1786) (often spelled Platzmann) was born in Muolaa, Carelian Isthmus, Finland (now Pravdino, Russia) and studied mathematical subjects in Åbo [81]. In 1777, after the death of his father, he applied for a position at the Petersburg Academy of Sciences; he was accepted as Lexell's student starting from the *Gymnasium*. In 1784, having shown a disposition to mathematics, he was appointed as Lexell's Adjunct. A year earlier he had already been given the responsibility for editing the German calendar for Saint Petersburg. However, due to a bad temper and contemptuous behaviour, he came increasingly into conflict with his mentor Lexell and his colleagues at the Academy (cf. Chap. 11).

in much the same way as the diurnal motion of the Earth. As we now understand, Herschel was in the course of discovering the motion of the Milky Way.

At the two following conferences Lexell delivered a number of astronomical papers:

— *Recherches sur la nouvelle planète, découverte par Mr Herschel et nommée Georgium Sidus* [Lexell 82], regarding the new planet,
— *Essai sur l'orbite elliptique de la comète observée en 1763* [Lexell 85], or an attempt to compute the elliptic orbit and the period of the comet discovered by Messier in 1763 (Comet "C/1763 S1"),
— *Solutio problematis geometrici in Actis Academiae Scientiarum Berolinensis pro anno 1776 a Cel. Castillon propositi* [Lexell 84], containing a response to the challenge of Castillon (see Sect. 8.3.5),
— *Solution d'une question astronomique* [Lexell 90], containing an analysis of the question which of the two fringes of the Moon, the upper one or the lower one, should be observed when the Moon passes through the meridian, and
— *Mémoire sur la comète observée en 1780*, a work which apparently was not published.

In a report to Catherine II, Director Dashkova also mentions that she had ordered academician Lexell to supervise the updating of the *Globe of Gottorf* with the most recent discoveries in geography and astronomy [104]. After the death of Lexell this commission was passed on to his astronomer colleague Stepan Rumovski and then to Friedrich Theodor Schubert.

However, under the placid surface a new conflict was brewing between the Director and the Academy. The crucial decision which seemingly unfolded the crisis was her dismissal of Pallas's Adjunct Vasily Zuyev, which did not seem entirely fair. At the conference session of 29 (18) March 1784, when Pallas had read out his response to the action, Princess Dashkova announced that she was seeking a vote of confidence by the Academy in respect of her position:

> [...] The Secretary collected the votes and all but two united in responding that they did not have the least thing to complain about Director Dashkova, whose directorship was as advantageous for them personally as for the progress of the sciences in general; that they on many occasions had reassured her of this, and consequently they requested Her Excellency not to doubt the sincerity of their sentiments. Only Academicians Pallas and Lexell were of a different opinion, which they promised to deliver in writing.
> [...] The latter limited himself to saying that he had many reasons to be dissatisfied with Madam the Princess, and if Her Excellency insisted, he would deliver his statement of objections in writing. Pallas confined himself mainly to the judgement against Zuyev and pretended that it was not only he who was dissatisfied with the Director's procedure [...]

Apparently, Lexell's discontentment concerned only his salary. At the academic conference on 2 April (22 March o.s.) 1784 [141, p. 734], the Secretary read a letter from the Director with the following content:

> Monsieur !
> I ask you to express to your colleague academicians how flattered I am to find in the Minutes of the meeting on Monday that, after the 14 months I have had the honour of presiding over the Academy, all academicians apart from Mr Pallas have proved to be content with the way

I have performed my duty. I say except Mr Pallas, because the discontent of Mr Lexell, of which he has informed me, was due to the fact that I have not accorded him an increase in salary, although he had never mentioned it before, despite having had (especially in view of the distinguished and cordial manner in which I have treated him) at least twenty times the opportunity to say a word.

Signed Princess Dashkova

According to [141, p. 735], the academicians and the adjuncts listened to the reading of the letter with feelings of the most perfect admiration and gratitude, to which Lexell, grateful for the kindness shown by the Director, read and inserted in this record the following confession:

Regarding the declaration of Academician Lexell in the Minutes of the Academy of 18 March, he believes it to be his duty to inform the academicians that having made his statement of complaint to Her Excellency the Princess, Director of the Academy, she has deigned to consent to his wishes in the most graceful manner, and he is forced to admit that due to a misunderstanding, he has chosen a less than suitable way of fulfilling his wishes.
21 March 1784 A. J. Lexell

It is clear from the formulation that Lexell regretted his tactless behaviour and the measures he had been forced to take against the Director, and indicated that it was due to a misunderstanding from his part. Lexell received his increase in salary, and peace was restored. At least so it appears. In a letter to Archbishop Mennander dated 23 April 1784 Henrik Gabriel Porthan in Åbo quotes the latest news from Lexell. Porthan writes that [132]

[...] their Madam Dashkova rules the Academy just as despotically as Domashnev; therefore they have once again declared war on their Director, who has tried to dismiss (without due inquiry and judgement) a certain Adjunct Zuyev as authoritatively as the previous Director tried to dismiss Kotelnikov from the supervision of the Natural Cabinet. She tries in particular to get at Pallas, who was on good terms with Domashnev, and who is now fighting against the same power of the Director, whom he had supported at the time. Zuyev is his adjunct and student. But thanks to the Empress's decision, the Academy has prevailed once again. I am really curious as to how Lexell dares to be as audacious as he has been again. But the result was that he was transferred to the mathematical class, after old Euler, with an increase of 300 Roubles in his salary, so that he now has 1,300 Roubles, in addition to his living quarters, light and heating. He is not entirely satisfied with the characters of the majority of his colleagues, but at least he is happy to have Ferber[7] as an honest compatriot and aid.

As a result of the crisis, it was not only Lexell's salary which was raised, but also his esteem and authority among the academicians of Saint Petersburg.

[7]The Swedish mineralogist Johan Jacob Ferber (1743–1790), a disciple of Linnaeus, had been invited to the Petersburg Academy of Sciences in 1783. In 1786, he left Saint Petersburg to join the Berlin Academy of Sciences.

Chapter 11
Standing on Euler's Shoulders?

11.1 Euler's Death in Lexell's Words

In the November issue of *The London Magazine or Gentleman's Monthly Intelligencer* of 1783, the following letter by Lexell dated 30 September 1783 and addressed to J. H. de Magellan, was published (translated into English probably by the Editor of the Magazine) [Lexell 100].

> My Dear Friend
> When I wrote to you last, I little thought it would so soon have fallen to my lot, to have announced to you the melancholy news of the death of our great and incomparable Euler. On the 16th of September he found himself much indisposed, and was taken with a giddiness in his head. On the 18th, at four o'clock in the afternoon, he was struck with an apoplectic stroke, which on a sudden deprived him of his senses. He lay until eleven o'clock the same evening, when he died. He retained all that presence of mind, and solidity of judgement, so natural to him, until the fatal moment that he was seised, as you will see by the conversation I had with him on the day of his death; and which I have the honour to send you in this letter.
> The life of our incomparable Euler had been one continual scene of the most sublime researches into every part of mathematics: even during the last days of his life, when the dizziness of his head prevented him from making calculations, his mind did not cease from being occupied in meditating on different subjects, and even the most delicate parts of mathematics, as I myself have been witness, in the conversation I had with this excellent mathematician a few hours before he was seised with the fatal stroke that put an end to a life so useful and so glorious to humanity. And as the last moments of the existence of great men do not fail to excite curiosity, I am persuaded that the recital of what passed in our conversation will give pleasure to those who knew the great merits of the deceased.
> After speaking concerning the state of his health, he began a conversation, by asking me if I had read the pieces which have been given in, relating to the astronomical question concerning the diurnal motion of the Earth[1]: and when I told him some things concerning

[1]The prize competition of the Petersburg Academy of Sciences for 1783 concerned the diurnal motion of the Earth; whether it is uniform or slows down, and in the latter case, for what reason.

J.C.-E. Stén, *A Comet of the Enlightenment*, Vita Mathematica 17,
DOI 10.1007/978-3-319-00618-5_11,

these memoirs, he assured me he was persuaded that the only circumstance capable of producing any change in the rotation of the Earth was the resistance of the ether, and as the effect of this resistance would lengthen the time of the diurnal revolution, it would consequently shorten the length of the year; and in comparing the ancient observation concerning the length of the year with the modern ones, he believed it might be nearly discovered if there was really any change in the duration of the diurnal revolution: for, if the duration of the diurnal rotation had suffered any alteration, it must appear in making these comparisons. When I observed to him, that much dependence could not be placed on the observations of the ancients, he replied, that from some ancient observations [Tobias] Mayer had found a secular equation of the Moon's motion with sufficient certainty.

Saying afterwards that he had understood that the trials I had made with Mr de Magellan's instrument, invented to measure the distance of the Moon from the fixed stars, &c. had been sufficiently correct, he desired a description of the construction of that instrument (the circular instrument) and asked what were the principal advantages to be derived from it, which gave him occasion to make some reflections on the use of instruments employed at sea.

I then spoke to him of the method of combining eye glasses[2] in a telescope, practiced by Mr Herschel; of which Mr de Magellan had sent me an account: he was very desirous to learn what effect these eye glasses had; and if, by magnifying three or six thousand times, it would not be impossible to discern anything distinctly for want of light.

Talking afterwards upon the principles on which the aerostatic globes [hot-air balloons] are constructed, he remarked that it was a curious mathematical problem to determine the motion of such a globe, from knowing the proportion between the density of the air contained in the globe and of the common air. He observed also, that supposing this proportion to be as *one* to *two*, the greatest velocity of the globe would be 41 feet in a second.[3]

During the time we were at table, he discoursed of the new planet discovered by Mr Herschel, and enquired if anybody had yet constructed tables of its motion.

Thus did the greatest and most illustrious mathematician of our age finish his course; having preserved, until the moment that he was struck with the apoplexy that terminated his existence, that strength of mind, and solidity of judgement which had always been so conspicuous in him; even his last moments were not unworthy of a life so illustrious and glorious! He has left a prodigious quantity of works, not yet printed, which the Imperial Academy of Sciences, at Petersburgh, mean to insert in their annual publications.

(signed) A. Lexell

This letter was followed in the same issue of the *London Magazine* by a letter from J. A. Euler to J. H. de Magellan, giving some more details of the events surrounding Euler's death. These descriptions, or summaries thereof, were presumably used in the often quoted description of the events in the Marquis de Condorcet's *Éloge de M. Euler* published in the *Histoire de l'Académie Royale des Sciences* pour l'année 1783 (1786) [175, Pt. I].

---

[2]The original French word was possibly 'lunette', lens, or 'verre oculaire', eyepiece. However, today the word 'eye glass' might be misunderstood as meaning spectacles.

[3]Footnote in the original: "He had, in the morning of the day on which he died, made calculations concerning the motion of the aerostatic globe, which a friend committed to writing". The text it refers to has been published by N. Fuss: "Sur les ballons aérostatiques, faits par le feu M. Léonard Euler, tels qu'on les a trouvés sur son ardoise, après sa mort arrivée le 7 Septembre 1783", *Mémoires de l'Académie Royale des Sciences*, 1784, pp. 264–268 (OO Ser. II, Vol. 16. E579).

Less than two weeks later, Lexell broke the sad news to Wargentin, who himself in his recent letters had complained about serious problems with his health (judging by Lexell's replies). In the letter, which turns out to be his last one to Wargentin, dated 10 October 1783, Lexell writes:

> The 18th (7th) of September the Academy of Petersburg had the misfortune to lose the greatest jewel in its crown, through the death of Euler, which took place after an cerebral stroke, which in the course of a few hours ended a life which has been so useful to humanity. Only two days before his death he had been troubled by giddiness, but nonetheless in those moments he felt himself clear in the head, attempted to calculate, and on the morning of the very day he died, he had written down on his slate some calculations of the motion of the aerostatic globe. It was a peculiar presentiment that drove me to visit him on that day, and I can hardly describe how moved I was, when he described his condition to me; that he had now lost all his eyesight and that he could no longer sense where he was, so that he was now a complete stranger in his own house. Notwithstanding he had retained the firmness of his mind, so that he talked to me about some mathematical subjects, on which discussion I have made a short essay, which I will have the honour to send by the next post. He dined with a rather good appetite, but when he after dinner smoked his pipe as usual, he started to feel faint and had to lie down, after which he slept well for a couple of hours. When he woke up, he came out to drink some tea. Having drunk a cup, he asked his wife if this was his first or second cup, to which she replied that it was the first, and he said: then I would like to drink another cup. No more than two minutes later he suffered a stroke, which in approximately one minute deprived him of all senses and consciousness, because while he still breathed and the pulse could be perceived until 11 that evening (it was 4 o'clock when he suffered the stroke), it was purely mechanical. All attempts to revive him, by cutting the veins, using enemas, Spanish flies, were useless. Although it has been three weeks since this happened, I cannot think about it without emotions, and I have difficulty believing that it really has happened. [...]
> I am with the most hearty affection [...][4]

[4]KVAC, Bergianska brevs. Vol. 20, pp. 34–35. "[...] Den $\frac{7}{18}$ september hade Petersburgska Academien den olyckan at förlora sin största prydnad, genom Eulers död, som timade efter en slag fluss, hwilken innom några timmar slutade et lif, som för menskligheten warit så nyttigt. Han hade allenast twänne dagar för sin död warit beswärad af swindel, men likafullt wid de ögnablick då han tyckte sig hafwa någorlunda klart hufwud, budit til at räkna, och hade han ännu om morgonen af den dagen han dog, uppå sin swarta tafla uptecknat några räkningar öfwer den aerostatiska kulans rörelse. Det war en besynnerlig aning som dref mig at besöka honom denna dag, och kan jag ei beskrifwa huru mycket jag blef rörd, då han berättade om sit tilstånd, at han nu mera förlorat synen aldeles och al sansning af hwar han war, så at han fann sig helt främmande i sit eget hus. Men likafullt hade han behållit al sin sinnes styrka, så at han talte med mig om åtskilliga mathematiska ämnen, öfwer hwilket samtal jag giort en liten upsats, den jag nästa påst skal hafwa den äran at öfwersända. Til middagen åt han ännu med tämmelig god appetit, men då han efter wanligheten efter måltiden rökte sin pipa, blef han swag och måste lägga sig, hwarefter han några timmar sof rätt sött. Sedan han waknat, kom han åter ut för at dricka thé. Sedan han druckit en kopp frågade han sin fru, om det war första eller andra koppen, och då hon swarade den första, sade han: så wil jag ännu dricka en til. Högst twå minuter derefter blef han rörd af slag, som ungefär innom en minut betog honom al känsla och sansning, ty ehuruwäl han ännu andades och pulsen gick til kl. 11 samma afton (klockan war 4 då han blef slagrörd) war det blott mecaniskt. Alla försök med åderns öpnande, clystirer, spanska flugor woro fruktlöse. Ehuru tre weckor redan äro förflutne sedan denna händelse, kan jag likwäl aldrig tänka derpå utan rörelse, och tycker jag mig ei rätt wilja förmås at tro, det så händt. Jag är med hiertelig tilgifwenhet [...]"

Wargentin passed away on 13 December 1783 and so these last scientific meditations of Euler were probably never communicated to Wargentin, perhaps not even written down. Thus, Lexell had lost two of his closest friends and scientific idols within a few months' time. It is clear, from the more frank letter to Wargentin, that he was still in a kind of shock from Euler's death, unwilling to believe what had happened.

However, life at the Academy went on as before, or so it seems. On 29 September 1783, according to the minutes of the conference [141, Vol. III, p. 698]:

> Professor Lexell read a letter from Mr de Magellan, dated in London on 29 August, communicating a new curious experiment and an announcement of new kitchens invented in London for ships; they are of iron and make soups, puddings and steaks, they also have two ovens for making bread and pastry and making fresh water out of salt water, and finally they circulate the air in the cargo deck and between the decks.

Further, on 9 October [p. 701]

> Her Excellency Madam Princess de Dashkova sent [...] by order of Her Majesty the Empress the new machine for engraving books, invented by the Abbé Rochon; it was exhibited at the Conference and then given to Professor Lexell, who is responsible for studying its use and to instruct the mechanic [...] whereupon the machine will be sent back to the Court.
> Professor Lexell returned to be sent to the Department of Geography a new map of the United States of North America printed in London.

And then, on 13 October, the anticipated decision came [p. 703]:

> Her Excellency Madam Princess the Director proposed Professor Lexell to replace the late Leonhard Euler in his function as Professor of Sublime Mathematics. The conference applauded the choice and gave it unanimous consent.

At the solemn assembly of the Academy on 3 November (23 October o.s.) 1783, Princess Dashkova and all the academicians were dressed in mourning. In the presence of the Archbishop of Mohilev,[5] academician Fuss read his eulogy to the deceased followed by a deep silence, "which more than the strongest words of rhetoric expressed how much and how widely the deceased was mourned". The feeling that pervaded the Academy is perceptible in Fig. 11.1, an image from 1784 which shows the "orphaned" academicians beside Euler's memorial monument.

11.2 Lexell's Unexpected Death

Given the lack of any personal letters from Lexell from the last year in Saint Petersburg it is difficult to assess how he and the Academy as a whole adapted to the new situation after Euler's death. However, from the activities at the academic

[5]Princess Dashkova's estate was situated near Mohilev in today's Belarus.

Fig. 11.1 Johann Friedrich Anthing's silhouette of the academicians of the Imperial Academy of Sciences of Saint Petersburg in 1784, which used to hang on the wall of the Secretary's office. From the left: Lexell, Fuss, Johann Albrecht Euler (standing by the memorial to his father), Pallas, Lepyokhin, Georgi and W. L. Krafft (Photo: Nikolai Nylander, Tallinn)

conferences it can be judged that at least Lexell had managed to maintain his capacity to work. His last scientific article [Lexell 105], presented at the conference in May 1784, concerned a remarkable theorem stated by Johann Heinrich Lambert in his book on orbit calculation (1761), namely how to determine the orbit of a comet, when its position is given at two moments in time.

All along, however, the minutes of the conferences are testimony to increasing tensions which must have hampered the working of the organisation. Besides making clear his lack of confidence in Princess Dashkova's leadership, as mentioned in Chap. 10, Lexell made an official complaint about the typographical quality of the Academy's publications, in particular regarding certain fonts for mathematical symbols [141, p. 764]. His Adjunct Martin Platzman also turned out to be a disappointment for the Academy; at the meeting of the conference on 27 September (o.s.) 1784, Lexell criticised Platzman in public for neglecting to communicate to him a dissertation he had written before presenting it to the Academy [141, p. 766]. Director Dashkova, surprised and furious about the news of yet another quarrel at the "sanctuary of science", wrote a serious letter [141, p. 773], which was read at the academic conference on 4 October (o.s.) 1784, reproaching not only Platzman but also Lexell for their actions in the matter.[6] How Lexell took the Director's criticism is unknown.

On 5 November (25 October o.s.) 1784, Lexell participated for the last time in the academic conference (Fig. 11.1). A month later, on 11 December 1784, as by a twist of fate, he passed away quite unexpectedly. He had been operated on for a tumour, but fell victim to complications, probably due to an infection causing severe fever. Two days after his death, on 13 December 1784 [141, Vol. III, p. 786], the Secretary of the Academy Johann Albrecht Euler

> ... announced with sorrow the death of Mr André Jean Lexell, ordinary member of the Academy, Professor of Mathematics, a distinguished celebrity, most worthy of being mourned by the Academy and by all who knew him. The deceased ended his career on the morning of 30 November (o.s.), after having been bedridden for only a few weeks. Her Excellency the Princess Dashkova testified at the Conference that she was deeply touched by this loss, and a profound silence was testimony to how everyone was struck with grief.

At the same conference session it was decided to charge Krafft with examining Lexell's manuscripts and scientific correspondence and Rumovski with ascertaining that all the instruments of the Academy were recovered from his apartment in the Academy building.

For the Swedish-Finnish community in Saint Petersburg, the loss of Lexell was also considerable. In the Book of Deceased (*Dödbok*) of Saint Catherine's congregation, the note of his death was supplemented with a short entry [86]: "A man of profound insights, a good Christian and an honest citizen. His departure is regretted with tears."[7]

Not long after Lexell's death, in a letter written in Åbo on 25 February 1785, Lexell's schoolmate and friend, Henrik Gabriel Porthan reports the sad news to Archbishop Mennander in Uppsala:

> Professor Lexell's death was not only a loss to the sciences, but also quite a considerable one to me personally. He was a hypochondriac, but an entirely honest man with a Finnish

---

[6]In [124], the Finnish mathematician P. J. Myrberg quotes a short biographical sketch of Platzman in Russian, saying that late in 1784, Platzman had presented to the Academy a paper entitled *Nova methodus integralia per series evolvendi*, without due approval by Lexell. On the following academic conference, held on 27 September (o.s.) 1784, Lexell criticised Platzman for his negligence, also pointing out the errors that needed to be corrected before publication. Princess Dashkova for her part reproached Platzman for being presumptuous towards his superior Lexell. After Lexell had passed away in November, Platzman formulated a response where he doubted the opinion of an "illustrious mathematician" (Lexell) and requested the judgement of the mathematical section of the Academy. Once a critical reply on the paper had been delivered by Nikolaus Fuss at a conference in January 1785, Platzman made a statement including some demeaning remarks on the academicians. Having written several defamatory critiques of Fuss's works he was cautioned. He died in Saint Petersburg in December 1786 after an illness of several months.

[7]En man af stora insigter, god Christen och rättskaffens medborgare. Hans förlust beklagas med tårar.

Fig. 11.2 The clock which was presumably the one Lexell employed in his astronomical observations in Saint Petersburg. The clock was designed by J. H. de Magellan (the inscription on the dial reads: "J. H. de Magellan invenit & fieri curavit Londini") and manufactured at the earliest in 1774 by Michael Ranger in Marylebone [163]. Unfortunately, the clock has been damaged and is not working. Museum of the University of Helsinki (Photo: the author)

> mindset.[8] First, a tumour which had to be removed deprived him of his strength. This was followed by a fever. Professor Ferber as his friend has written to us about the circumstances surrounding his illness and death; in his last letter, which was almost a formal goodbye, he advised his relatives to address their affairs concerning him to Ferber.

At the academic conference on 13 December 1784 [141, Vol. III, p. 788] it was decided that Lexell's treatise on the biharmonic wave equation governing the vibrations of a flexible strip in the form of a circular annulus [Lexell 92] was to be published in the Academy's proceedings, the *Acta Academiae Scientiarum Imperialis Petropolitanae*. Moreover, Ferber, who was commissioned to take care of Lexell's correspondence, "gave the Secretary a letter addressed to Lexell by Mr Arnold written in London on 22 November. Mr Arnold tells that the four timepieces

---

[8] [. . .] hypochondriacus, men en ganska redelig man med Finskt tänkesätt.

ordered from him have been completed, and that he still wishes to examine their working [. . .]." On 10 January 1785 [141, p. 790], Ferber delivered 40 papers on mathematics and physics, which Krafft had chosen among Lexell's manuscripts and which he considered worth preserving at the Academy. From this, the Secretary initially separated those which had been in the Archives, from the remainder, which was found to contain several papers, calculations and observations, which in part had already been published and presented at the conferences. On 24 January 1784 [141, p. 794], Rumovski delivered a report on the state of the astronomical instruments Lexell had received from the Academy. Having compared them with a list he found nothing missing, except a copper cap belonging to a 3 ft achromatic refractor, some keys of different cases, and an astronomical clock by le Paute,[9] which was meant for Golovin [. . .] In exchange, at the Observatory there was a clock marked Magellan belonging to Lexell, which was given to the care of Ferber. The University of Helsinki owns a clock attributed to J. H. de Magellan, which possibly is the pendulum clock which after Lexell's death was delivered from Saint Petersburg to Åbo by J. J. Ferber [14, pp. 36–38] (see Fig. 11.2). All the instruments were found to be in good condition, except the Gregorian telescope made by Short,[10] "whose mirrors were damaged by rust, an inevitable effect of humidity and temperature inconsistency especially during the winter, and for which Mr Rumovski has no other remedy than to build another observatory which can be heated and constantly maintained at an even temperature".

These are the final notes concerning Lexell in the minutes of the conferences of the Petersburg Academy of Sciences [141].

[9]Company of Jean-André Lepaute, French royal clockmaker.

[10]James Short (1710–1768), Scottish-born instrument maker [139].

Chapter 12
A Sketch of Lexell's Personality

In this book, different traits of Lexell's personality have come to light. We have seen how Lexell was regarded by his contemporaries, but also how he judged himself and how he *characterised himself* to others. All these assessments are subjective and influenced by the personality of the observer as well as the person described. The main sources for the present assessment of Lexell's personality are his letters and the contents of his library [47].

Lexell's life can be divided roughly into three periods: (1) his Finnish-Swedish formative years, (2) his professional career in Saint Petersburg, and (3) his great journey to Europe and his final mature years in Saint Petersburg. These periods differ from each other in many ways. The time in Åbo was dominated by studies in humble but familiar surroundings and a growing ambition to accomplish something significant in the mathematical sciences. The second period was, especially at the beginning, a fulfilment of Lexell's dreams: to be accepted by and allowed to work alongside one of the greatest heroes of mathematics and to profit from the stimulating intellectual atmosphere at one of the world's most prestigious research institutions of the time, the Imperial Academy of Sciences. On the other hand, he was increasingly concerned about the future of his career: should he occupy the Chair of Mathematics which had been offered to him in his *Alma Mater* in Åbo, Finland, or should he give priority to the stimulating, but at the same time exacting, environment in Saint Petersburg? Questions of loyalty and honesty concerned Lexell deeply.

The "Academic Journey" in 1780–1781 put a stop to all these speculations. He decided to stay in Saint Petersburg for good. Not only had the journey improved his own self-esteem; but the esteem he enjoyed among his colleagues also seems to have increased. In the course of events he became more audacious in that he openly contested the directors of the Academy, and that in a way which seems discourteous and unwise. On the other hand, in his weakest moments, especially when he compared himself to his super-prolific Master, Leonhard Euler, he felt inadequate and played down his own importance. In comparison with his mathematician

J.C.-E. Stén, *A Comet of the Enlightenment*, Vita Mathematica 17,
DOI 10.1007/978-3-319-00618-5_12,

colleagues in Sweden at the time, he was undoubtedly the most celebrated scientist in his field, but in Saint Petersburg and the rest of Europe, he realised that he enjoyed a somewhat lesser status, namely as one among a number of peers. At the same time, one must remember the unique conditions in which Lexell lived and worked: a backward, autocratic Russia of the eighteenth century, the Imperial Academy of Sciences consisting largely of learned foreigners, and among few friends and compatriots. To survive in such conditions, Lexell searched for guidance and support from his closest friend in Sweden, Pehr Wargentin.

12.1 A Melancholic Disposition

A feature of Lexell's personality, particularly noticeable in his more intimate letters to Wargentin, is his susceptible and introspective character. In fact, a conspicuous number of contemporary Swedish scientists mentioned in this book (e.g. Carl Linnaeus, Johan Peter Falck, Fredrik Mallet and Martin Johan Wallenius) seem to have been troubled spirits. This may have been due to various personal circumstances and consequently a mere coincidence or, alternatively, a more serious symptom of the troubled frame of mind of the nation.

If the latter were the case, there were good external reasons for it: First of all, there had been two wars lost in the first half of the eighteenth century, with ensuing poverty and deprivation. Social problems, such as vagrancy, alcoholism and child labour were common, and incurable diseases and premature death were an everyday reality. Scientists, presumably more susceptible than the majority of the population living in simple, rural conditions, were perhaps more easily subject to melancholy and hypochondria. We know, for instance, that Lexell's private library contained a number of books on medicine and the description of illnesses [47].

The greatest interpreter of the eighteenth-century Swedish mindset is without a doubt the celebrated poet and song-writer Carl Michael Bellman (1740–1795), whose poetry combines the great joys of the senses with their fragility in the face of imminent death [6]. An example of this dichotomy in Bellman's poetry is given by song number 30 of his *Fredman's Epistles*, "To Father Movitz, during his sickness, elegy", whose first verse reads:

> Drick ur ditt glas,
> se döden på dig väntar,
> slipar sitt svärd
> och vid din tröskel står.
> Bliv ej förskräckt,
> han blott på gravdörrn gläntar,
> slår den igen kanske än på ett år.
> Movitz din lungsot den drar dig i graven.
> Knäpp nu oktaven;

Stäm dina strängar,
sjung om livets vår![1]

In Sweden, the eighteenth century was a time of discord and conflict. The "Era of Liberty", as the period 1718–1772 is called in Swedish-Finnish history, was not quite as free as the name may suggest. Freedom of religion, press and speech were limited and strictly supervised. Expensive imported products (e.g. coffee) were rationed for economic reasons, since the Crown was seriously in debt. While it is true that the King's role was at best only nominal and true power lay in the hands of the Secret Committee (*Sekreta Utskottet*), Swedish politics was dominated by ruthless struggles between the "Hats" and the "Caps", where bribes and corruption were rife. On the other hand, the Church's grip on society tightened, at least temporarily: the calamities of the wars with ensuing famine and plague were pictured as God's punishments for sins committed by the leaders. Religious dissenters, in particular pietists, were severely sanctioned by an edict (*Konventikelplakatet*, 1726), which banned religious sects and enforced religious orthodoxy, but the attitude was to soften towards the mid-eighteenth century.

As to the state of learning in Sweden, there were only a few university positions available in the sciences, and they were usually poorly salaried. The Gustavian era (1771–1792) did not improve the situation: with the lack of resources, Swedish scientists started to realise that, in spite of the great promises of prosperity and power which the sciences and crafts were supposed to entail, not much had been gained. Instead, the circumstances created a good breeding ground for superstition, mysticism and fanaticism.[2] In short, as far as the sciences are concerned the general outlook was far from promising, and from this point of view the mere existence of a "Swedish Enlightenment" may be questioned (as it indeed has been, cf. [51]).

As to Lexell's personality, his correspondence demonstrates his earnestness and his total commitment to his scientific mission. Although he has a penchant for telling disparaging anecdotes about his adversaries, he rarely makes up stories for mere amusement, but often ironically reproaches others for not taking him seriously. Sarcasm seems to have been the only kind of humour Lexell knew. In a moment of hesitant reflection in 1780, agonising over the more or less unpleasant alternatives for his future career, he confesses to Wargentin (letter dated 31 March 1780, KVAC):

> My knowledge as well as the services I have rendered to this Academy are so mediocre, that I in no way have claimed to be indispensable, or even been able to make myself indispensable here, which is why I have been silent about my resignation until the moment

---

[1]Drain off your glass! See death upon you waiting, sharpening his sword, standing on your threshold. Do not be frightened: he only opens the door of the tomb, shuts it again maybe for a year. Movitz, your consumption will be the death of you. Pluck the octave; tune your chords, sing of the springtide of life!

[2]In this atmosphere of despair, the Swedish poet and journalist Johan Henrik Kellgren (1751–1795) wrote a satirical poem "The enemies of light" (*Ljusets fiender*; 1792) about the superstitious mob that in his eyes threatened the ideals of the Enlightenment. Kellgren's worst fears became reality after the murder of Gustav III in 1792, which was followed by a brief reign of dictatorship marked by secret societies, obscurantism and terror.

> when I have taken my decision to leave the Academy and I could not stay here any longer, if I was to be regarded as an honest man in my Fatherland.
> It is sad for me that this event may bring upon me a reputation for being irresolute and inconstant, but then I console myself by thinking that if I make a mistake in this affair, my error will at least be smaller than the one I committed in not resigning long ago. A man of melancholic temperament causes himself problems where there are none, and so it has been in this matter.

On the other hand, Lexell seems to have formed a fairly accurate idea of his talent and his insights in comparison with his contemporary colleagues. As the honest, frank and at the same time extremely sensitive personality he was, he was prone to dismiss the slightest signs of flattery or exaggerated praise. He resented people with ulterior motives or false conceptions, as he explains to Johann III Bernoulli (17 June 1774, UBB):

> Maybe you are right, Monsieur, to conclude that I am a man filled with truth and that I am very eager for praise, also a passage in your last letter made me believe that you have formed this disapproving opinion of me since you have inserted there the praise for the Theory of the Moon published in the *Göttingische Anzeigen von gelehrten Sachen* as if it would affect me considerably. But let me tell you, Monsieur, that I am not quite so weak on this matter. I am eager for praise, but only from those great men that I think of being much greater than myself, because I am convinced I have some reason for self-esteem, when I see that I have deserved the praise and approbation of men which I am sure surpass me much in knowledge and experience. So I was extremely flattered by the praise I received from Mr Euler, your Uncle[3] [Daniel Bernoulli] and Mr La Grange, of which the latter expressed himself quite favourably about me in a letter to Mr Euler.[4] In general I like to be praised when I think it is coming from a man who can appreciate what he praises, but if not, you can be sure that I would be the first to mock the compliments made to me. *Voilà, mon caractère Monsieur, tracé au naturel.*[5]

---

[3]In the correspondence between Daniel Bernoulli and the Eulers (to appear in: OO Ser. IVA, Vol. 3, eds. Emil Fellmann and G.K. Mikhailov, quoted here by permission of the Bernoulli-Euler-Zentrum, Basel), we find:

— Johann Albrecht Euler to Daniel Bernoulli (St. Petersburg, 14 (3) May 1776): "...Mr Lexell est Suédois, natif d'Abo en Finlande: il peut avoir tout au plus quarante ans, et c'est lui qui a fait dans nos commentaires les extraits [de] Vos mémoires."
— Daniel Bernoulli to J. A. Euler (Basel, 6 July 1776) "...Je suis charmé de connoître un peu plus particulièrement Mr Lexell; la Suède a toujours eté féconde en grands hommes et Mr Lexell soutient, on ne peut pas mieux, la gloire de sa Patrie. [...]"

[4]In Lagrange's scientific correspondence with Laplace and Euler ([85, Volumes XIII & XIV] and [76]) Lexell is frequently mentioned and praised. A letter by Lexell dated 5 March 1772, is also included (Vol. XIV, pp. 228–235).

[5]Vous avez peut être raison Monsieur de juger, que je sois un homme rempli de vérité et qui est fort avide des louanges, même un endroit de Votre dernière lettre m'a fait croire que Vous avez cette opinion désavantageuse de moi puisque Vous avez inséré les éloges qu'on à donné à la Théorie de la Lune dans les Gazettes Littéraires de Göttingue comme si cela m'affecteroit fort vivement. Mais permettez moi de Vous dire Monsieur, que je ne suis pas tout à fait si foible sur ce sujet. Je suis avide des louanges mais de ces grands hommes que je crois de beaucoup supérieurs à moi, puisque je me persuade d'avoir quelque raison de m'estimer moi-même, quand je vois que j'ai mérité les louanges et les adprobations des hommes, dont je suis persuadé qu'ils me surpassent beaucoup en

However, although Lexell was appreciated for his knowledge, talent and good manners, there is another side of his personality: he could be a rather hasty judge of character, which could manifest itself in unpleasant ways, as was the case with his compatriot Laxman, whom he slandered quite inappropriately. The sometimes exaggerated distrust he showed for instance towards his colleagues Lalande, Hell, Planman and Johann Bernoulli, which is revealed in his correspondence, seems to have been provoked by his own lack of self-confidence in difficult situations rather than by true ill-will or negligence on the part of his adversaries. Occasionally, his suspiciousness grew into pathological dimensions, perhaps bordering on paranoia. As his letters testify, he did not fear to engage in a battle when the cause seemed just. And when it came to it, he could certainly show a temperamental side. Afterwards, he often suffered from a guilty conscience with respect to his words and actions and regretted them deeply.

12.2 Religion and Ethics

Lexell's scientific ideals were embodied by his colleagues Leonhard Euler and Pehr Wargentin, whose perfect commitment and honesty in the pursuit of truth he never doubted. More important than finding the absolute truth was for him the fact that its discovery was guided by clear and reasonable logic.

Also on the personal side, the most important human traits Lexell appreciated, and expected, from his colleagues and friends were honesty and sincerity. For Lexell, these were essentially Christian virtues. In his opinion, God was the unadulterated source and basis of everything true and good,[6] and since God is the judge of what is just and appropriate for every man, no one can attempt anything he is unworthy of without incurring the anger of God. A wise man knows himself, the limits of his capacity, and what is appropriate in the eyes of God. Frequently, Lexell labelled as "charlatans" all those scientists who did not stick to solid facts and flawless reasoning.

In general terms, the traditionally strong ties between theology and jurisprudence gradually loosened during the seventeenth and eighteenth century. The foundation for human laws was no longer to be based solely on the Bible, but on rational natural principles. Notwithstanding, the belief in God's vengeance on those who have transgressed was still not uncommon in the eighteenth century: an unconcealed

sçavoir et en expérience. Ainsi j'ai été infiniment flatté des louanges que j'ai reçu de Mr Euler, Mr Votre Oncle et M^r^ la Grange, dont la dernier s'a exprimé fort avantageusement à mon sujet, dans une lettre adressée à Mr Euler. En général j'aime d'être loué, quand je crois que ces louanges viennent d'un homme qui sçait apprécier ce qu'il loue, mais si cela n'est pas, Vous pouvez être sûr que je serois le premier qui se moquera des complimens qu'on me fait.

[6]Cf. letter from Paris to J. A. Euler, dated 10 November 1780 (АРАН Фонд 1, опись 3, дело № 65, л. 109).

expression of this belief is Linnaeus's treatise *Nemesis divina*,[7] which is filled with deterrent moral examples and speculations. However, being the sceptical and rational man he was, it seems unlikely that Lexell would have embraced Linnaeus's or Johann Esaias Silberschlag's brand of natural theology (see Sect. 9.2). Rather, for him, religion and scientific knowledge seemed to be connected on a more subtle and indefinite level.

The fact that Lexell had been brought up next to the Lutheran cathedral in Åbo, where many of his teachers and friends were priests, notably Jacob Gadolin and Carl Fredrik Mennander, certainly influenced his view of religion and its relation to scientific knowledge. He was also influenced by the views and sentiments of Leonhard Euler, known as a devout Christian of the Reformed faith.[8] For Lexell it seemed utterly impossible to conceive the moral philosophy and incentives of an atheist, as he writes to Johann Albrecht Euler from Paris [35]: Why should an atheist strive to be good and decent, unless there is a reward for it (namely in the afterlife)? In his letters to Wargentin and J. A. Euler, Lexell describes in very lively terms how deeply he suffered when Lalande, apparently without being contradicted, declared to him his disbelief in the existence of God.

As a Lutheran, brought up in a country where the practice of other religions was prohibited by law, Lexell was certainly not insusceptible to the wide-spread spiteful rumours and prejudices regarding the Catholics and the Jesuits in particular. However, as was the case with the Jesuit Father Mayer, or with the "atheists" of Paris [35], Lexell did not allow these differences to prevent co-operation, when it was considered appropriate.

12.3 Political Views and Patriotism

Once Lexell's religious stance is known, it is not difficult to surmise his political viewpoint: God has bestowed on the monarchs of the world all the wisdom and necessary ability to rule their countries to the benefit of their nations. The fact that wars and misrule, seen as manifestations of God's anger, nevertheless existed, was not the fault of the Almighty, but that of the monarchs who defied the His Holy Word.

Like many of his learned compatriots, Lexell did not have a high opinion of the parliamentary rule in Sweden starting from 1719, when the two "parties",

[7] As a physico-theologian Linnaeus continually searched for meaning in God's Creation, which is visible e.g. in his dissertations *Œconomia naturæ* (1749) and *Politia naturæ* (1760). Linnaeus's "Divine retribution" (*Nemesis divina*) extended these thoughts to the human sphere, but due to its controversial nature, the book was published as late as the 1960s (a more recent English translation with commentaries is [102]). The book contains Linnaeus's famous dictum, *Innocue vivito, numen adest*—"Live unimpeachably, God is present".

[8] Euler's father, Paul Euler (1670–1745), was a minister at the village of Riehen near Basel [43].

the Caps and the Hats, bitterly strove for supremacy. These political factions were sworn enemies, unable to agree on practically anything: whilst the Hats advocated an assertive policy against Russia, a strong alliance with France and a rigid mercantilism, the Caps were for friendly relations with Russia and more liberal foreign trade. However, in Lexell's eyes, these endless struggles hampered wise rule of the country.[9] Nevertheless, the age of parliamentarianism also brought about many reforms which considerably promoted the economy, the sciences and industry. In general, the sciences and in particular the Royal Academy of Sciences gained from the rule of the Hats.

The Age of Liberty ended in 1772 by the *coup d'état* of Gustav III, by which monarchical autocracy was restituted in Sweden. Contrary to expectations, the sciences did not particularly benefit from the change of constitution and Gustav's reign. Whereas Gustav's mother, Queen Lovisa Ulrika (a sister of Frederick II) had already become familiar with scientists and philosophers at the Berlin court, Gustav's taste lay more in theatre and the arts [10]. However, initially Lexell saw Gustav's revolution as good news and so in a letter to Wargentin dated 7 September 1772, he writes enthusiastically:

> I cannot well enough describe the joy with which most of our fellow-countrymen at this place think about the change that has taken place in Sweden. For my part I am convinced that it was the only way to save Sweden from the total collapse which would sooner or later have overtaken her. God, who has endowed the King of Sweden with such outstanding qualities to undertake this great task, will also support Him in order to accomplish this job to the end, which I and every upright Swede wish from the heart. Every foreigner at this place thinks the same of this change.
> Perhaps I would not have dared to send this by post, but as the letter goes with Baron Taube,[10] I have wished to take the opportunity to assure you, Herr Secretary, that I am in my heart and thoughts entirely Swedish and that, with the grace of God, I hope to remain so, as long as a single warm drop of blood moves in my veins.[11]

This forceful declaration of patriotism may have been a response to those who, on the occasion of Lexell's move to Russia, might have doubted his loyalty to his King and country. In later years, Lexell would never return to the issue.

---

[9]For instance, in a letter to Wargentin dated 10 December 1771, Lexell comments regarding a nomination process which had been submitted to the Swedish Diets, that soon "...even the truth of two times two is four has to be considered doubtful, unless the Estates of the Swedish Nation can agree on it".

[10]Baron Evert Wilhelm Taube (1737–1799), Swedish nobleman and favourite of Gustav III.

[11]Jag kan ei beskrifwa med hwad fägnad de fleste af wåra Landsmän här å orten anse dem i Swerige föregångna förändringen. För egen del är jag öfwertygad, at det war det enda medel til Sweriges räddning ifrån det totala undergång som snart eller sent detta Land förestod. Gud som utrustat Swea Konung med så förträfflige egenskaper, at detta stora wärf företaga lär ock understöda Dess Krafter, at det fulkomligen til ända föra, hwilket jag och war redlig Swensk med mig af hiertat önskar. Alla utlänningar som här äro, tänka likaledes om denna förändring. Detta hade jag kanske med Påsten ei tordt skrifwa, men efter brefwet afgår med Baron Taube, har jag dock welat nyttia tilfället at försäkra Herr Secreteraren, at mit hierta och tankesätt ä[f]wen är aldeles Swenskt och at jag med Guds nåd hoppas deruti kunna fortfara, så länge någon warm blodsdroppa i mig röres.

Lexell's admiration of monarchs was by no means limited to Swedish royalty. Having visited the Royal Palace at Versailles in 1781 and witnessed the royal family dining, he delivers in a letter to Johann Albrecht Euler a long report on the physiognomies and appearances of the members of the French royal family (АРАН Фонд 1, опись 3, дело № 65, л. 175–181. Paris, 4 March 1781 — for the French original, see Appendix section "Letters"):

> The King [i.e. Louis XVI] and his two brothers have the traits of the family, that is to say, they resemble each other enough to see that they are related. Their appearance is also reminiscent of their father's, in particular *Monsieur* [the future Louis XVIII]. The King's physiognomy is slightly indolent and phlegmatic, he is quite strongly built and he sways when he walks. I saw him dining; he seemed to have a healthy appetite and to eat very heartily. It is said that he does not have wit, but until now he has been only beneficial for France. Being well informed, he has a lot of confidence in his ministers, who for the most part are decent people. The favourite pastimes of the Bourbon family are women and hunting; till now the King has loved no one else than his Spouse, and alas for France if he changes his thoughts. He loves hunting very much. Besides, he wants only what is best for his country, that is to say he has a love of saving money and therefore he has gracefully accepted all the reforms of Mr Necker,[12] which, if Mr Necker is spared, can make France the most fortunate country in the world. As to the character of *Monsieur*, sentiments are very much divided, some say that he has no wit, that he is superstitious and a bigot, and that he loves to drink; others have assured me that of the three brothers, it is he who is the wittiest, that he devotes himself to reading and to developing his talents. The *Comte d'Artois* [the future Charles X], although otherwise good-looking, has the defect of eyebrows meeting too closely. He is known to love the amusements, women and gambling. The Queen [Marie Antoinette] is very good-looking and resembles her brother the Emperor,[13] but she has not the same gloomy and forbidding look. She has a cheerful temperament, but it is said that she has somewhat liberal passions; at least for the sake of her compatriots she ought to restrain her passion for gambling, which has caused here much unhappiness among families. In the Parliament the question of prohibiting gambling with unequal odds has been discussed, and it was even said that a ban will soon be issued, but a few days later it was said that there had been gambling at the Court and that the Queen had lost a huge sum of money. Almost everywhere it is said that the Queen is expecting. *Madame* and the *Comtesse d'Artois* are rather like each other. Even if they are both good-looking, they do not have the same gracefulness as the Queen, they are both short in stature. Madame Élisabeth,[14] the King's sister has a very pretty face, the most beautiful nose, a beautiful mouth and very beautiful eyes; but she is very strong, she has almost no neck at all and she is extremely strongly built for her age. She does not make herself up like the Queen and her two sisters-in-law."

The Soviet-Russian historian of science Inna Lyubimenko characterised Lexell's implicit admiration of the royal family and their qualifications as expressions of his political naivety, and so it indeed may appear, if one ignores Lexell's convictions

[12]Jacques Necker (1732–1802), statesman and French finance minister. Necker's daughter Anne Louise Germaine married the Swedish ambassador Erik Magnus Staël von Holstein and became famous as an author under the name Madame de Staël.

[13]Emperor Joseph II visited St. Petersburg in July 1780 under the alias of "Count of Falkenstein". Thus, in principle, Lexell could have seen him at the Petersburg Academy of Sciences.

[14]Princess Élisabeth of France, sister of Louis XVI and the future kings Louis XVIII and Charles X. She was guillotined in 1794.

[106, 108]. In truth, few could foresee the imminent collapse of the *Ancien Régime* in 1789.

12.4 The Private Library

As the proverb says, you can tell a man from his library. Nevertheless, the list of Lexell's books [47][15]–around 900 titles bound in more than a thousand volumes—which was compiled in 1785 for the purpose of an international auction, hardly surprises in any other way than by showing the remarkable breadth of his philosophical, cultural and literary interests. Obviously, his books mostly deal with the mathematical sciences, but there were also numerous books on geography, exploration and navigation, philosophy, logic, theology and natural law.

It is impossible to present here but a selection of the books. Besides the main works of Euler, the books on mathematical and physical subjects included:

- Newton's *Philosophiae naturalis principia mathematica* (1760 Cologne edition by the Minim friars and mathematicians Thomas Le Seur and François Jacquier), *Arithmetica universalis* (Leiden, 1732) and *Opticks* (Latin edition, Lausanne & Geneva, 1740).
- Descartes's *Principia philosophiae* (Amsterdam, 1644) and *Geometria* (Latin edition 1659), William Oughtred's *Clavis Mathematicae* (1631), Jacques Ozanam's *Traité des lignes* (Paris, 1687), *Traité de la construction des équations* (1687) and *Traité des lieux géométriques* (1687).
- Jakob Hermann's *Phoronomia* (Amsterdam, 1716), Daniel Bernoulli's *Hydrodynamica* (Strasbourg, 1738), Pierre Varignon's *Nouvelle mécanique ou statique* (Paris, 1725), and J. H. Lambert's *Hinterlassene Schriften* (in 5 volumes, Berlin, 1782).
- The collected works of Jakob Bernoulli, Tycho Brahe, Albrecht Dürer and Robert Boyle, as well as the philosophical writings of Francis Bacon (in three volumes).
- d'Alembert's *Opuscules* (8 volumes; 1761–1780), *Traité de l'équilibre & du mouvement des fluides* (Paris, 1754), *Traité de dynamique* (1748) and *Réflexions sur la cause générale des vents* (1747).
- Gabriel Cramer's *Introduction à l'analyse des lignes courbes algébriques* (Geneva, 1750), Brook Taylor's *Methodus incrementorum* (London, 1717), Condorcet's *Essais d'analyse* (1768), and Étienne Bézout's *Théorie générale des équations algébriques* (Paris, 1779).
- J. A. Segner's *Rechenkunst und Geometrie* (1747), Ignace-Gaston Pardies's *Œuvres de mathématiques* (Amsterdam, 1725), Roger Cotes's *Harmonia*

[15] I am indebted to Staffan Rodhe for discovering this rare publication in the Uppsala University Library (UUB).

mensurarum (Cambridge, 1722), Colin Maclaurin's *Traité d'algèbre* (Paris, 1753) and *Traité des fluxions* (1749).
- Clairaut's *Théorie de la Lune* (Paris, 1752), as well as the translated *Anfangsgründe der Algebra* (Berlin, 1752) and *Inledning til geometrien* (Stockholm, 1744), Daniel Melander's *Fundamenta astronomiae* (1779), Jacques Cassini's *Élémens d'astronomie* (Paris, 1740), La Caille's *Astronomiae fundamenta* (Paris, 1757), Edmond Halley's *Tabulae astronomicae* (London, 1749) and Tobias Mayer's *Tabulae motuum Solis & Lunae* (London, 1770).
- Tiberius Cavallo's *Complete treatise on electricity* (London, 1782), Franz Aepinus's *Tentamen theoriae electricitatis et magnetismi* (St. Petersburg, 1759), and William Gilbert's *De magnete* (1628).

Lexell also owned an important collection of terrestrial and celestial maps, planispheres and logarithmic, trigonometric and sexagesimal tables by the most celebrated astronomers. Among the mathematical classics of antiquity we find the *Ten books on architecture* by Vitruvius, the works of Archimedes, the *Conics* by Apollonius, the *Elements* of Euclid and the *Spherics* by Theodosius.

Of Swedish mathematical authors we may mention Mårten Strömer's *Spheriska trigonometrien* (Stockholm, 1759), Fredric Palmquist's[16] *Inledning til algebra* (Stockholm, 1748), *Coniske sectionerne* (1752), and *Grunderna til mechaniken* (1756), Johan Bilberg's[17] *Elementa geometriae* (1691) as well as Kexlerus's *Arithmetica* (Åbo, 1658).[18]

The classics of antiquity naturally found a place in Lexell's library: the complete works of Homer, Cicero, Virgil, Tacitus, Seneca, Ovid, Lucian and Suetonius, the Tragedies of Sophocles, the Comedies of Terence, the Epigrams of Martial, the *Satyricon* by Petronius, Quintus Curtius' *Historia Alexandri Magni* and Caesar's *De bello Gallico et civili*. From the Modern Era we find Machiavelli's *Opere*, Boccaccio's *Decamerone*, the *Stultitiae laus* by Desiderius Erasmus (also in a French transation: *L'Éloge de la folie*), as well as *La Gerusalemme liberata* by Torquato Tasso. From the Age of Reason we find e.g. Leibniz's *Theodicaea*, Thomas Hobbes's *Leviathan* (London, 1651), Locke's *Essay concerning human understanding* (Latin edition, Leipzig, 1701), the complete works of Montesquieu, Voltaire and Alexander Pope, *La révolution de l'Amérique* by the Abbé Raynal and *La nouvelle Héloïse* by J. J. Rousseau.

Of books in the genre of *belles lettres* we may mention the complete works of La Fontaine and Corneille, François Fénelon's *Télémaque*, John Bunyan's *The pilgrim's progress* in German, Daniel Defoe's *Robinson Crusoe* likewise in a German adaptation, Sheridan's *Art of reading*, Tobias Smollett's *The adventures of Roderick Random* and Laurence Sterne's *Yorick's sentimental journey* and *Letters*

---

[16]Baron Fredric Palmquist (1720–1771) was a fortification engineer and mathematician, who wrote a number of textbooks on mathematics and its applications.

[17]Johan Bilberg (1646–1717) was an ecclesiastic and a scholar, Professor of Mathematics in Uppsala, later Professor of Theology, and eventually Bishop in Strängnäs.

[18]Simon Kexlerus (1602–1669) was the first *Matheseos* professor at Åbo [81, 153].

to Eliza. The library also contained the epic *Works of Ossian* (1765), collected and published by James Macpherson; among female authors, we may mention *Evelina* by Fanny Burney and *Memoirs of a certain island, adjacent to Utopia* by Eliza Haywood. On the whole it seems that, especially after his visit to England, Lexell favoured English literature more than French or German. Fiction in the Swedish language was very rare at the time.

Lexell's books on natural history were almost exclusively travelogues by scientific explorers, primarily by the Russian academicians, i.e. the Gmelins, Pallas, Georgi, Rumovski and Lepyokhin (the last two in Russian), who traveled to different parts of the Russian Empire. The accounts of the voyages of Captain Cook and Constantine John Phipps, as well as the expeditions of Maupertuis and La Condamine were found on his bookshelves. Somewhat surprisingly there is only one work by Linnaeus, written in Swedish, and containing his observations from his expedition to the islands of Öland and Gotland in 1741. Numerous volumes concerned history, especially of Sweden, the German-speaking countries, England and America, as well the lives of kings and emperors. Engineering technology was represented by books on naval architecture, such as F. H. Chapman's *Tractat om skeppsbyggeriet* (Stockholm, 1775) as well as warfare and fortification, e.g. François Blondel's *Die Kunst Bomben zu werffen* (Nürnberg, 1686), G. F. von Tempelhof's *Le bombardier Prussien* (Berlin, 1781), G. A. Böckler's *Neue Kriegs-Schule* (Frankfurt, 1665). Numerous books were devoted to scientific instruments.

Lexell owned Bibles in several languages: Swedish, Finnish, Dutch, Latin and Greek. He also had several dictionaries of Greek and Hebrew. Not surprisingly, Lexell's theological literature was in the tradition of German Lutheranism of Philipp Melanchthon, Johann Franz Buddeus and Johann Arndt, but did not exclude occasional works by mystics such as Emanuel Swedenborg's *Appendix ad veram Christianam religionem* (London, 1771) or the Jewish platonist Moses Mendelssohn's *Phédon* (French edition, Berlin, 1772), nor even the symbolism of the Freemasons (William Preston's *Illustrations of Masonry*, 1775). Among the classics of natural law, Hugo Grotius's *De jure belli ac pacis* (edition Amsterdam, 1720), Samuel von Pufendorf's *Droit de la nature* (Amsterdam, 1706) and Christian Thomasius's *Institutiones jurisprudentiae divinae* (Halle, 1717) may be mentioned as the most notable.

12.5 On Women

As to Lexell's relations with the fair sex, there is not much to tell, except for a passage of a letter quoted below. As is known, Lexell remained unmarried all his life. He never mentions having thought of marrying, not even desiring anyone. However, in one of his letters from Paris to J. A. Euler, dated 15 January 1781, he turns sentimental and writes openly about his dilemma (АРАН Фонд 1, опись 3, дело № 65, л. 156–160):

I have always been a great admirer of beautiful forms, and as among all beautiful things, that of a feminine face is the most exciting, I have not been insensitive on that point. Whenever I see a beautiful face I cannot help wishing that I was a painter and that I was allowed to paint a copy.

The population of Paris being very large, it is quite natural to find there many beautiful female faces, many mediocre and also many ugly ones, although it appears to me that the fair sex is very pretty here, by comparison with what I have seen in Russia, in my home country and in Germany. I have seen all sorts of very beautiful women here from Duchesses to daughters of working people. An Englishman who has published letters on France maintains that French women are not beautiful, but pretty.[19] As for me, I remark first that it may be rare to find a large number of perfect beauties in any country whatsoever; maybe also the fair sex is more beautiful there than here; but I think it is an injustice against the French to say that they are not beautiful. Those too, who are but pretty, are more exciting than women of other countries, because they have a physiognomy which is more suggestive of intelligence and wit. French women have more colour than the women of the northern countries, which accentuates their facial features and makes them interesting. Yet we must confess that the beauty of the French women extends only to the neck, that it is rare to find waists of great beauty. In fact, some dancers from the Opera are very slender and still young, but this is all gone after a few years. In general I think it is a fair comment that the further south one goes, the more bodies become squarely built, at least it is certain that almost all peasant women in France are very strongly built.

It is easily supposed that if the people of this country are amiable in society, the women are even more so. They are generally wittier and more educated than the women in many other countries. I am not talking about women scientists, or those who profess to be such, although in our times, they are more friendly, as long as they are not those ridiculous ones Molière had in mind in his Comedy.[20] But what I have said concerns women in general, at least those of a certain standing. I have also noticed that in the society in which I moved, the manners of women were very decent, but what shocked me somewhat is that in conversation with women we speak as naturally, as if they were not there. Maybe we are convinced that honesty is in the actions and not in the words, yet it would be better not to be too accustomed to certain ideas.

As regards the dresses of women, so much has been invented thereupon that there remains little to add to it, except perhaps a few small changes. Besides, there is no country where we copy the French so well in this regard as in Russia; I even observed here several bourgeois women more modestly dressed than they would be in Petersburg. One does make oneself up here, since the Queen paints her lips, but those who are beautiful enough without do not care for spoiling their complexions.

Whilst on this subject, I ought to name all those Academicians who, as far as I know, are married. They are Mr Cassini, le Monnier, Montigny,[21] le Roy, Bezout, Daubenton, Lavoisier, Le Gentil, Monge and Cassini the younger. You can see that they are not numerous and the reason for it is clear. I have often thought that only rich people's children ought to go to school, because a poor man choosing this way is also obliged to renounce all the pleasures of life. We preach against being single and it is right to condemn those who are so without any necessity, but there are many in number who are so because they would render themselves and their families unhappy by taking a wife. Though I probably spent the

[19]Here, Lexell might be alluding to Laurence Sterne's (1713–1768) *A sentimental journey through France and Italy* (1768). Lexell owned a copy of the book [47].

[20]Probably alluding to Molière's play "Les femmes savantes".

[21]Étienne Mignot de Montigny (1714–1782), engineer and official, known as Voltaire's nephew [72].

> biggest and the best part of my life single, I only can think with regret about the necessity for me to be so even in the future, as much from consideration as for economic reasons. Now finishing this article on women, as I suppose you will read this letter in the presence of your worthy wife, I must apologize that I could not render my description as elegant as I had hoped, considering the feelings of respect and affection that I have for the fair sex.

12.6 Maybe a Finn, After All

There have always been people who are difficult to place in a national framework, and Lexell, with his cosmopolitan career, is certainly among them. The concept of nationality is a controversial and sensitive one: it is so today, and it was so in the eighteenth century. Depending on how the question is asked, the view of nationality may vary according to citizenship, ethnicity, language, culture and religion. Of Lexell's citizenship there is no doubt: he was born a Swedish citizen, but after his promise to remain in Russian service in 1780, it is technically possible that he also changed his citizenship to Russian. In the light of the present knowledge, this seems not to be the case, however. Being of Lutheran religion was virtually a part of Swedish citizenship at the time, and Lexell showed no signs of wanting to abandon this part of his cultural heritage.

Lexell's contemporary, the philosopher Johann Gottfried von Herder (1744–1803), envisioned nationality as being based on cultural and linguistic affinities rather than ethnicity and origin. Herder's views were embraced during the nineteenth century, when national-romantic poets and philosophers tried to grasp the essentials of a nation's "soul" and identity.

During the Enlightenment the concept of national character was a subject of debate and development. It was generally agreed that each nation has its special characteristics, but opinions varied as to what constitutes national character. In the debate on nationality and national character during the eighteenth century, where Montesquieu played an important role, the crucial question was whether the differences in character and temperament of different nations as a whole were due to the physical environment, including the climate, the length of the seasons, the geography, vegetation and so on, or on the spiritual state, the morals, religion, and political rule [49].

Despite the difficulties in defining nationality and national character, I am tempted to take a view on Lexell's personality—admittedly a subjective one—which I take as an indication of his "Finnish mindset", as Porthan described it. In the patriotic epic *Fänrik Ståls sägner*,[22] the national poet of Finland, Johan Ludvig Runeberg (1804–1877) described the "classic Finnish virtues" in a way that seem to fit Lexell perfectly: honesty, modesty and loyalty—in particular obedience to the sovereign and to God—as well as courage to resist evil.

[22]"Tales of Ensign Stål" (1848, 1860), an epic of the events and the personalities of the Finnish war in 1808–1809.

In his sonnet *Finnische Landschaft* (1940), Bertolt Brecht described the Finns as a nation which remains silent in two languages: *ein Volk, das in zwei Sprachen schweigt*. This tallies well with the description of Lexell given in the obituary (most probably written by Johann Albrecht Euler) [41]:

> Mr Lexell spoke only sparingly without being embarrassed in the circles where he frequented: he loved, he even looked for good company, and there he was perfectly compensated.[23]

[23]M. Lexell parloit peu sans être embarassé dans des cercles où il se trouvoit: il aimoit, il recherchoit même la bonne compagnie, et il étoit payé d'un parfait retour.

Chapter 13
Conclusion

What is life but a series of fortuitous events? For a mathematical talent with sufficient ambition and capacity, such as Anders Johan Lexell, to have access to appropriate education and to find a suitable opportunity to develop his skills to the fullest may seem unlikely. Yet, sometimes things fall into place. Looking back at the present biography, we find that Lexell had recognised his vocation early in life and only waited for a chance to fulfil his duty in the sciences. We have seen how he, with a rather modest background, in a small town in Finland, at the time a province of Sweden, and with no obvious family connections or intellectual role-models to follow, could have the career he did. In spite of being distantly related to Pehr Wargentin, the Secretary of the Royal Academy of Sciences in Stockholm and a pre-eminent figure in the exact sciences in Sweden of the mid- and late eighteenth century, Lexell's relation to him was more professional than familial.

We have briefly summarised Lexell's efforts in mathematics and theoretical astronomy and examined how these were received by his contemporaries. Whereas his most original contributions stem from his research in celestial mechanics and astronomy, in which he was initially guided by Leonhard Euler, he also made some independent advances in the theory of differential equations and geometry; his most significant contributions in mathematics concern spherical trigonometry and the properties of general polygons. The fact that he simultaneously provided Swedish naturalists, such as Carl Linnaeus, with information and research material, suggests that he was regarded as competent even outside his own field of research. We have presented excerpts from his personal and professional correspondence, thereby using Lexell as a "witness of his time" from the Republic of Letters in the second half of the eighteenth century.

The fact that it was the greatest mathematical scientist of the century, Leonhard Euler, who provided an opening for Lexell's abilities was undoubtedly the most remarkable circumstance in his whole life. But as we have seen, this was not the only fortunate coincidence. At the time Lexell joined the Petersburg Academy of Sciences, the institution was well-connected and open to international cooperation. This had not always been so, however: in the years 1741–1766, when Euler was in Berlin, the Petersburg Academy of Sciences stagnated and lost much of its original

J.C.-E. Stén, *A Comet of the Enlightenment*, Vita Mathematica 17,
DOI 10.1007/978-3-319-00618-5_13,

attraction and prestige. The remarkable recovery began with Euler's return in 1766. The preparations for the great international campaign to observe and analyse the results of the transit of Venus in 1769 were in full swing. At the same time, natural historians were commissioned to different parts of Russia to explore the resources of the empire. However, the real reason for Lexell's employment at the Academy was Euler's need for competent associates to cope with the numerous scientific topics at hand. On his arrival to Saint Petersburg, Lexell was smoothly and successfully integrated into Euler's circle of collaborators, but, and perhaps more importantly, he was also granted intellectual freedom to mature as an independent researcher.

The known fact that Euler did not give regular lectures does not mean that he did not have students. Joseph Louis Lagrange is often mentioned as Euler's student and de facto successor in Berlin, although the two never met. Nevertheless, as we know, Lagrange contributed to many of the topics Euler had initiated. From time to time, Euler instructed students by correspondence; some of them he could even lodge temporarily in his house [43]. During the two periods Euler spent in Saint Petersburg, the mathematical section and indeed the whole Academy was strongly influenced by his presence. He trained the scientists and supervised the mathematical work done at the Academy and thus he contributed to its high standard. Among his disciples and collaborators in Saint Petersburg, no one was as productive and at the same time as much appreciated internationally as Anders Johan Lexell.

Lexell ultimately became known as Euler's closest associate and one of the few at the time who could actually understand his reasoning. The *grand tour* Lexell undertook in 1780–1781 to the great centres of learning in Europe made him internationally known and set a definitive seal of approval on his talents. For instance Pierre Simon de Laplace, in his letters to the Petersburg Academy, henceforth addressed his discoveries to both Euler and Lexell. Due to his poor eyesight, Euler entrusted his correspondence with Daniel Bernoulli, Lagrange and Lambert to the hands and eyes of Lexell. Most colleagues could nevertheless perceive that Lexell was more than a mere amanuensis or even a sincere imitator of his Master. In scientific matters, Euler clearly had confidence in Lexell's opinion and judgement.

Lexell had the misfortune of not being fully understood and appreciated in his native country, Finland. Although Lexell was throughout his life a Swedish citizen, some Swedish historians of science label him as a Finnish Swede and, especially after his removal to Russia, as an expatriate. In Russian historiography, on the other hand, Lexell is presented as a Russian academician, whereas on the Finnish side very little has been written on Lexell before the present study and very few today would even recognise his name. He is often mentioned as an expatriate Finn who, despite occupying the Chair of Mathematics in his home town of Åbo, did not exert any significant influence on conditions of learning and science in Finland. However, since Lexell was in constant correspondence with scientists in Sweden and Finland, this view can be contested. The contents of the few surviving letters to Planman and Gadd suggest that Lexell was aware of his unique position as a go-between for the Academies of Saint Petersburg and Stockholm, and even the continental centres of science. He also felt a strong responsibility to keep his colleagues in Åbo, as well as in Stockholm and Uppsala, updated on his activities and concerns.

In particular, Lexell's locum tenens and successor in Åbo, Professor Johan Henrik Lindquist, a student of Wallenius like himself, pursued in his dissertations several problems initially addressed by Lexell. Lexell also made an effort to improve the conditions of learning in Åbo by offering one half of his salary for the acquisition of modern astronomical instruments for the university. These are now among the oldest scientific instruments of the University of Helsinki, the historical successor of the University at Åbo.

Nevertheless, Lexell spent most of his active career in Russia, where he also died and was buried.[1] His scientific legacy of books and papers remained there.[2] His closest colleagues and students[3] also pursued their careers in Saint Petersburg. Hence, many have viewed Lexell as being exclusively a figure in the Russian history of science.

As has been pointed out in this biography, however, a narrow nationalistic view of Lexell is misleading. Undoubtedly Lexell had warm feelings towards his home country, family and friends, but his dedication to science was even stronger. Disregarding the political tensions between Sweden and Russia, the Petersburg Academy of Sciences, in the presence and guidance of Euler, provided him the best available platform to be useful to science, he seems to have reasoned. He might well have endorsed Cicero's contented statement in his *Tusculan disputations* (Book 5):

Patria est ubicumque est bene.[4]

Nevertheless, it took years for Lexell to decide—and not without considerable hesitation and anguish—whether he should move to Åbo or pursue his work in Saint Petersburg. Unfortunately, there are no clues as to how Lexell may have reassessed his situation after Euler's death. So far, only a few letters from Lexell have been discovered from the approximately 14 months separating Euler's passing away in 1783 and his own untimely demise in 1784.

With the growing interest in the history of the Petersburg Academy of Sciences, in particular its most remarkable years in the eighteenth century, it is possible that additional archive material may still surface and shed new light on the working dynamics of Euler's closest circle, on which Lexell generally remains markedly silent. Despite the extensive project of publishing Leonhard Euler's books, papers and correspondence in the *Opera Omnia*, much work remains to be done to complete the picture of Euler's impact on the sciences and the scientists surrounding him. We believe that a part of this rewarding, multifaceted puzzle has been completed by the present study of the life and science of Anders Johan Lexell.

[1]Lexell is likely to have been buried in the Lutheran part of the *Smolensky cemetery* (Смоленское кладбище) on the Vasilyevsky Island. The site of his grave is unknown.

[2]Lexell bequeathed his belongings to the children of his late sister Ulrika. His books were auctioned in Saint Petersburg [47]. The fate of his scientific correspondence is unclear: it is possible that it was transported to Åbo with his belongings, either to the family or to the academic library, in which case it may have been destroyed in the fire of Åbo in 1827.

[3]Lexell's research topics were pursued in Saint Petersburg by Nikolaus Fuss and Friedrich Theodor Schubert [109].

[4]Where it is well with me, there is my country.

Appendices

Letters

Letter to Lambert

UBB Ms L I a 703, f° 206–209 (edited by Karl Bopp in the *Abhandlungen der Preussischen Akademie der Wissenschaften*, Physikalisch-Mathematische Klasse. Nr. 2. Berlin, 1924, pp. 38–40). After having communicated Euler's greeting and thanks for Lambert's recent letter (dated 18 October 1771; edited in the same volume on pp. 36–37), Lexell takes the opportunity to discuss two of his latest results in calculus. The first one concerns his research into the theorem due to Lagrange and the second one the integrability of differential equations. As an example of the generalisation of his research to multivariate calculus, he gives the characteristic criteria for the integrability of $V\mathrm{d}x\mathrm{d}y$: In the resulting expression, the sum of the terms on the horizontal lines, as well as those along the diagonals, must be zero. As to the discovery of the criteria, Lexell gives the priority to Euler.

Monsieur

Monsieur Euler m'ayant chargé de faire une réponse à cette obligeante lettre, qu'il vient de recevoir de Votre part et de Vous présenter ses remercîmens pour les sincères preuves d'attachement dont Vous l'avez honoré; j'ai cru être d'autant plus obligé d'obéir à ses commandemens, que je dois regarder cette occasion comme la plus favorable de Vous témoigner, Monsieur, les sentiments d'affection et de respect, que Vos solides connoissances et rares mérites m'ont inspiré depuis longtemps.
Comme la dernière partie de Votre lettre roule sur une matière, à laquelle j'ai travaillé un peu à l'occasion de ce remarquable théorème: ayant supposé $x = y + \varphi x$, il sera

$$\psi x = \psi y + \frac{\varphi y \mathrm{d}\psi y}{\mathrm{d}y} + \frac{\mathrm{d}((\varphi y)^2 \mathrm{d}\psi y)}{2\mathrm{d}y^2} + \text{etc.},$$

vous me permettez, Monsieur, de Vous communiquer mes sentiments là dessus. Il est bien vrai, que lorsque une fonction quelconque de x et y est égale à zéro, on peut déterminer telle fonction de x ou de x et y, qu'on voudra, par la seule variable y; mais que le calcul

J.C.-E. Stén, *A Comet of the Enlightenment*, Vita Mathematica 17,
DOI 10.1007/978-3-319-00618-5,

deviendroit plus prolixe, quand des quantités intégrales entrent dans cette fonction, ce c'est que je ne comprends pas. Dans un mémoire, qui sera imprimé dans le tome XVI de nos Commentaires, j'ai considéré cette équation $x = y + P$, P étant supposé être une fonction quelconque de x et y, et j'ai montré, qu'on peut exprimer telle fonction qu'il plait de x ou de x et y, par la seule variable y, au moyen d'une suite. Or je pense que cette équation est la plus générale pour le cas des fonctions de deux variables. Néanmoins j'ose assurer que la suite à laquelle la résolution de cette question se réduit, est presque aussi simple, que celle, qu'on vient de trouver pour le cas, où P est une fonction de x seulement, ou une somme des fonctions de x et y. Quand il s'agit des fonctions de trois ou plusieurs variables, je suis bien convaincu, qu'il est possible de trouver pour une fonction quelconque de toutes ces variables, une expression, dont une est excluse; mais en même temps je doute fort si les formules qu'on viendroit de trouver par ces recherches, seroient assez élégantes pour mériter quelque attention. L'expression

$$\psi x = \psi y + \frac{\varphi y \cdot \mathrm{d}\psi y}{\mathrm{d}y} + \text{etc.}$$

peut à la vérité se déduire de celle ci

$$\psi' x = \psi' y + \frac{\varphi x \mathrm{d}\psi' y}{\mathrm{d}y} + \frac{\varphi x^2 \mathrm{dd}\psi' y}{2\mathrm{d}y^2} + \text{etc.}$$

au moyen de la reversion des suites, mais Vous conviendrez sans doute, Monsieur Vous-même, que cette réversion est fort compliquée et qu'il y a raison de croire, que la démonstration de Votre expression

$$\psi x = \psi y + \frac{\varphi y \mathrm{d}\psi y}{\mathrm{d}y} + \text{etc.}$$

se peut trouver d'une manière beaucoup plus aisée.

Dans le mémoire dont je viens de parler, j'en ai donné une démonstration assez simple et Mr Euler, qui m'avoit proposé de chercher cette démonstration, a depuis trouvé plusieurs autres, dont quelqu'unes sont fort élégantes, lesquelles seront aussi insérées dans nos Commentaires.

Voici une formule assez élégante, pour exprimer la variation d'une fonction quelconque de x et y. Soit $x' = x + P$ et $y' = y + Q$, que Z désigne une fonction quelconque de x et y et Z', une semblable fonction de x', y', je dis, qu'il sera:

$$\begin{aligned} Z' = Z + P\Big(\frac{\mathrm{d}Z}{\mathrm{d}x}\Big) + \frac{1}{2}P^2\Big(\frac{\mathrm{dd}Z}{\mathrm{d}x^2}\Big) + \frac{1}{6}P^3\Big(\frac{\mathrm{d}^3Z}{\mathrm{d}x^3}\Big) &+ \text{etc.} \\ +Q\Big(\frac{\mathrm{d}Z}{\mathrm{d}y}\Big) + \frac{2}{2}PQ\Big(\frac{\mathrm{d}^2Z}{\mathrm{d}x\mathrm{d}y}\Big) + \frac{3}{6}P^2Q\Big(\frac{\mathrm{d}^3Z}{\mathrm{d}x^2\mathrm{d}y}\Big)& \\ +\frac{1}{2}Q^2\Big(\frac{\mathrm{dd}Z}{\mathrm{d}y^2}\Big) + \frac{3}{6}PQ^2\Big(\frac{\mathrm{d}^3Z}{\mathrm{d}x\mathrm{d}y^2}\Big)& \\ +\frac{1}{6}Q^3\Big(\frac{\mathrm{d}^3Z}{\mathrm{d}y^3}\Big)&. \end{aligned}$$

Il est fort évident, qu'on peut pousser ces recherches encore plus loin et même à tel nombre des variables qu'on veut. Dans le tome XV de nos Commentaires, qui vient d'être imprimé, il y a un mémoire dans lequel j'ai traité des charactères d'intégrabilité. Il s'agit principalement d'une démonstration de ce remarquable théorème de Mr Euler. Supposant

$\mathrm{d}y = p\mathrm{d}x, \mathrm{d}p = q\mathrm{d}x$ etc., $\mathrm{d}V = M\mathrm{d}x + N\mathrm{d}y + P\mathrm{d}p + Q\mathrm{d}q +$ etc., la formule $V\mathrm{d}x$ sera intégrable lorsque

$$N - \frac{\mathrm{d}P}{\mathrm{d}x} + \frac{\mathrm{dd}Q}{\mathrm{d}x^2} - \frac{\mathrm{d}^3 R}{\mathrm{d}x^3} + \text{etc.} = 0.$$

Mr Euler avoit lui-même démontré ce théorème par les principes du calcul de la variation et par les Mémoires de Paris l'an 1765 j'ai appris, que M^r^ le Marquis de Condorcet a donné une exposition détaillé de ce théorème, mais il m'est encore inconnu, de quels principes il a déduit sa démonstration. Pour ce qui est de la mienne elle est déduite seulement des principes du calcul différentiel, qui me paroissent beaucoup plus analogues à ce sujet que ceux, qui se tirent du calcul de la variation. J'ai encore poussé la recherche des charactères d'intégrabilité plus loin, car j'ai même considéré des formes intégrales doubles ou triples comme $\int\int V\mathrm{d}x\mathrm{d}y; \int\int\int V\mathrm{d}x\mathrm{d}y\mathrm{d}z$. Pour Vous donner un échantillon de mes recherches, j'ajouterai l'expression dans laquelle les charactères d'intégrabilité pour la formule $V\mathrm{d}x\mathrm{d}y$ sont compris. Posant donc

$$\begin{aligned} \mathrm{d}z = p\mathrm{d}x + p'\mathrm{d}y; \qquad & \mathrm{d}p = q\mathrm{d}x + q'\mathrm{d}y; \\ & \mathrm{d}p' = q'\mathrm{d}x + q''\mathrm{d}y; \quad \text{etc.} \end{aligned}$$

$$\begin{aligned} \mathrm{d}V = L\mathrm{d}x + M\mathrm{d}y + N\mathrm{d}z + P\mathrm{d}p + Q\mathrm{d}q & \quad + \text{etc.} \\ +P'\mathrm{d}p' + Q'\mathrm{d}q' & \\ +Q''\mathrm{d}q'', & \end{aligned}$$

la formule $V\mathrm{d}x\mathrm{d}y$ sera intégrable doublement lorsque dans l'expression suivante:

$$\begin{aligned} N - \left(\frac{\mathrm{d}P}{\mathrm{d}x}\right) + \left(\frac{\mathrm{dd}Q}{\mathrm{d}x^2}\right) - \left(\frac{\mathrm{d}^3 R}{\mathrm{d}x^3}\right) + \text{etc.} = 0 \\ -\left(\frac{\mathrm{d}P'}{\mathrm{d}x}\right) + \left(\frac{\mathrm{dd}Q'}{\mathrm{d}x^2}\right) - \left(\frac{\mathrm{d}^3 R'}{\mathrm{d}x^3}\right) \\ +\left(\frac{\mathrm{dd}Q''}{\mathrm{d}x^2}\right) - \left(\frac{\mathrm{d}^3 R''}{\mathrm{d}x^3}\right) \\ -\left(\frac{\mathrm{d}^3 R'''}{\mathrm{d}x^3}\right), \end{aligned}$$

tous les termes qui sont placés ensemble dans les mêmes lignes horizontales et diagonales, font des expressions évanouissantes. Je sçais que Mr Euler avoit envoyé un mémoire sur son théorème à Berlin et que Mr de la Grange avoit fait quelque difficulté de la faire insérer dans les Mémoires de Votre Académie, parce que cette matière étoit déjà traité de M^r^ le Marquis de Condorcet, et qu'il pourroit en exister des différends sur la gloire de l'invention de ce théorème. Je pense que cette crainte est très mal fondé, M^r^ la Grange étant lui-même convaincu, que l'invention de ce théorème appartienne uniquement à Mr Euler. Pour moi je sçais que Mr Euler avoit communiqué à quelques uns de ces disciples ce même théorème, il y a déjà quinze ou seize ans, et Mr Euler lui-même m'a dit que ce théorème lui est connu depuis vingt années. La première fois que Mr Euler l'a publié, c'est dans nos Commentaires Tome X, qui est imprimé avant les Mémoires de Paris l'an 1765, et peut être en même temps que le livre de M^r^ le Marquis de Condorcet.

La nouvelle théorie de la Lune de Mr Euler est actuellement sous la presse. Pour ce qui regarde la partie théorique, c'est sans doute l'ouvrage le plus complet, qu'on peut désirer,

mais que les nouveaux tables de la Lune, surpasseront celles de Mr Mayer en exactitude, c'est dont je ne suis pas encore persuadé. Mr Euler avoit aussi commencé des recherches sur la perturbation des planètes, qu'il reprendra sans doute, lorsqu'il sera un peu mieux rétabli. Mais comme je dois craindre, que ma lettre par sa longueur, ne Vous fasse plus de dégoût, que de plaisir, je vais finir en Vous assurant du profond respect avec lequel je suis
Monsieur

Votre

très humble et très obéissant
serviteur
A. J. Lexell

St. Pétersbourg ce 22 Janvier 1772
P:S: Ce seroit pour moi le plus sensible plaisir, si Vous voudriez bien, Monsieur, m'honorer quelque fois de Vos lettres. Oserois je Vous prier de présenter mes très humbles complimens à M[r] la Grange et à Mr Bernoulli?

Extracts from Letters to J. A. Euler Quoted in the Text

Göttingen, 27 September 1780 (АРАН Фонд 1, опись 3, дело № 65, л. 101–103)

[...] En allant de Berlin à Göttingen, j'ai préféré le chemin, qui passe par la Saxe, à celui qui va par Magdebourg et Halberstadt, tant pour voir la Saxe, qui est un assez beau païs, que pour faire connoissance avec quelques sçavans à Leipzig et principalement un certain Monsieur Hindenburg, qui est un très habile mathématicien. Étant donc parti de Potsdam comme j'ai eu l'honneur de marquer dans ma dernière lettre, le 15 Septembre de Potsdam, je suis arrivé à Leipzig le 17, et m'étant arrêté à Leipzig jusqu'au 19, je n'ai pû écrire que le 22, à Göttingue. À Leipzig il n'y a presque rien a faire pour l'astronomie et même on n'y trouve des instruments astronomiques, que le reste de ceux, qui ont appartenu à Monsieur Heinsius et dont M[r] [B]orz, Professeur en Mathématique, est en possession. Les deux principales Bibliothèques de Leipzig sont celles de l'Université et du Magistrat, dont la première ne paroît pas être très considérable et contient pour la pluspart des livres de théologie, l'autre est beaucoup plus belle et contient quarante mille livres, la plus grande collection est celle de l'histoire. Cette Bibliothèque est aussi placée dans une très belle maison qui appartient à la ville. Les environs de Leipzig sont très beaux, tant par la nature que par le moyen de l'art, étant environné de plusieurs beaux jardins, dont le meilleur appartient à un négociant nommé Richter. Un autre négociant nommé Winckler a une très belle collection de tableaux au nombre de 900 et même des meilleurs maîtres Italiens, Flamands et Allemands. [...]
[...] L'Observatoire de Göttingue est bâti sur une tour, qui appartient aux vieilles fortifications de cette ville et contient 18 jusqu'à 22 pieds en diamètre, la hauteur n'étant que de dix pieds tout au plus. On voit donc que la place est d'autant plus étroite, que l'escalier par où on monte est précisément au milieu de l'Observatoire. Au reste l'horizon de Göttingue est borné de tous côtés par des montagnes, en sorte qu'il faut renoncer aux observations, qui passent à 3° près de l'horizon. La collection des instruments est assez belle et les principaux en sont: 1° Un quart de cercle de Bird de 6 pieds en rayon, qui est placé vers le sud, mais dont la muraille, à laquelle il est attaché, a subi quelque changement, en sorte que l'instrument décline un peu du plan vertical, ce qui est embarrassant pour les observations et demande qu'à tout moment on tienne compte de cette déviation. Vis à vis de la muraille, à laquelle l'instrument est attaché, il y a encore une muraille pour placer l'instrument au nord. 2° Un quart de cercle Anglois de Sisson de 2 pieds. [3°] Un quart de cercle de 2 $\frac{1}{2}$ pieds, fait à Göttingue par Campe à la manière des quarts de cercle François, mais pour la lunette la construction ressemble à celle qu'on emploie pour les quarts de cercle muraux, ce qui pour un quart de cercle mobile est sans aucune utilité. 3° trois lunettes achromatiques, dont la

principale, avec un double objectif, est fabriquée à Londres par Dollond: cette lunette n'est que de cinq ou six pieds, mais d'un très bel effet et surpasse beaucoup les deux autres, qui sont faites à Göttingue par un artiste ici établi, nommé Bauman[n]. 4° Cinq pendules, entre lesquelles il y a une avec la verge composée, faite à Londres par Shelton, les autres sont faites à Göttingue. À l'horloge de Shelton appartient encore un compteur du même artiste. 5° Une construction d'horloge, montrant les demi-secondes, les quarts et les huitièmes parties des secondes. Cet instrument est un présent à l'Observatoire de la part de la Reine d'Angleterre. 6° Une horloge qui montre les tierces par Campe à Göttingue. [...]

Paris, 20 November 1780 (АРАН Фонд 1, опись 3, дело № 65, л. 131–134)

Monsieur
et très honoré Confrère.

Dans ma première lettre, qui étoit écrite il y a huit jours, je Vous ai marqué avec quel respect Messieurs de l'Académie des Sciences d'ici, qui sont mathématiciens, parlent de Monsieur Votre Père, mais pour Vous en donner la preuve la plus décisive, je Vous envoye le N° 323 du *Journal de Paris*, qui contient le détail de l'entrevüe entre le Prince de Prusse et Votre Père, d'après la lettre que Vous aviez la bonté de m'écrire. L'article du Journal est certainement du Marquis de Condorcet et quoique il n'est pas tout à fait exact, je crois cependant qu'il Vous fera plaisir d'apercevoir qu'à Paris tout ce qui est mathématicien, sçait rendre justice aux grands talens et aux mérites immortelles de Monsieur Votre Père. J'aurois bien souhaité, que le Marquis de Condorcet ne s'auroit exprimé si précisément par rapport à l'ouïe de Monsieur Votre Père et cela n'est pas conforme à ce que je lui en ai dit; car quoique en parlant de Monsieur Votre Père j'ai dit qu'il entendoit difficilement, je n'ai pas prétendu dire, qu'il étoit sourd. Par rapport aux nombres des mémoires, qu'il a donné pour nos actes, j'en ai parlé en nombre rotonde et peut être qu'il y en a encore plus que celles, que j'ai nommé; aussi n'ai je pas déterminé en combien de temps que ces mémoires sont faites, mais j'ai dit en général qu'on pouvoit compter que Votre Père donnoit un mémoire par semaine. Vous sçavez Monsieur et cher Confrère que les François vont un peu vite, et qu'à l'ordinaire ils ne se font pas scrupule de changer un peu les circonstances. Mais tout ceci me paroît de peu d'importance, l'essentiel est, qu'on vient de donner un hommage public à Monsieur Votre Père, qu'il a si bien mérité et que personne, qui sçait estimer les mathématiques, ne désavouera. Je suis assuré que Monsieur le Marquis de Condorcet a agi de concert avec Monsieur d'Alembert; voici au moins ce qui est arrivé le même jour, que ce N^{o} du Journal paroissoit.

Il y avoit Assemblée de l'Académie des Sciences ce jour [là] et après la séance, comme c'étoit un des jours, que Monsieur d'Alembert tient ses assemblés, j'y allois aussi bien pour entendre ce qu'il disoit sur cet article, que parce que je n'avois pas été chez lui depuis quelque temps. Aussitôt que j'étois entré, il m'a demandé si j'avois lû cet article, et il m'en a nommé l'auteur, quoique moi-même je l'avois déjà conjecturé. Ensuite comme il y avoit quelques personnes à l'assemblée, à qui cet article n'étoit pas connu, M^{r} d'Alembert en a fait la lecture à l'assemblée et lorsque je m'en allois, il m'a chargé de faire ses compliments à Vous et à Monsieur Votre Père. Voilà qui est bien honnête. Aussi puis je Vous assurer, que tous les autres mathématiciens que j'ai vû ce jour là et qui avoient lû le Journal, en étoient très contents ainsi qu'il n'y [a] qu'une voix par rapport aux mérites de Mr Votre Père et que le Marquis de Condorcet a interprété très bien.

La semaine passée, j'ai assisté à deux fêtes littéraires: la première étoit au Collège Royal, où M^{rs} les Professeurs célébrèrent leur rentrée, et la seconde à l'Académie des Sciences. Au Collège Royal quatre Professeurs ont lû. M^{r} le Monnier commença la séance par un mémoire sur les différentes densités de l'air, selon les observations barométriques; je suis persuadé qu'il ne sçavoit bien lui-même ce qu'il disoit. Ensuite Mr Vauvilliers a lû une traduction d'un morceau de Thucydides, qui contenait l'oraison funèbre prononcée par Périclès à l'honneur des guerriers, qui périrent dans la première guerre de Péloponnèse. Après lui Mr de la Lande a donné un extrait de son traité sur le flux et reflux de la mer, où

il a dit des choses très singuliers; mais il faut voir son traité avant que d'en pouvoir juger avec précision. M^r l'Abbé Delille a terminé la séance, en récitant de très beaux vers, dans lesquels il a donné l'art de peindre en vers ou la description d'une campagne.
La séance de l'Académie des Sciences a été commencée par l'éloge de Mr Bucquet, Adjoint, prononcé par M^r le Marquis de Condorcet. Ensuite Mr de la Lande a lû un mémoire, où il prétend prouver que l'obliquité d'écliptique ne diminue que 35$''$ par siècle et que la durée moyenne de l'an est à présent 365J. 6h. 48$'$.48$''$, qu'elle a diminuée de 4$''$ depuis 2000 ans; il prétendait même en déduire, que la masse de Vénus n'est que la moitié de celle de la Terre. Il y a bien de gens ici qui pensent, Mr de la Lande auroit fort bien fait de ne pas se mêler de cette question et qu'il n'a pas ni assez de patience, ni assez de discernement d'en juger. En général, il n'est pas beaucoup aimé de ses confrères; c'est, je crois, parce qu'il est trop charlatan. Après la Lande, Mr Messier a donné une notice sur la comète, observée en dernier lieu. M^r l'Abbé Rochon a continué la séance par la lecture d'un mémoire sur un nouveau instrument pour mesurer les hauteurs solsticiales; c'est une mire qui est placée dans le méridien et laquelle réfléchit l'image du Soleil. Il me paroît que l'Abbé Rochon s'est trop hâté de publier son instrument, avant qu'il en a fait l'épreuve, et de le préférer aux grands instruments, placés dans le méridien. Mr Vandermonde a lû un mémoire sur la musique, duquel je ne sçaurois prononcer, n'étant pas musicien. Enfin Mr Cornet[te] a lû le résultat des éxpériences qu'il a fait avec Mr Lassone pour faire des phosphores des os. Samedi passé, à la séance de l'Académie, Mr Lavoisier a lû deux mémoires sur les phosphores. La chimie est à présent la science la plus à la mode.
L'autre jour étant allé voir la Bibliothèque du Roi, le portier m'a conduit dans quelques apartements, qui contenoient la collection des ouvrages de Mr Houdon célèbre sculpteur dans cette ville. Je ne me rappelle plus, tout ce qu'il y avoit à voir, mais je donnerai pourtant les pièces les plus principales. 1° Il s'y trouve une statue de Mr de Voltaire assis dans un fauteuil en robe de chambre et la tête nue, ouvrage très parfait et excellent. C'est l'Impératrice qui a ordonné cette pièce-là, pour la placer dans la Bibliothèque de Mr de Voltaire. L'auteur en fera encore deux exemplaires, l'un pour l'Académie Françoise et l'autre pour la Comédie Françoise. 2° Diane marchant, elle se tient sur le bout d'un des pieds, ouvrage hardi et beau. Cette statue est pour le Roi de Pologne et pour le Duc de Saxe-Gotha. Elle coûte en marbre 15000 Livres et en bronze 10000. L'image de Voltaire coûtera 25000 Livres. 3° Une najade pour le Duc de Chartres. 4° Bustes de plusieurs personnes distingués parmi les philosophes et les gens en charge. L'Imperatoire de la Russie, Monsieur et Madame de France, Mesdames les tantes du Roi, Mr Voltaire à tête nue et avec la perruque, M^r d'Alembert qui lui ressemble très bien, J. J. Rousseau, Franklin, Le Duc de Praslin, Mr Turgot, le Garde des sceaux, M^r le Noir, le Chevalier Gluck, Mr Palissot, Paul Jones, Mr de la Lande aussi très ressemblant, Mr Vitinghöf de Riga et encore plusieurs autres. Il est encore jeune ce Mr Houdon et s'il continue à cultiver son art il ne manquera pas à devenir le plus grand sculpteur de l'Europe.
Je n'ai pas encore beaucoup profité de mon séjour ici, c'est qu'en parti le temps a été très mauvais, en parti aussi je me suis occupé à calculer la comète d'à présent, que Mr Messier n'a pas vû depuis. Pendant qu'elle a été observée elle n'a décrit qu'un angle de quelques degrés autour le Soleil, ainsi les élémens de son orbite sont extrêmement difficiles à déterminer et la moindre erreur dans les observations y produit des changemens considérables.
Mes plus respectueux compliments à Madame Votre digne Épouse, Monsieur Votre Père et tout Votre chère famille, aussi bien que Messieurs les Académiciens de celui, qui est avec le plus sincère attachement

Monsieur et cher Confrère

Votre très humble
et très obéissant serviteur
Lexell

Paris, ce $\frac{20}{9}$ de Nov: 1780

Paris, 15 January 1781 (АРАН Фонд 1, опись 3, дело № 65, л. 156–160)

[...] J'ai toujours été un grand amateur des belles formes, et comme parmi toutes les beautés, celle d'un visage féminin est la plus touchante, je n'ai pas manqué d'être très attentif sur ce point. Aussi souvent que je vois un beau visage je ne puis m'empêcher de souhaiter que je fusse peintre et qu'il me seroit permis d'en prendre une copie.
Comme la population de Paris est très grande, il est bien naturel d'y trouver beaucoup de beaux visages des femmes, beaucoup de médiocres et encore des laides, cependant il m'a paru que le sexe est très bien fait ici, en proportion de ce que j'ai vû en Russie, dans ma Patrie et en Allemagne. J'ai trouvé des femmes de toutes sortes de condition très belles depuis des Duchesses jusqu'à des filles des petits ouvriers. Un Anglais qui a publié des lettres sur la France prétend que les Françoises ne sont pas belles, mais jolies; pour moi je fait d'abord la remarque qu'il est peut-être rare de trouver beaucoup de parfaites beautés dans quelque païs que se soit; peut-être aussi qu'en Angleterre le sexe est plus beau qu'ici; mais en tout cas il me paroit qu'on fait injustice aux Françoises lorsque on dit, qu'elles n'ont pas de beauté. Celles même, qui ne sont que jolies, sont souvent plus touchantes, que les femmes les plus belles des autres païs, c'est qu'elles ont des physionomies plus spirituelles. Les Françoises ont en général plus de couleur que les femmes des païs du nord, c'est ce qui relève leurs physionomies et les rend intéressantes. Cependant il faut que j'avoue que la beauté des Françoises ne s'étend que jusqu'à la gorge, c'est qu'il est rare de trouver des tailles d'une grande beauté. On voit en effet quelques danseuses de l'Opéra très sveltes, encore jeunes, mais après quelques ans, cela change bien. En général je crois la remarque juste, que plus on avance vers le sud, plus les corps deviennent quarrés et ramassés, au moins il est sure, que presque toutes les paysannes de France sont bien fortes de taille.
Il est aisé de supposer que si les hommes de ce pays sont aimables dans la société, les femmes le doivent être encore davantage. Elles sont en général plus spirituelles et plus instruites que dans les autres païs. Je ne parle pas des femmes sçavantes, ou qui font profession de l'être, quoique dans le temps à présent, celles ci sont d'autant plus aimables, qu'elles n'ont ces ridicules, que Molière a eu en vüe dans sa comédie. Mais ce que je dis se rapporte aux femmes au moins d'une certaine condition, en général. Dans les sociétés que j'ai fréquenté, j'ai aussi remarqué, que les manières des femmes étoient très décentes, mais ce qui m'a un peu choqué, c'est que dans la conversation avec les femmes on parle tout aussi naturellement, que si elles n'y étoient pas. Peut-être qu'on est persuadé, que l'honnêteté consiste dans les actions et non dans les paroles, cependant il vaudroit mieux de ne pas trop se familiariser avec certaines idées.
Pour ce qui regarde la parure des femmes, on a tant inventé là dessus, qu'il reste fort peu d'y ajouter, excepte peut-être quelques petits changements. D'ailleurs il n'y a aucun païs, où on copie les François si bien, par rapport à cet article, qu'en Russie et même j'ai observé plusieurs bourgeoises ici plus modestement habillées, qu'elles le seroient à Pétersbourg. On se farde ici, puisque la Reine met du rouge, mais celles qui sont assez belles sans cela, ne se soucient pas de gâter leur teint.
À propos de cet article, je devrois Vous encore nommer ceux des Académiciens qui sont mariés, il s'entend parmi ceux que je connois. Ce sont Mrs Cassini, le Monnier, Montigny, le Roy, Bézout, Daubenton, Lavoisier, le Gentil, Monge et Cassini fils. Vous voyez qu'ils ne sont pas en grand nombre et la raison en est claire. J'ai souvent pensé, qu'il n'y a que les gens riches, dont les enfans devroient s'appliquer aux études, car un homme pauvre en prenant ce parti est en même temps obligé de renoncer à tous les agrémens de la vie. On prêche contre les célibataires et on a raison de condamner ceux, qui le sont sans aucune nécessité; mais il y en a bien en grand nombre, qui le sont parce qu'ils se rendroient eux et leurs familles malheureuses en prenant des femmes. Quoique j'ai sans doute passé la plus grande et la meilleure partie de ma vie solitaire, je ne puis penser qu'à regret à la nécessité où je me trouve de la faire à l'avenir autant par réflexion que par considération d'économie.
En finissant l'article concernant les femmes, comme je suppose que Vous ferez la lecture de cette lettre en présence de Votre digne Épouse, je dois faire mes excuses, que je n'ai pas pû

mettre tant d'élégance dans ma description, que j'aurai bien souhaité et qui serait en rapport avec les sentiments de respect et de tendresse, que j'ai pour le beau sexe. [. . .]

Paris, 4 March 1781 (АРАН Фонд 1, опись 3, дело № 65, л. 175–181)

[. . .] Le Roi et ses deux frères ont une physionomie de famille, c'est à dire ils se ressemblent assez pour qu'on puisse voir, qu'ils sont frères; aussi ont ils beaucoup de ressemblance avec leur Père, principalement Monsieur. Le Roi a la physionomie un peu indolente et phlegmatique, il est assez fort et marche en se balançant. Je l'ai vû dîner, il paroissoit manger avec beaucoup d'appétit et mettre à cette occupation assez d'intérêt. On dit qu'il n'a pas beaucoup d'esprit, cependant il n'en a résulté jusqu'à présent que du bien pour la France. Comme il se connoît soi-même il a beaucoup de confiance à ses Ministres, qui pour la pluspart sont des gens du bien. Les inclinations favorites dans la famille de Bourbon, ce sont les femmes et la chasse; jusqu'à présent le Roi n'a aimé que son Épouse, et malheur à la France, s'il changeroit de sentiment. Il aime beaucoup la chasse. Il a d'ailleurs une qualité très salutaire pour son pays, c'est d'aimer l'épargne et par cette raison il s'est prêté à toutes les réformes de Mr Necker, ce qui peut encore, si Mr Necker est conservé rendre la France le pays le plus heureux du monde. Pour ce qui regarde le caractère de Monsieur les sentiments sont très partagés, quelques uns disent qu'il a peu d'esprit, qu'il est superstitieux et bigot, qu'il aime à boire; d'autres m'ont assuré que parmi les trois frères, c'est celui qui a le plus d'esprit, qu'il s'occupe à lire et à cultiver ses talents. Le Comte d'Artois, quoique d'ailleurs pas mal fait, a le défaut que ses sourcils s'approchent trop. Il aime comme on sçait les plaisirs, les femmes et le jeu. La Reine est certainement d'une figure prévenante et ressemble beaucoup à son frère l'Empereur, mais elle n'a pas ce regard morne et rebutant. Elle est d'une humeur très enjouée, mais on dit qu'elle a des passions un peu libres; au moins pour le bonheur de ses sujets elle devroit réprimer son inclination pour le jeu, qui a causé ici des grands malheurs parmi des familles. Il y a eu des délibérations au Parlement pour défendre les jeux aux chances inégales, et on disoit même qu'un arrêt de Parlement alloit paroître, cependant quelques jours après on a dit, qu'on avoit joué à la Cour et que la Reine avoit perdu une grosse somme. On dit presque partout, que la Reine est grosse. Madame et la Comtesse d'Artois se ressemblent beaucoup; quoique pas mal faites, elles n'ont pas la même grâce que la Reine, elles sont toutes deux petites. Madame Élisabeth, sœur du Roi a une très belle tête, le plus beau nez du monde, une belle bouche et des beaux yeux; mais elle est extrêmement puissante, elle n'a presque aucun cou et sa taille est extrêmement forte pour son âge. Elle ne se farde point comme fait la Reine et ses deux autres belles sœurs. [. . .]

London, 15 June 1781 (АРАН Фонд 1, опись 3, дело № 65, л. 208–210об)

Je ne sçaurois décider si c'est effet du climat, ou de l'éducation, qui fait qu'on remarque chez les enfans en Angleterre un caractère bien différent de celui qu'ils ont dans d'autres païs, peut être que toutes les deux causes concourent à former cet esprit de hardiesse, de courage, d'amour pour la liberté et l'indépendance, qu'on observe chez les plus petits enfans ici. J'ai été surpris de voir des enfans de trois ou quatre ans, qui montoient des échelles avec autant de hardiesse, que des garçons de quinze ans dans d'autres païs. Même le sexe ne fait aucune différence car les filles aussi bien que les garçons courent partout dans les rües, et même dans les meilleures maisons, on est accoutumé de faire promener les enfans quelque fois par jour, ce qui est bon et utile pour leur santé. L'abord des Anglois est en général très ressemblant à celui des autres peuples du nord, c'est à dire, ils semblent très posés et froids et ils n'ont pas le souplesse des François; mais on auroit tort de juger, qu'ils ne sont pas aussi serviables que les François; car au contraire, je suis persuadé, qu'ils se font un honneur de faire toutes les civilités possibles aux étrangers, et même quand on sçait gagner leur confiance, ils sont extrêmement francs et sans aucune réserve. J'oserois presque soutenir, qu'il n'y a aucune nation si instruite que les Anglois, car on trouve dans toutes les conditions des gens, qui ont quelque teinture des plusieurs sciences; mais pour des véritables sçavans, c'est à dire tels qui sont capables de faire des découvertes, leur nombre est ici extrêmement borné et principalement pour les mathématiques, où il n'y a pas plus que deux ou trois, qui ont quelque réputation. Et même je suis surpris, qu'ils ne sont pas tous extrêmement stupides, puisque leur principale nourriture, c'est la viande, qu'ils ont plus excellente, que dans aucun païs du monde. En remarquant que les Anglois mangent beaucoup de viande, boivent des vins très forts, comme le vin de Port, respirent un air extrêmement épais et rempli de la fumée de charbon, il n'est pas surprenant, qu'ils deviennent mélancoliques, tristes et rêveurs.

[...] Exceptée les Hollandois, il n'y a peut être aucune nation qui affecte tant de propreté, que les Anglois, principalement pour le linge et si je ne me trompe, c'est la nécessité, qui leur a fourni ce sentiment, puisque à cause de la fumée, on est obligé de changer le linge tous les jours. Les femmes sont ici presque en général habillées en blanc en été, et comme leur habits sont d'une blancheur éblouissante, cela fait un très bel effet. Je suis assez content de la manière de vivre des Anglois, excepté qu'ils restent si longtemps à la table, et qu'ils se remplissent de leur vins forts; car quoique on n'est pas forcé de boire avec les autres, si on se croit indisposé, si on ne veut pas passer pour singulier, on ne peut pas éviter de boire autant des santés des 'toastes' que les autres convives. Voici quelques coutumes singuliers des Anglois, qu'on ne trouve pas dans d'autres païs. En entrant dans une maison, on laisse le chapeau dans l'anti-chambre. Si on est à table avec des femmes, c'est une politesse d'un homme de demander, si quelqu'une des dames veut bien accepter de boire un verre de vin avec lui. Après le dessert les femmes sortent et c'est alors, que les bouteilles commencent à faire leur tour. Une des singularités les plus extraordinaires des Anglois, c'est de laisser leurs ongles pousser à une longueur presque égale à celle des Chinois.

[...] Quoique je n'ai vû que fort peu de chose du païs, tout ce que j'en ai vû étoit très beau et excellent, et il m'a paru, que tout le païs depuis Londres jusqu'à Oxford, n'étoit qu'un jardin bien cultivé. La situation de Oxford est charmante, et je n'ai jamais vû aucune ville, où il y a une Université, dont la situation est si belle. Leipzig en approche le plus, mais Oxford m'a paru avoir la préférence. L'Observatoire d'Oxford est le plus magnifique temple, qui jamais a été consacré au service de la Vénus Uranie, tant par rapport à l'élégance, que l'utilité. Il y a outre l'Observatoire plusieurs autres bâtiments très magnifiques à Oxford, comme le Collège de Christ, dans le quel la façade de la bibliothèque est certainement un très beau morceau d'architecture, la Bibliothèque de Ratclif, qui a un très beau dôme et celle du Collège de toutes les âmes, qui est aussi très beau. Mr Hornsby m'a fait beaucoup de civilité pendant mon séjour à Oxford, qui étoit fort court, et il m'a montré l'Observatoire et la construction de ses instruments, dont j'ai eu lieu d'être extrêmement content. Il a entre autres instruments une lunette achromatique de Dollond, avec un appareil de six oculaires, qui agrandissent depuis 50 jusqu'à 300 fois, toujours en augmentant de 50 à 50. J'ai regardé Saturne avec l'oculaire qui agrandit 200 fois, et je suis obligé d'avouer que je ne l'ai jamais vû si distinctement quoique il n'étoit qu'à 2° de hauteur.

London, 4 May 1781 (АРАН Фонд 1, опись 3, дело № 65, л. 195–197об)

Il y a ici un grand charlatan, qui se nomme le Doct. Graham et qui a un grand apparatus d'électricité avec des conducteurs très considérables. Il donne des représentations deux fois par semaine, il débite les choses les plus extravagants et très contraires à la modestie. Cependant pour ne pas exposer la pudeur des femmes qui y assistent, il fait éteindre les chandelles lorsque il commence a déclamer sur l'influence de l'électricité par rapport à la génération. Il a acheté une maison dans le Pall Mall, où il prétend établir ce qu'il appelle des *celestial beds* et il a fait mettre sur la porte l'inscription *Sacred to Hymen* avec bien d'autres extravagances des statues et des emblèmes conformes à ce sujet. Il a cependant été obligé d'ôter quelques statues nües qu'il avoit placé sur le frontispice de cette maison, mais on auroit aussi-bien eu raison de lui enjoindre de faire rayer l'inscription qui selon ses propres avertissements dans les gazettes est très équivoque et suspecte. Ce n'est pas un amusement des plus ordinaires d'entendre ce charlatan déclamer et de le voir. Sa manière de s'habiller répond à tout le reste, il étoit lorsque je le voyois dans un habit de velours, des dentelles très fines et des boucles des cheveux d'une grosseur énorme. Il va aussi dans une voiture qui est décorée et tout couverte des figures emblématiques.

Åbo, 8 November 1781 (АРАН Фонд 1, опись 3, дело № 66, л. 161–162об)

[...] Mon voyage par la Suède à Stockholm a été bien fatiguant et pénible, pendant un temps extrêmement pluvieux; mais j'en ai été amplement dédommagé par la reception la plus amicale de plusieurs de mes compatriotes. J'ai été présenté au Roi à Drotningholm; mais j'ai aperçu très clairement, que les études en général et les mathématiques en particulier, ne sont pas les choses les plus distinguées, ni à la cour de Suède, ni chez les personnes les plus distinguées de ce royaume. Mr Wargentin a beaucoup vieilli, il a l'oui foible. Mr Wilke Votre ancien ami est aussi maladif, il étoit indisposé de la goutte. Il se souvenoit toujours avec reconnoissance de Mr Votre Père et de Vous. Après avoir quitté Stockholm, j'ai fait une petite excursion à Upsal, où j'ai trouvé tous mes amis bien portans, excepté Mr Bergman, qui après une maladie très critique est encore extrêmement foible et dont la vie ne m'a paru être tout à fait hors du danger. Ce seroit dommage pour la Chemie, s'il viendroit de mourir, étant certainement un des chemistes les plus distingués. Mr Melanderhielm est toujours le même, extrêmement paresseux et si fort par la commodité, qu'il se remue avec une peine excessive de sa chaise à la porte. Mr Mallet, Professeur de Géométrie a pris d'embonpoint et se porte beaucoup mieux depuis qu'il s'est marié. Mr Prosperin est un très honnête homme, mais hypochondre au dernier degré. Le voyage d'Upsal jusqu'ici étoit aussi fort peu agréable, mais j'ai eu le plus sensible plaisir de trouver ma sœur et sa famille en bonne santé. Mon cœur a été vivement affecté de voir ma sœur environnée de tant de jolis enfans, elle en a dix, et je souhaiterois avec la meilleure grâce du monde d'être puissamment riche, pour faire la fortune de toutes mes nièces. Je Vous supplie cher Confrère de faire mes excuses à l'Académie, si je suis obligée de rester ici une quinzaine de jours. Il est encore incertain, quand je puisse avoir le plaisir de revoir ma sœur, on ne trouvera donc étrange, si cédant aux mouvements de la nature, je ne sçaurais si tôt m'arracher des bras de ma famille. [...]

Lexell's Publications

List of Lexell's publications arranged chronologically.

1. *Dissertatio sistens animadversiones subitaneas circa principium universae opticae Leibnitianum, quatenus idem in catoptrica adhibetur, praeside Martino Johanne Wallenio, respondens Andreas Johannes Lexell.* Holmiae (Stockholm): Lorentz Ludvig Grefing, 1759. (23p.)
2. *Aphorismi mathematico-physici, dissertatio praeside Jacobo Gadolin, respondens Andreas Johannes Lexell.* Aboa (Åbo): Jacob Merckell, 1760. (16p.)
3. *Dissertatio de methodo inveniendi lineas curvas, ex datis radiorum osculi proprietatibus, dissertatio praeside Andreas Johannes Lexell, respondens Ericus Östling.* Upsaliae (Uppsala): 1763.
4. "De integratione aequationis differentialis: $a^n \mathrm{d}^n y + ba^{n-1}\mathrm{d}^{n-1}y\mathrm{d}x + ca^{n-2}\mathrm{d}^{n-2}y\mathrm{d}x^2 + \cdots + ry\mathrm{d}x^n = X\mathrm{d}x^n$", NCASIP, Tomus XIV, Pars I, pp. 215–237, 1769.
5. "Methodus integrandi, nonnullis aequationum differentialium exemplis illustrata", NCASIP, Tom. XIV, Pars I, pp. 238–246, 1769.
6. *Recherches et calculs sur la vraie orbite elliptique de la comète de l'an 1769 et son tems périodique, executées sous la direction de M. Léonard Euler, par les soins de M. Lexell.* Saint Pétersbourg: Academiae Scientiarum, 1770. [OO Ser. II, Vol. 28. E389]
7. "Solutio problematis algebraici, de investigatione numerorum continue proportionalium, quorum datur summa a et summa quadratorum b", NCASIP, Tom. XV, pp. 107–126, 1770.
8. "De criteriis integrabilitatis formularum differentialium", NCASIP, Tom. XV, pp. 127–194, 1770.
9. "Determinatio longitudinis geographicae plurimorum locorum, in quibus eclipsis Solis a. 1769 observata fuit", NCASIP, Tom. XV, pp. 588–644, 1770.
10. "Longitudo observatorii Petropolitani ex observatione eclipsis Solis a. 1769 determinata", NCASIP, Tom. XV, pp. 645–654, 1770.

J.C.-E. Stén, *A Comet of the Enlightenment*, Vita Mathematica 17,
DOI 10.1007/978-3-319-00618-5,

11. "Lärda nyheter från Petersburg" ("Learned news from Petersburg", extract from a letter to H. G. Porthan), *Tidningar utgifne af et Sällskap i Åbo*, N° 7, 15 April 1771, pp. 54.
12. "Lärda nyheter från Petersburg". *Tidningar utgifne af et Sällskap i Åbo*, N° 21, 15 November 1771, p. 166.
13. "Uträkning öfver solens parallaxis, i anledning af observationer, som blifvit gjorde öfver Veneris gång genom solen år 1769" ("Calculations of the parallax of the Sun made upon observations of the transit of Venus across the disc of the Sun in 1769"), KVAH, pp. 220–234, 1771.
14. "Uträkning öfver solens parallax, i anledning af de uppå King Georg Eyland gjorde observationer öfver Veneris gång genom solen, år 1769" ("Calculations of the parallax of the Sun made upon observations on King George Island [Tahiti] of the transit of Venus across the disc of the Sun in 1769"), KVAH, pp. 301–310, 1771.
15. "De criteriis integrabilitatis formularum differentialium, dissertatio secunda", NCASIP, Tom. XVI, pp. 171–229, 1771.
16. "Demonstratio theorematis analytici a celeb. La Grange inventi", NCASIP, Tom. XVI, pp. 230–254, 1771.
17. "De parallaxi Solis conclusa ex transitu Veneris per Solem a. 1769 in insula Regis Georgii observato", NCASIP, Tom. XVI, pp. 586–648, 1771.
18. "De latitudine Veneris geocentrica tempore transitus anno 1769", NCASIP, Tom. XVI, pp. 586–648, 1771.
19. *Theoria motuum Lunae, nova methodo pertractata una cum tabulis astronomicis, unde ad quodvis tempus loca Lunae expedite computari possunt incredibili studio atque indefesso labore trium academicorum: Johannis Alberti Euler, Wolffgangi Ludovici Krafft, Johannis Andreae Lexell, opus dirigente Leonhardo Eulero*. Petropoli: Typis Academiae Scientiarum, 1772. [OO Ser. II, Vol. 22. E418]
20. "Om geographiska longituden för Åbo stad" ("On the geographical longitude of the city of Åbo [Turku]"), *Tidningar utgifne af et Sällskap i Åbo*, N° 21, 21 Maji 1772, pp. 165–168.
21. "Solutio problematis analytici", NCASIP, Tom. XVII, pp. 155–172, 1772.
22. "Disquisitio de investiganda parallaxi Solis ex transitu Veneris per Solem anno 1769", NCASIP, Tom. XVII, pp. 609–672, 1772.
23. *Disquisitio de investiganda vera quantitate parallaxeos Solis ex transitu Veneris ante discum Solis anno 1769, cui accedunt animadversiones in tractatum Rev. Pat. Hell de parallaxi Solis*. Petropolis: Typis Academiae Scientiarum, 1772.
24. "Formule pour calculer la parallaxe d'un astre, sa distance vraie du Zénith étant donnée". *Recueil pour les astronomes* (edited by Johann III Bernoulli), Tome II, pp. 311–314, Berlin, 1772.
25. "Några svenska orters geographiska longituder, uträknade af observationer öfver sol-förmörkelserne, åren 1764 och 1769" ("Geographical longitude of some Swedish towns computed from the observations of the solar eclipses in 1764 and 1769"), KVAH, pp. 44–66, continuation on pp. 117–136, 1773.

"Rättelser vid uträkningen öfver några svenska orters longituder" (Corrections to the former article), KVAH, pp. 170–172, 1774.

26. "Utdrag ur et bref ifrån Petersburg, dat. den 30 April 1773" ("Extract from a letter from Petersburg"), *Tidningar utgifne af et sällskap i Åbo*, N° 13, 15 Junii 1773, pp. 97–99.
27. "Epistola celeberrimi Lexellii, data 22. Febr. 1773. Petropoli", *Supplementum ad ephemerides astronomicas anni 1774 ad meridianum Vindobonensem.* Viennae, 1773, pp. 15–62.
28. "Observationes variae circa series, ex sinibus vel cosinibus arcuum arithmetice progredientium formatas", NCASIP, Tom. XVIII, pp. 37–70, 1773.
29. "Comparatio inter theoriam Lunae Illustris Euleri et tabulas recentiores Celeberrimi Mayeri", NCASIP, Tom. XVIII, pp. 537–567, 1773.
30. "Observatio eclipsis Solis facta Petropoli die $\frac{12}{23}$ Martii 1773", NCASIP, Tom. XVIII, pp. 571–601, 1773.
31. "Observationes astronomicas ab astronomis Academiae Imperialis Scientiarum Stephano Rumovski et And. I. Lexell anno 1773 institutas, recensuit And. Joh. Lexell", NCASIP, Tom. XVIII, pp. 602–630, 1773.
32. "Eine neue Methode, die Sonnenfinsternisse zu berechnen" (from a letter to Johann Bernoulli), AJEB für das Jahr 1776, pp. 174–180, 1774.
33. "Beobachtung der Sonnenfinsterniss vom 23 März 1773 zu St. Petersburg" (from a letter in French to Johann Bernoulli), AJEB für das Jahr 1776, 180–184, 1774.
34. "Tafeln der stündlichen heliocentrischen Bewegung der Planeten", AJEB für das Jahr 1776, 187–195, 1774.
35. "De resolutione polygonorum rectilineorum. Dissertatio prima", NCASIP, Tom. XIX, pp. 184–236, 1774.
36. "De differentia inter parallelum Lunae verum et apparentem", NCASIP, Tom. XIX, pp. 549–579, 1774.
37. "Nonnulla loca Lunae ex observationibus circa occultationes fixarum a Luna, anno 1774 Petropoli, et alibi institutis, determinata", NCASIP, Tom. XIX, pp. 580–609, 1774.
38. "Eclipses satellitum Jovis anno 1774, Petropoli in specula astronomica observatas, recensuit And. Joh. Lexell", NCASIP, Tom. XIX, pp. 636–638, 1774.
39. "Uplösning på ett astronomiskt problem" ("Resolution of an astronomical problem"), KVAH, pp. 87–93, 1775.
40. "Extract of a letter from Mr Lexel to Dr. Morton. Dated Petersburg, June 14, 1774", *Philosophical Transactions of the Royal Society of London*, Vol. 65, pp. 280–282, January 1775.
41. "Beobachtungen von Finsternissen der Jupiters Trabanten, zu St. Petersburg", AJEB für das Jahr 1777, pp. 126–127, 1775.
42. "Formeln, verschiedene Parallaxen zu berechnen", AJEB für das Jahr 1777, pp. 152–157, 1775.
43. "Eine kurze Methode, die Wirkungen der Parallaxe bey Durchgängen von Planeten vor der Sonne, zu berechnen", AJEB für das Jahr 1777, pp. 157–162, 1775.

44. "De resolutione polygonorum rectilineorum. Dissertatio secunda", NCASIP, Tom. XX, pp. 80–122, 1775.
45. "Theoremata nonnulla generalia de translatione corporum rigidorum", NCASIP, Tom. XX, pp. 239–270, 1775.
46. "Observationes astronomicas pro determinando situ geographico variorum per imperium Russicum locorum a Nob. Christophoro Eulero, annis 1769 et 1770 factas, recenset A. I. Lexell", NCASIP, Tom. XX, pp. 541–576, 1775.
47. "De observatione eclipseos Solis Petropoli die $\frac{15}{26}$ Augusti anno 1775 instituta", NCASIP, Tom. XX, pp. 577–592, 1775.
48. "Lettre de M. Lexell à M. le Marquis de Condorcet". In: *Théorie complète de la construction et de la manoeuvre des vaisseaux* par Léonard Euler. Paris: 1776, pp. 257–265. [OO Ser. II, Vol. 21. E426]
49. "Recherches sur la Période de la Comète, observée en 1770, d'après les observations de M. Messier", *Mémoires de l'Académie Royale des Sciences de Paris pour l'Année 1776*, pp. 638–651, 1779.
50. "Auszug aus einem französischen Schreiben des Herrn Lexell's an Herrn Bernoulli, dat. St. Petersburg $\frac{13}{24}$ Dec. 1775", AJEB für das Jahr 1778, pp. 65–68, 1776.
51. "Untersuchungen über die Länge der königl. Sternwarte zu Berlin" (translated by Johann Bernoulli), AJEB für das Jahr 1778, pp. 154–158, 1776.
52. "Anmerkungen über einige Widersprüche, die in den aus den Beobachtungen der Sonnenfinsternissen gezogenen Schlüssen gefunden worden" (translated by Johann Bernoulli), AJEB für das Jahr 1778, pp. 161–163, 1776.
53. "De methodis quae adhiberi possunt, ad integrandas aequationes differentiales lineares, quas differentialia plurium variabilium ingrediuntur", AASIP, Tom. I, Pars I, pp. 61–90, 1777 (impr. 1778).
54. "Solutio problematis astronomici, de inveniendo loco heliocentrico cometae ex dato loco eius geocentrico, si pro cognitis habeantur locus nodi et inclinatio orbitae, in qua cometa movetur", AASIP, Tom. I, Pars I, pp. 317–331, 1777 (impr. 1778).
55. "Tentamen astronomicum de temporibus periodicis cometarum et speciatim de tempore revolutionis cometae, anno 1770 observati", AASIP, Tom. I, Pars I, pp. 332–369, 1777 (impr. 1778).
56. "Conjectura de locis coeli, in quibus cometa anni 1770, in proximo suo ad perihelium reditu, e Tellure nostra conspici debet", AASIP, Tom. I, Pars II, pp. 328–342, 1777 (impr. 1780).
57. "Observationes circa methodum inveniendi longitudinem loci ex observata distantia Lunae a stella fixa", AASIP, Tom. I, Pars II, pp. 343–358, 1777 (impr. 1780).
58. "Formeln, die horizontale Strahlenbrechung aus der scheinbaren Weite der Sterne zu finden" (from a letter to Mr Bernoulli in French dated $\frac{1}{12}$ September 1776), AJEB für das Jahr 1779, pp. 33–37, 1777.
59. "Uplösning af det så kallade omvända centripetalkrafternas problem" ("Solution of the inverse problem of the so-called centripetal forces"), KVAH, pp. 55–59, 1778.

60. "En märkvärdig lärosats, om planernas vinklar uti triangulaira pyramider" ("A remarkable theorem concerning the dihedral angles in triangular pyramids"), KVAH, pp. 228–234, 1778.
61. "De reductione formularum integralium ad rectificationem ellipseos et hyperbolae", AASIP, Tom. II, Pars I, pp. 58–101, 1778 (impr. 1780).
62. "Ulteriores disquisitiones de tempore periodico cometae anno 1770 observati", AASIP, Tom. II, Pars I, pp. 317–352, 1778 (impr. 1780).
63. "Supplementum ad dissertationes, Novis Commentariis insertas, de eclipsibus solaribus annis 1769 et 1773 observatis, ut et occultationibus fixarum a Luna", AASIP, Tom. II, Pars I, pp. 353–393, 1778 (impr. 1780).
64. "Réflexions sur le temps périodique des comètes en général et principalement sur celui de la comète observée en 1770", AASIP, Tom. II, Pars II, *Histoire de l'Académie*, pp. 12–34, 1778 (imprimé en 1781). (et separatim)
65. "Ad dissertationem de reductione formularum integralium ad rectificationem ellipseos et hyperbolae additamentum", AASIP, Tom. II, Pars II, pp. 55–84, 1778 (impr. 1781).
66. "De eclipsi Solis anno 1778 die 24 Junii st. nov. observata", AASIP, Tom. II, Pars II, pp. 303–331, 1778 (impr. 1781).
67. "Supplementum ad dissertationem, de eclipsi Solis anno 1778 observata", AASIP, Tom. II, Pars II, pp. 332–344, 1778 (impr. 1781).
68. "Über die Umlaufszeit des Cometen vom Jahr 1770" (from a letter to Johann Bernoulli in French dated $\frac{5}{16}$ March 1778), AJEB für das Jahr 1781, pp. 21–31, 1778.
69. "De epicycloidibus in superficie sphaerica descriptis", AASIP, Tom. III, Pars I, pp. 49–71, 1779 (impr. 1781).
70. "Disquisitio de tempore periodico cometae anno 1770 observati" (communicated by Nevil Maskelyne), *Philosophical Transactions of the Royal Society of London*, Vol. 69, pp. 68–86, January 1779.
71. "De aestimando tempore, quo diameter Solis per circulum quendam sive verticalem, seu horizonti parallelum, transire videtur", AASIP, Tom. III, Pars I, pp. 279–299, 1779 (impr. 1783).
72. "Observationes de problemate, quo quaeritur elevatio poli, ex observata altitudine Solis, et observato quoque tempore, quo diameter Solis filum aliquod, sive verticaliter, sive horizontaliter dispositum, pertransit", AASIP, Tom. III, Pars I, pp. 300–309, 1779 (impr. 1783).
73. "Continuatio dissertationis de methodis integrandi aequationes differentiales lineares in tomo actorum I", AASIP, Tom. III, Pars II, pp. 52–80, 1779 (impr. 1783).
74. "De elementis orbitae cometae a. 1773 observati, ubi praeprimis disquiritur, utrum huius cometae tempus periodicum assignare liceat?", AASIP, Tom. III, Pars II, pp. 335–358, 1779 (impr. 1783).
75. "De perturbatione in motu Telluris ab actione Veneris oriunda", AASIP, Tom. III, Pars II, pp. 359–392, 1779 (impr. 1783).

76. "Aus einem französischen Schreiben des Herrn Prof. Lexell an Hrn. Bernoulli, dat. St. Petersburg, den $\frac{3}{14}$ Februar 1780", AJEB für das Jahr 1783, p. 62, 1780.
77. "Betrachtungen über die von Herrn Bode in die Ephemeriden auf das J. 1780 (Samml. 164 S) eingerückte Berechnung, um die geographische Länge von Mannheim vermittelst der Bedeckung des Aldebaran am 29. Januar. 1776 zu bestimmen", AJEB für das Jahr 1783, pp. 63–68, 1780.
78. "Auflösung der Astronomischen Aufgabe: Für eine gegebene Zeit den heliocentrischen Ort eines Cometen zu finden, wenn für diese Zeit sowohl die geocentrische Länge und Breite, als die Länge des Knoten und die Neigung der Bahn bekannt sind", AJEB für das Jahr 1783, pp. 68–73, 1780.
79. "Betrachtungen über die Bahn des im Jahr 1773 erschienenen Cometen", AJEB für das Jahr 1783, pp. 73–78, 1780.
80. "Quantité de l'eau de pluie observée à Saint Pétersbourg en 1778, 1779 et 1780", AASIP, Tom. IV, Pars I, *Histoire de l'Académie*, pp. 17–18, 1780 (impr. 1783).
81. "Hauteurs de l'eau dans la Néva observées en 1778, 1779 et 1780", AASIP, Tom. IV, Pars I, *Histoire de l'Académie*, pp. 19–20, 1780 (impr. 1783).
82. "Recherches sur la nouvelle Planète, découverte par M. Herschel et nommée Georgium Sidus", AASIP, Tom. IV, Pars I, pp. 303–329, 1780 (impr. 1783). (avec un supplément)
83. "Solutiones quorundam problematum astronomicorum ad doctrinam de motu planetarum et cometarum in sectionibus conicis pertinentium", AASIP, Tom. IV, Pars I, pp. 330–369, 1780 (impr. 1783).
84. "Solutio problematis geometrici, in Actis Academiae Scientiarum Berolinensis, pro anno 1776 a celeb. Castillon propositi", AASIP, Tom. IV, Pars II, pp. 70–90, 1780 (impr. 1784).
85. "Essai sur l'orbite elliptique de la comète de 1763", AASIP, Tom. IV, Pars II, pp. 324–346, 1780 (impr. 1784).
86. "Mémoire sur les élémens de la comète de l'année 1780", AASIP, Tom. IV, Pars II, pp. 347–358, 1780 (impr. 1784).
87. "Des Herr Prof. Lexell Nachtrag zu seinem in den Ephemeriden für 1783, Seite 66 und folg. stehenden Aufsatze über die Parallaxen-Berechnung" (from a letter to Johann Bernoulli dated 7 July 1780), AJEB für das Jahr 1784 (Berlin), pp. 187–189, 1781.
88. "Solutio problematis geometrici ex doctrina sphaericorum", AASIP, Tom. V, Pars I, pp. 112–126, 1781.
89. "Solutio problematis mechanici", AASIP, Tom. V, Pars I, pp. 196–208, 1781, (impr. 1784).
90. "Solution d'une question astronomique", AASIP, Tom. V, Pars I, pp. 351–366, 1781, (impr. 1784).
91. "Integratio formulae cuiusdam differentialis per logarithmos et arcus circulares", AASIP, Tom. V, Pars II, pp. 104–117, 1781 (impr. 1785).
92. "Meditationes de formula, qua motus laminarum elasticarum in annulos circulares incurvatarum, exprimitur", AASIP, Tom. V, Pars II, pp. 185–218, 1781 (impr. 1785).

93. "Examen criticum observationum a celeb. Messier circa cometam anni 1770 institutarum", AASIP, Tom. V, Pars II, pp. 351–372, 1781 (impr. 1785).
94. "Untersuchungen über die Bahn des neuen Planeten" (from a letter in French to Johann Bernoulli dated St. Petersburg $\frac{4}{15}$ April, 1782), AJEB für das Jahr 1785 (Berlin), pp. 201–204, 1782.
95. "De proprietatibus circulorum in superficie sphaerica descriptorum", AASIP, Tom. VI, Pars I, pp. 58–103, 1782 (impr. 1786).
96. "De motu corporis ad duo centra virium fixa attracti", AASIP, Tom. VI, Pars I, pp. 157–190, 1782 (impr. 1786).
97. "Determinatio errorum qui in longitudines et latitudines alicuius cometae geocentricas inducuntur, ex commissis erroribus in elementis orbitae", AASIP, Tom. VI, Pars I, pp. 281–311, 1782 (impr. 1786).
98. "Demonstratio nonnullorum theorematum ex doctrina sphaerica", AASIP, Tom. VI, Pars II, pp. 85–95, 1782 (impr. 1786).
99. "De occultationibus quibusdam singularibus, sive stellarum fixarum a planetis, seu etiam planetarum a se invicem", AASIP, Tom. VI, Pars II, pp. 291–320, 1782 (impr. 1786).
100. "Extract of a letter from Mr A. J. Lexell", *The London Magazine, or Gentleman's Monthly Intelligencer*, pp. 446–448, 1783.
101. "Recherches sur la nouvelle planète, découverte par M. Herschel et nommée par lui Georgium Sidus", NAASIP, Tom. I, *Histoire de l'Académie*, pp. 69–82, 1783. (et separatim) (Lues à l'assemblée publique de l'Académie Impériale des Sciences de Saint Pétersbourg, le 11 de Mars 1783)
102. "Rapport au sujet d'un nouvel instrument du Capitaine Burdett, nommé compas optique", NAASIP, Tom. I, *Histoire de l'Académie*, pp. 111–114, 1783.
103. "Rapport fait à l'Académie au sujet d'un Ouvrage de M. l'Abbé Rochon, qui a pour titre: Recueil de Méchanique et de Physique" (par Mrs. Roumovski, Fuss et Lexell), NAASIP, Tom. I, *Histoire de l'Académie*, pp. 115–119, 1783.
104. "Rapport au sujet d'un nouvel instrument nautique envoyé et soumis à l'approbation de l'Académie par M. de Magellan" (signé par Mrs. Roumovski, Krafft et Lexell), NAASIP, Tom. I, *Histoire de l'Académie*, pp. 141–150, 1783.
105. "Disquisitio de theoremate quodam singulari Cel. Lamberti, pro aestimandis temporibus, quibus arcus sectionum conicarum describuntur a corporibus, quae ad alterutrum focum attrahuntur viribus reciproce proportionalibus quadratis distantiarum", NAASIP, Tom. I, *Histoire de l'Académie*, pp. 140–183, 1783.
106. *Polygonometrie, oder Anweisung zur Berechnung jeder gradlinichten Figur* (translated by J. F. Lempe). 2 Theile. Leipzig: Johann Jacob Kindervater, 1783.
107. "Integration af en differential-formel" ("Integration of a differential formula"), KVANH, pp. 197–204, 1784.

Source Material in Saint Petersburg

Contents of the Lexell files at the Saint Petersburg Branch of the Archives of the Russian Academy of Sciences (АРАН).

—Разряд I, опись 91. Рукописи трудов акад. А. И. Лекселя по астрономии и математике. Manuscripts of academician A. I. Lexell's research in astronomy and mathematics.

—Разряд V, оп. L, № 14. Краткие биографические сведения. Список трудов в 1769–1789 гг. Дело об отпуске его за границу сроком на один год (1780–1782 гг.) Brief biographical data. Work from 1769–1789. Mission abroad for one year during 1780–1782.

—Фонд 3, оп. 1, кн. 313, л. 302–304. 11 августа 1768 г. О приглашении его на службу в АН астрономической обсерватории с жалованием по 400 руб. в год и о высылке 100 руб. на проезд в Санкт-Петербург. Concerning the invitation to work at the Academy's astronomical observatory for a salary of 400 Roubles/year and the expense of 100 Roubles for the move to Saint Petersburg.

—Фонд 3, оп. 1, № 314, л. 309–312. Фонд 3, оп. 1, № 539, л.684. 20 октября 1768 г. О его прибытии в Петербург. Рапорт архивариуса Унгебауера. On his arrival to Saint Petersburg. Note by archivist Ungebauer.

—Фонд 3, оп. 1, кн. 540, § 231. 1769 г. Маг., произв. в адъюнкты. Magister appointed Adjunct.

—Фонд 3, оп. 1, кн. 323, л. 339–341. 1770 г., сент. 10. О напечатании Recherches et calculs sur la vraie orbite elliptique de la comete de l'an 1769 et son tems périodique, executées par les soins de M. Lexell sous la direction de M. L. Euler. Образцы рисунков. On the printing of the mentioned work, including plates.

—Фонд 3, оп. 20, № 19. 1771 г. Записка об организации Астрономического Кабинета в нижнем этаже "Волкова дома". Correspondence regarding the organisation of an astronomical study in the ground floor of Volkov's home (possibly the lodgings of the Academy's former translator).

—Фонд 3, оп. 1, № 344, л. 530, 4 мая 1780 г. Выписка из журнала заседания комиссии АН об увольнении академика Лекселя. Extract from the minutes of the session of the Academy of Sciences concerning the resignation of academician Lexell.

—Фонд 3, оп. 1, кн. 331, л. 320–323. 27 марта 1783 г. Переписка о напечатании и продаже его речи, произнесенной в общем собрании АН. Correspondence regarding the printing and selling of his talk at the public assembly of the Academy of Sciences.

J.C.-E. Stén, *A Comet of the Enlightenment*, Vita Mathematica 17,
DOI 10.1007/978-3-319-00618-5,

—Фонд 3, оп. 1, кн. 335, л. 189. 1783 г. Рапорт акад. А. Протасова о поручении акад. Лекселю описания типографической машины. Academician Protasov's report on academician Lexell being charged with delivering a description of a typographical machine.
—Фонд 3, оп. 1, № 339, л. 301–308, апрель 1784 г. Списки книг, потребных Лекселю, Крафту и Ферберу. Lists of books that are required for Lexell, Krafft and Ferber.
—Фонд 136, оп. 1, № 155, л. 115–158. Auflösung der geradlichten Vielecke (*Novi Comm.* Tom. XIX, pp. 184–236, Tom. XX, pp. 80–122).
—Фонд 3, оп. 1, кн. 344, л. 158–159. О смерти ординарного академика Лекселя 30 ноября 1784 г. On the death of ordinary member Lexell 30 November 1784.
—Фонд 1, оп. 3, № 66. Фонд: Конференция АН. Eingekommene Briefe 1780–1781 (520 листов). Microfilmed on 23 September 1954.
№ 52: л. 101, 101 об.
№ 68: л.130, 130об, 131.
№ 84: л.161, 161об, 162, 162об.
—Лубименко И. И.: Заграничная командировка акад. А. И. Лекселя. Lexell's foreign mission.
—Разряд V. оп. L-14.
1. 1769 г. Выписки из протокола заседания конференции от 20 марта 1769 г. О назначении его адъюнктом. На нем. яз. и перевод на русский яз. 2 л. Extract of the minutes of the session of the conference held on March 20, 1769, regarding his appointment as adjunct. In German, with a Russian translation. 2 pages.
—2. 1772 г. Копия с журнала комиссии АН от дек. 21 1772 г. Об отпуске Лекселя в г. Або с 30 XII 1772 по 28 I 1773. К отпускн. отношению акад. комиссии в Ямскую канцелярию от 29 дек. 1772 г. 2 л. Copy of the journal of the commission of the Academy of Sciences of 21 December 1772. Concerning a leave of absence (mission) for Lexell to travel to Åbo from 30 Dec. 1772 to 28 Jan. 1773. 2 pages.
—3. 1779 г. Список работ Лекселя 1769–1779 гг. Написан неизв. рукой. 2 л. A list of Lexell's works 1769–1779. Unknown author. 2 pages.
—4. 1780–1782 гг. Дело об отпуске акад. Лекселя в чужие края на один год с произвождением ему жалования. 35 л. Regarding the foreign mission of Academician Lexell and grant of salary. 35 pages.
—5. 1784 г. Краткая биография А. Лекселя, составленная Юст. Иоахимом Леппингом. На нем. яз. 2 л. Short biography of A. Lexell compiled by Just. Joachim Lepping. In German. 2 pages.
—6. 1770 г., авг. 20/21. Фотокопия письма Лекселя к Варгентину, находящегося в библ. Стокгольмской АН. На швед. яз. Прил. письмо д-ра Гольмберга от 22 окт. 1952 г. из Стокгольма. Photocopy of a letter from Lexell to Wargentin (KVAC, Stockhom). In Swedish. Attached is Dr. Holmberg's letter from Stockholm on 22 October 1952.
(Material concerning Kepler: Фонд 285, оп.1, № 39. Выдержки из отзывов академиков Г. В. Крафта и А. И. Лекселя о собрании рукописей Кеплера и об их издании. На франц. яз. 8 л. Отзывы доложены в заседании конференции АН 16. окт. 1775 г. Excerpts from the review of the academicians G. W. Krafft and A. J. Lexell on the Kepler manuscript collection and its publication. 8 pages in French. Delivered at the Academic Conference on 16 October 1775 [o.s.])

List prepared by M. V. Krutikova, 26 October 1953. Recorded by Carl-Fredrik Geust, 17–19 February 2008.

Bibliography

1. Altmann, Simon L. *Rotations, Quaternions, and Double Groups*. Oxford: Clarendon, 1986.
2. Artemyeva, Tatiana. "The Status of Intellectual Values in the Russian Enlightenment." In *The Northern Lights, Facets of the Enlightenment Culture*, ed. Tatiana Artemyeva, Mikhail Mikeshin, and Vesa Oittinen. The Philosophical Age, Almanac 36, 25–41. St. Petersburg: Saint Petersburg Centre for the History of Ideas, 2010.
3. Aspaas, Per Pippin. "Le Père Jésuite Maximilien Hell et ses relations avec Lalande." In [17, pp. 129–148].
4. ——. *Maximilianus Hell (1720–1792) and the Eighteenth-Century Transits of Venus: A Study of Jesuit Science in Nordic and Central European Contexts*. Thesis. University of Tromsø, 2012.
5. Aspaas, Per Pippin, and Truls Lynne Hansen. "The Role of the Societas Meteorologica Palatina (1781–1792) in the History of Auroral Research". *Acta Borealia : A Nordic Journal of Circumpolar Societies* 29, N° 2 (2012): 157–176.
6. Austin, Paul Britten. *The Life and Songs of Carl Michael Bellman. Genius of the Swedish Rococo*. Malmö: Allhem, 1967.
7. Badinter, Élisabeth. *Les passions intellectuelles III. Volonté de pouvoir (1762–1778)*. Paris: Fayard, 2007.
8. Baker, Keith Michael. *Condorcet: From Natural Philosophy to Social Mathematics*. Chicago: The University of Chicago Press, 1975.
9. Barrow-Green, June. "The Dramatic Episode of Sundman." *Historia Mathematica* 37, N° 2 (2010): 164–203.
10. Barton, H. Arnold. "Gustav III of Sweden and the Enlightenment." *Eighteenth Century Studies* 6, N° 1 (1972): 1–34.
11. Bergquist, Olle. *Det skvallrande ansiktet: Fysiognomikens historia* [History of Physiognomy]. Lund: Sekel, 2009.
12. Bernoulli, Johann (III). *Reisen durch Brandenburg, Pommern, Preussen, Curland, Russland und Pohlen in den Jahren 1777–1778*. Leipzig: Caspar Fritsch, 1779.
13. ——. *Liste des astronomes connus, actuellement vivans, par ordre alphabétique des lieux de leur demeure*, avec supplémens au *Recueil pour les astronomes*. Berlin: Haude & Spener, 1776.
14. Betts, Jonathan. "John Hyacinth de Magellan (1722–1790), Part 3: The Later Clocks and Watches." *Antiquarian Horology* 30, N° 1 (2007).
15. Birembaut, Arthur. "L'Académie royale des Sciences en 1780 vue par l'astronome suédois Lexell." *Revue d'Histoire des Sciences* 10 (1957): 148–166.

J.C.-E. Stén, *A Comet of the Enlightenment*, Vita Mathematica 17,
DOI 10.1007/978-3-319-00618-5,

16. Bogolyubov, N. N., G. K. Mikhailov, and A. P. Yushkevich. *Euler and Modern Science*. Translation from the Russian by R. Burns. Washington, DC: Mathematical Association of America, 2007.
17. Boistel, Guy, Jérôme Lamy, and Colette Le Lay, eds. *Jérôme Lalande (1732–1807). Une trajectoire scientifique*. Rennes: Presses Universitaires de Rennes, 2010.
18. Bradley, R. E., and C. E. Sandifer, eds. *Leonhard Euler: Life, Work and Legacy*. Amsterdam: Elsevier, 2007.
19. Braunmühl, Anton von. *Vorlesungen über Geschichte der Trigonometrie*. Leipzig: Teubner, 1903.
20. Calinger, Ronald S., and Elena N. Polyakhova. "Princess Dashkova, Euler, and the Russian Academy of Sciences." In [18, pp. 75–96].
21. Callmer, Christian. "Svenska studenter i Göttingen under 1700-talet" [Swedish Students in Göttingen During the Eighteenth Century]. *Lychnos årsbok* (1956): 1–29.
22. Candaux, Jean-Daniel, Sophie Capdeville, Michel Grenon, René Sigrist, and Vladimir Somov, eds. *Deux astronomes genevois dans la Russie de Catherine II, Journal de voyage en Laponie russe de Jean-Louis Pictet et Jacques-André Mallet pour observer le passage de Vénus devant le disque solaire 1768–1769*. Ferney-Voltaire: Centre International d'Étude du XVIIIe Siècle, 2005.
23. Cantor, Moritz. *Vorlesungen über Geschichte der Mathematik*. Vierter Band, 1759–1799. Leipzig: Teubner, 1908.
24. Carpelan, Tor. *Åbo i genealogiskt hänseende på 1600- och början af 1700-talen* [Åbo of the Late 17th and Early 18th Century in a Genealogical Perspective]. Helsinki: Finska Litteratur-Sällskapets tryckeri, 1890.
25. Cederberg, Arno Rafael. *Pehr Wargentin als Statistiker. Untersuchungen in der Geschichte der Bevölkerungsstatistik während der zweiten Hälfte des 18. Jahrhunderts*, Annales Academiae Scientiarum Fennicae, Series B, vol. 4/4, 1–185. Helsinki: Suomalaisen tiedeakatemian toimituksia, 1919.
26. ——. "Muistiinpanoja suomalaisista aineksista Bergiuksen kopiokokoelmassa Ruotsin tiedeakatemian arkistossa" [Material Related to Finnish Scientists in the Bergius Collection at KVA in Stockholm]. *Historiallinen arkisto* XXX, N° 4 (1923). Helsinki.
27. Chandler, Seth Carlo. "On the Action of Jupiter in 1886 upon comet d 1889, and the Identity of the Latter with Lexell's Comet of 1770." *Astronomical Journal* 9, N° 205 (1889): 100–103.
28. *Collectio omnium observationum quae occasione transitus Veneris per Solem A. MDCCLXIX iussu Augustae per Imperium Russicum institutae fuerunt una cum theoria indeque deductis conclusionibus*. Petropoli: Academia Scientiarum, 1770.
29. Crilly, Tony. "Review of: James Joseph Sylvester: Jewish Mathematician in a Victorian World by Karen Hunger Parshall." *The Mathematical Intelligencer* 29, N° 2 (2007): 72–75.
30. Darrigol, Olivier. *A History of Optics from Greek Antiquity to the Nineteenth Century*. Oxford: Oxford University Press, 2012.
31. Daschkoff, Ekaterina Romanovna. *Furstinnan Dasjkovs memoarer: efter originaltexten i Vorontsovska arkivet*. Stockholm: Wahlström & Widstrand, 1916. (Swedish translation of *Mémoires de la Princesse Daschkoff*. Paris, 1859)
32. Dauben, Joseph W., and Christoph J. Scriba. *Writing the History of Mathematics: Its Historical Development*. Basel: Birkhäuser, 2002.
33. Dawson, John W. "In Quest of Kurt Gödel: Reflections of a Biographer." *Notices of the AMS* 53, N° 4 (April 2006): 444–447.
34. Donner, Anders. *Den astronomiska forskningen och den astronomiska institutionen vid det finska universitetet* [Astronomical Research and the Astronomical Institution in the Finnish University]. Del I. Akademisk inbjudningsskrift. Helsinki, 1907.
35. Dulac, Georges. "L'astronome Lexell et les athées Parisiens (1780–1781)." *Dix-Huitième Siècle*, N° 19 (1987): 347–361.
36. ——. "Un nouveau La Mettrie à Pétersbourg: Diderot vu de l'Académie impériale des Sciences." *Recherches sur Diderot et sur l'Encyclopédie*, N° 16 (1994): 19–43.

37. ——. "La vie académique à Saint-Pétersbourg vers 1770 d'après la correspondance entre J. A. Euler et Formey." In *Académies et Sociétés Savantes en Europe (1650–1800)*, ed. D.-O. Hurel and G. Laudin, 221–263. Paris: Honoré Champion, 2000.
38. Dumont, Simone. *Un astronome des Lumières: Jérôme Lalande*. Paris: Observatoire de Paris, Vuibert, 2007.
39. *Encyklopädie der Mathematischen Wissenschaften mit Einschluß ihrer Anwendungen*. 6 Bände. Leipzig: Teubner, 1904–1935.
40. Enneper, Alfred. *Elliptische Functionen. Theorie und Geschichte*. Academische Vorträge. Halle a/S: Nebert, 1876.
41. Euler, Johann Albrecht. "Précis de la vie de M. Lexell" [Lexell's Obituary], NAASIP, Tom. II, *Histoire de l'Académie de l'an 1784*, pp. 12–15, 1788.
42. Fejes Tóth, László. *Lagerungen in der Ebene auf der Kugel und in Raum*. Die Grundlehren der mathematischen Wissenschaften, Band LXV. Berlin: Springer, 1953.
43. Fellmann, Emil A. *Leonhard Euler*. Reinbek bei Hamburg: Rowohlt Taschenbuch Verlag, 1995. In English: Basel: Birkhäuser, 2007.
44. Ferreiro, Larrie D. *Measure of the Earth. The Enlightenment Expedition that Reshaped Our World*. New York: Basic Books, 2011.
45. Fitzpatrick, Martin, Peter Jones, Christa Knellwolf, and Iain McCalman, eds. *The Enlightenment World*. London: Routledge, 2004.
46. Forsman, Juho Rudolf. *Kaarle Fredrik Mennander ja hänen aikansa* [Carl Fredrik Mennander and His Times]. Turku: Turun suomalainen kirjapaino, 1900.
47. *Förtekning på [A.J. Lexells] böcker som i St. Pettersburg komma at försäljas på öpen auction den ... 1785* [List of Lexell's Books that are to be Auctioned in St. Petersburg]. Åbo: Joh. C. Frenckells Enka, 1785.
48. Francesco, Grete de. *Die Macht des Charlatans*. Basel: Benno Schwabe, 1937.
49. Frängsmyr, Carl. *Klimat och karaktär: Naturen och människan i sent svenskt 1700-tal* [Climate and Character: Nature and Man in the Late 18th-Century Sweden]. Stockholm: Natur och Kultur, 2000.
50. Frängsmyr, Tore. "The Mathematical Philosophy." In *The Quantifying Spirit in the Eighteenth Century*, ed. T. Frängsmyr, J. L. Heilbron, and R. E. Rider. Berkeley: University of California Press, 1990: 27–44.
51. ——. *Sökandet efter upplysningen: Perspektiv på svenskt 1700-tal* (Rev. ed., in Swedish). Stockholm: Natur och Kultur, 2006. (French translation: *À la recherche des Lumières: une perspective suédoise*. Bordeaux: Presses Universitaires de Bordeaux, 1999)
52. Fumaroli, Marc. *Quand l'Europe parlait français*. Paris: de Fallois, 2001.
53. Gardberg, Carl Jacob, and Pentti Virrankoski. "Lexell, Jonas (1699–1768)." Biographical entry in [78].
54. Gilain, Christian. "Sur la correspondance de Condorcet avec Euler et ses disciples de Pétersbourg." *Mélanges de l'École française de Rome. Italie et Méditerranée*, Tom. 108, N° 2 (1996): 517–531.
55. Gillispie, Charles Coulston. *Dictionary of Scientific Biography*. New York: Charles Scribner's Sons, 1970–1980.
56. Grate, Pontus, ed. *Solen och Nordstjärnan. Frankrike och Sverige på 1700-talet*. Stockholm: Nationalmuseum, 1993. French edition: *Le Soleil et l'Étoile du Nord. La France et la Suède au XVIIIe siècle*. Paris: Réunion des Musées Nationaux, 1994.
57. Grigorian, A. T., and A. P. Yushkevich. "Anders Johan Lexell." Biographical entry in [55].
58. Grosser, Morton. *The Discovery of Neptune*. Cambridge: Harvard University Press, 1962.
59. Hakfoort, Casper. *Optics in the Age of Euler: Conceptions of the Nature of Light 1700–1795*. Cambridge: Cambridge University Press, 1995.
60. Häkli, Esko, ed. *Gelehrte Kontakte zwischen Finnland und Göttingen zur Zeit der Aufklärung, Ausstellung aus Anlass den 500-jährigen Jubiläums des finnischen Buches, Universitäts-Bibliothek Helsinki und Universitäts-Bibliothek Göttingen*. Göttingen: Vandenhoeck und Ruprecht, 1988.

61. Hankins, Thomas L. *Jean d'Alembert. Science and the Enlightenment*. New York: Gordon and Breach, 1970.
62. Hebbe, P. M. "Svensk-ryska vetenskapliga förbindelser under 1700-talets senare hälft" [Swedish-Russian Scientific Contacts in the Latter Part of the 18th Century]. *Lychnos årsbok* (1938): 388–392.
63. Heikel, Ivar A. *Helsingfors Universitet 1640–1940*. Helsinki, 1940.
64. Heilbron, John L. *Elements of Early Modern Physics*. Berkeley: University of California Press, 1982.
65. Henkel, Thomas. "August Ludwig Schlözers Russlandbeziehungen—Briefwechsel, Wissenstransfer, Spätwerk." In [118, pp. 200–249].
66. Hintikka, Toivo Johannes. "A. J. Lexellin elämänvaiheista. Hänen syntymäpäivänsä 200-vuotispäivän johdosta" [The Life of A. J. Lexell on the Occasion of his Bicentenary]. Helsinki: Suomalaisen Tiedeakatemian esitelmät ja pöytäkirjat, 1940: 101–129.
67. Holmberg, Peter. "Anders Planman och solens parallax." *Arkhimedes* [Periodical of the Finnish Mathematical Society and the Physical Societies in Finland], N° 5 (2008): 15–21.
68. Holmberg, Peter, and Tapio Markkanen. "Jacob Gadolin, en mångfasetterad vetenskapsman i Åbo på 1700-talet." *Nordenskiöld-samfundets tidskrift* 69 (2010): 33–60. Helsinki.
69. Home, Roderick Weir. *Electricity and Experimental Physics in Eighteenth-Century Europe*. Hampshire: Variorum, 1992.
70. Houzel, Christian. *Fonctions elliptiques et intégrales abéliennes: Abrégé d'histoire des mathématiques 1700–1900* (éd. J. Dieudonné). Paris: Hermann, 1978, Vol. II, pp. 1–113. Reprinted in Ch. Houzel, *La géométrie algébrique: Recherches historiques* (Collection Sciences dans l'Histoire). Paris: Blanchard, 2003, pp. 81–190.
71. Im Hof, Ulrich. *Das Europa der Aufklärung*. München: Beck, 1993.
72. Institut de France. *Index Biographique de l'Académie des Sciences*. Paris: Gauthier-Villars, 1979.
73. Israel, Jonathan I. *Radical Enlightenment. Philosophy and the Making of Modernity*. Oxford: Oxford University Press, 2001.
74. Jangfeldt, Bengt. *Svenska vägar till S:t Petersburg* [Swedish Ways to Saint Petersburg]. Stockholm: Wahlström & Widstrand, 1998.
75. Юшкевич, Адольф Павлович. *История Математики в России до 1917 года* [A. P. Yushkevich: History of Mathematics in Russia Before 1917]. Москва: Академия Наук СССР, Наука, 1968.
76. Juškevič, A. P. et R. Taton (réd.). *Leonhardi Euleri Opera Omnia, Series 4A, Volumen 5, Commercium epistolicum: Correspondance de Leonhard Euler avec A. C. Clairaut, J. d'Alembert et J. L. Lagrange*. Basel: Birkhäuser, 1980.
77. Kallinen, Maija. *Change and Stability. Natural Philosophy at the Academy of Turku 1640–1713*. Studia Historica, N° 51. Helsinki: Suomen historiallinen seura, 1988.
78. *Kansallisbiografia* [Finnish National Biography]. Helsinki: Suomalaisen Kirjallisuuden Seura, 2007.
79. Kline, Morris. *Mathematical Thought from Ancient to Modern Times*, Vol. 2. New York: Oxford University Press, 1972.
80. Klinge, Matti, Rainer Knapas, Anto Leikola, and John Strömberg. *Helsingfors Universitet 1640–1990*. Första delen: Kungliga Akademien i Åbo 1640–1808. Helsinki: Otava, 1988. Abridged German translation: *Eine nordische Universität: die Universität Helsinki, 1640–1990*. Helsinki: Otava, 1992. Updated English edition: *A European University. The University of Helsinki 1640–2010*. Helsinki: Otava, 2010.
81. Kotivuori, Yrjö. *Ylioppilasmatrikkeli 1640–1852* (Register of students at the Finnish University. In Finnish). Helsinki: University Press, 2005. (Available online at http://www.helsinki.fi/ylioppilasmatrikkeli/)
82. Kronk, Gary W. *Cometography: A Catalog of Comets*, Vol. 1, Ancient–1799. Cambridge: Cambridge University Press, 1999.
83. Lacroix, Sylvestre-François. *Traité du calcul différentiel et du calcul intégral*, 3 Vols. Paris: Duprat, 1797–1798.

84. ——. *Traité des différences et des séries*. Paris: Duprat, 1800.
85. Lagrange, Joseph Louis. *Œuvres de Lagrange* (14 Vols. ed. J.-A. Serret). Paris: Gauthier-Villars, 1867–1892.
86. Lagström, Hugo. "Lexell." *Genos* (Periodical of the Genealogical Society of Finland) 32 (1961): 61–69.
87. Lagus, Wilhelm. *Erik Laxman, hans lefnad, resor, forskningar och brefvexling* [Erik Laxman, His Life, Travels, Researches and Correspondence]. Bidrag till kännedom af Finlands natur och folk, 34. Helsingfors: Finska Vetenskaps-Societeten, 1880.
88. ——. *Album studiosorum academiae Aboensis MDCXL–MDCCCXXVII. Åbo akademis studentmatrikel å nyo upprättad*. Förra afdelningen 1640–1740. Helsingfors: Svenska Litteratursällskapet i Finland (skrifter 11:1–3). 1889–91. — Senare afdelningen 1740–1827. Helsingfors: Svenska Litteratursällskapet i Finland (skrifter 11:4–6). 1892–95.
89. Laine, Tuija. *Carl Fredrik Fredenheim — en nyhumanist och hans klassiska bibliotek* [C. F. Fredenheim and His Classical Library]. Helsinki: Svenska Litteratursällskapet i Finland, 2010.
90. Lamm, Martin. *Upplysningstidens romantik. Den mystiskt sentimentala strömningen i svensk litteratur. I & II* [Romanticism of the Swedish Enlightenment]. Stockholm: Hugo Gebers, 1920.
91. Lehti, Raimo, and Tapio Markkanen. *History of Astronomy in Finland 1828–1918*. The History of Learning and Science in Finland 1828–1918 4b. Helsinki: Societas Scientiarum Fennica, 2010.
92. Lehto, Olli. *Tieteen aatelia. Lorenz Lindelöf ja Ernst Lindelöf* [Double Biography of L. and E. Lindelöf]. Helsinki: Otava, 2008.
93. ——. "Lexell, Anders Johan (1740–1784)." Biographical entry in [78].
94. Leikola, Anto. "Learned Men of Finland in the 17th and 18th Century." *Books from Finland* XXII (1988): 38–41.
95. Levasseur-Regourd, Anny-Chantal, and Philippe de La Cotardiere. *Halley, le roman des comètes*. Paris: Denoël, 1985.
96. ——. *Comètes et astéroïdes*. Paris: Le Seuil, 1997.
97. Lindberg, Sten G. *Bengt Ferrner. Resa i Europa 1758–1762* [Bengt Ferrner's Travels in Europe in 1758–1762 According to His Journal]. Lychnos Bibliotek. Uppsala: Almquist & Wiksell, 1956.
98. Lindroth, Sten. *Les chemins du savoir en Suède. De la fondation de l'Université d'Upsal à Jacob Berzelius*. Archives internationales d'histoire des idées, N° 126. Dordrecht: Martinus Nijhoff, 1988.
99. ——. "Svensk-ryska vetenskapliga förbindelser under 1700-talet" [Swedish-Russian Scientific Relations in the Eighteenth Century]. *Nordisk tidskrift för vetenskap, konst och industri. Utgiven av Letterstedtska föreningen*, Vol. 43, 1967, pp. 24–48. Reprinted in *Löjtnant Åhls äventyr, svenska studier och gestalter*. Stockholm: Wahlström & Widstrand, 1967.
100. ——. *Kungl. Svenska Vetenskapsakademiens historia 1739–1818. Tiden intill Wargentins död (1783)* [History of the Royal Academy of Sciences. First Part: 1739–1783]. Stockholm: Almquist & Wiksells, 1967.
101. Lindroth, Sten, and Gunnar Eriksson. *Svensk lärdomshistoria. Gustavianska tiden* [Swedish History of Learning: The Gustavian Era]. Stockholm: P. A. Norstedt & Söners Förlag, 1986.
102. Linné, Carl von. *Nemesis Divina*, ed. and trans. M. J. Petry. Archives internationales d'histoire des idées, N° 177. Dordrecht: Kluwer, 2001.
103. *List of Fellows of the Royal Society 1660–2007*. Library and Information Services. The Royal Society of London, 2007. (Available online at http://royalsociety.org/)
104. Лозинская, Лия Яковлевна: *Во главе двух академий*. [L. Ya. Lozinskaya: At the Head of Two Academies, Catherine Dashkova]. Москва: Наука, 1983. (Electronic version http://n-t.ru/ri/lz/da.htm)
105. Lubet, J.-P. "De Lambert à Cauchy: La résolution des équations littérales par le moyen des séries." *Revue d'Histoire des Mathématiques* 4 (1998): 73–129.

106. Lyubimenko, Inna. "Un académicien russe à Paris (d'après ses lettres inédites 1780–1781)." *Revue d'Histoire Moderne* Tome 10, N° 20, (Nov.–Déc. 1935): 415–447.
107. ——. "Заграничная командировка акад. А. И. Лекселя в 1780–1781 гг." [The Foreign Commission of Academician A. J. Lexell in 1780–1781]. *Архив Истории Науки и Техники* Том. 8, стр. 327–358, 1936.
108. ——. *Ученая корреспонденция Академии наук XVIII века*. 1766–1782 гг. *La correspondance scientifique de l'Académie de Pétersbourg dans le XVIIIe siècle*. Ленинград: Академия Наук СССР, 1937.
109. Лысенко, Валентин Иванович. "О работах Петербургских академиков А. И. Лекселя, Н. И. Фусса, и Ф. И. Шуберта по сферической геометрии и сферической тригонометрии". [V. I. Lysenko: The Work of Lexell, Fuss and Schubert in Spherical Geometry and Trigonometry], История Физико-Математических Наук. *Труды Института Истории Естествознанния и Техники*. Том. 34, стр. 384–414, Москва, 1960.
110. ——. "О математических работах А. И. Лекселя". [Lysenko: Lexell's Mathematical Work]. История и Методология Естественных Наук. Вып. XXV. Москва: Издательство Московского Университета, 1980.
111. ——. "Работы по полигонометрии в России" XVIII века. [Lysenko: Polygonometric Work in Russia in the Eighteenth Century]. *Историко-Математические Исследования*, 1959, стр. 161–178.
112. ——. *Николай Иванович Фусс (1755–1826)*. [Lysenko: Nikolai Ivanovich Fuss]. Москва: Наука, 1975.
113. ——. "Дифференциальные уравнения в работах А. И. Лекселя". [Lysenko: Differential Equations in the Work of Lexell]. *Историко-Математические Исследования*, 1990, стр. 39–52.
114. Mädler, Johann Heinrich von. *Geschichte der Himmelskunde von der ältesten bis auf die neueste Zeit*. Band I. Braunschweig: George Westermann, 1873.
115. Marshall, William. *Peter the Great*. London and New York: Longman, 1996.
116. McClellan, James E. *Science Reorganized: Scientific Societies in the Eighteenth Century*. New York: Columbia University Press, 1985.
117. Miller, Arnold. "The Annexation of a *Philosophe*: Diderot in Soviet Criticism, 1917–1960." In *Diderot Studies XV*, ed. Otis Fellows and Diana Guiragossian. Genève: Librairie Droz, 1971.
118. Mittler, Elmar, and Silke Glitsch, eds. *300 Jahre St. Petersburg. Russland und die "Göttingische Seele"*. Göttingen: Niedersächsische Staats- und Universitätsbibliothek, 2004.
119. Модзалевскій, Борисъ Львовичъ. *Списокъ членовъ Императорской Академіи Наукъ* 1725–1907 [Boris L. Modzalevski: List of Members of the Imperial Academy of Sciences]. С.-Петербургъ: Академія Наукъ, 1908.
120. Moutchnik, Alexander. *Forschung und Lehre in der zweiten Hälfte des 18. Jahrhunderts. Der Naturwissenschaftler und Universitätsprofessor Christian Mayer SJ (1719–1783)*. Augsburg: Dr. Erwin Rauner Verlag, 2006.
121. Mustelin, Olof. "En finländsk astronom på resa i Europa 1780–1781" [The Journey of a Finnish Astronomer in Europe]. *Finsk Tidskrift* Nr. 1 (1963): 147–157.
122. ——. "Anders Johan Lexell." *Svenskt biografiskt lexikon*, Band 22, (1977–1979): 670.
123. Myrberg, Pekka Juhana. "Matemaattiset tieteet vanhassa Turun akatemiassa" [The Mathematical Sciences at the old Academy of Åbo]. *Suomalainen Tiedeakatemia: esitelmät ja pöytäkirjat 1950*. Helsinki: Academia Scientiarum Fennica, 1951.
124. ——. "Martin Platzman (1760–1786)." *Arkhimedes* [Periodical of the Finnish Mathematical Society and the Physical Societies in Finland], N° 1, 1963.
125. Nordenmark, Nils Viktor Emanuel. *Pehr Wilhelm Wargentin. Kungliga Vetenskapsakademiens sekreterare och astronom 1749–1783*. Uppsala: Kungliga Svenska Vetenskapsakademien, Almquist & Wiksell, 1939.
126. ——. *Fredrik Mallet och Daniel Melanderhjelm, två Uppsala-astronomer*. Uppsala: Kungliga Svenska Vetenskapsakademien, Almquist & Wiksell, 1946.

127. ——. *Astronomiens historia i Sverige intill år 1800* [History of Astronomy in Sweden Until 1800]. Uppsala: Almquist & Wiksell, 1959.
128. Nordenmark, Nils Viktor Emanuel, and Johan Nordström. Om uppfinningen av den akromatiska och aplanatiska linsen: med särskild hänsyn till Samuel Klingenstiernas insatser [On the Invention of the Achromatic and the Aplanatic Lenses, Especially with Regard to the Contribution of Klingenstierna]. *Lychnos årsbok* (1938): 1–52, and ibid. (1939): 313–384.
129. *Nordisk Familjebok – Konversationslexikon och realencyklopedi* [Swedish Encyclopedia]. Stockholm, 1876–1899.
130. Nyström, Eva. "Naturalhistoriens tillstånd i Ryssland. Johan Anders Lexells brev till Linné 1772–1776" [On the Natural History of Russia. Correspondence Between Lexell and Linnaeus]. *Svenska Linnésällskapets årsbok*. Uppsala, 2004–2005, pp. 7–56.
131. Oittinen, Vesa. "Between Radicalism and Utilitarianism: On the Profile of the Finnish Enlightenment." In *The Northern Lights, Facets of the Enlightenment Culture*, ed. Tatiana Artemyeva, Mikhail Mikeshin, and Vesa Oittinen. The Philosophical Age, Almanac 36, 10–24. St. Petersburg: Saint Petersburg Centre for the History of Ideas, 2010.
132. Österbladh, Kaarlo, ed. *K. F. Mennanderin lähettämiä ja saamia kirjeitä* [Letters Sent and Received by Bishop C. F. Mennander]. Suomen historian lähteitä, Vols. 1–3. Helsinki, 1939–1942.
133. Palais, Bob, Richard Palais, and Stephen Rodi. "A Disorienting look at Euler's Theorem on the Axis of Rotation." *American Mathematical Monthly* 116, N° 10 (2009): 892–909.
134. Пекарский, Пётр Петрович. *Исторія Императорской Академіи Наукъ въ Петербургѣ* [P. P. Pekarsky: History of the Imperial Academy of Sciences, 2 Vols.]. Санкт Петербургъ: Академия Наукъ, 1870–1873.
135. Pekonen, Osmo. *La rencontre des religions autour du voyage de l'abbé Réginald Outhier en Suède en 1736–1737*. Rovaniemi: Lapland University Press, 2010.
136. ——. *Salaperäinen Venus* [The Mysterious Venus]. Ranua: Mäntykustannus, 2012.
137. Pfrepper, Regine, and Gerd Pfrepper. "Georg Moritz Lowitz (1722–1774) und Johann Tobias Lowitz (1757–1804) – zwei Wissenschaftler zwischen Göttingen und St. Petersburg." In [118, pp. 163–182].
138. Pihlaja, Päivi Maria. *Tiedettä Pohjantähden alla: Pohjoisen tutkimus ja Ruotsin tiedeseurojen suhteet Ranskaan 1700-luvulla* [Science Under the North Star: Nordic Research and the Relations of the Swedish Scientific Societies with France in the Eighteenth Century]. Bidrag till kännedom av Finlands natur och folk, 181. Helsinki: Finska Vetenskaps-Societeten, 2009.
139. Pipping, Gunnar. *The Chamber of Physics. Instruments in the History of Sciences Collections of the Royal Swedish Academy of Sciences*. Uppsala: Stiftelsen Observatoriekullen, 1991.
140. Poggendorff, J. C. *Biographisch-literarisches Handwörterbuch zur Geschichte der Exacten Wissenschaften*. Leipzig: J. A. Barth, 1863.
141. *Протоколы засѣданій конференціи Императорской Академіи Наукъ с 1725 по 1803 года = Procès-verbaux des séances de l'Académie Impériale des Sciences depuis sa fondation jusqu'à 1803*. Tomes II & III. Saint Pétersbourg, 1897–1911. (All volumes available electronically at http://www.ranar.spb.ru/rus/protokol1/cat/232/)
142. Reich, Karin. "Euler's Contribution to Differential Geometry and Its Reception." In [18, pp. 479–502].
143. Ribenboim, Paulo. *Fermat's Last Theorem for Amateurs*. New York: Springer, 1999.
144. Rodhe, Staffan. *Matematikens utveckling i Sverige fram till 1731* [The Development of Mathematics in Sweden Until 1731]. Department of Mathematics, Uppsala University, 2002.
145. Rosenfeld, Boris A. *A History of Non-Euclidean Geometry*. Translated from the Russian. New York: Springer, 1988.
146. Rosenhane, Shering. *Anteckningar hörande till Kongl. Vetensk. Akademiens historia* [Notes on the History of the Royal Academy of Sciences]. Stockholm: Joh. P. Lindh, 1811.
147. Sarton, George. *Das Studium der Geschichte der Naturwissenschaften*. Frankfurt am Main: Vittorio Klostermann, 1965. [Originally in English: *The Study of the History of Science*. Harvard University Press, 1936]

148. Shank, J. B. *The Newton Wars and the Beginning of the French Enlightenment*. Chicago: The University of Chicago Press, 2008.
149. Sigrist, René. "Quand l'astronomie devint un métier: Grandjean de Fouchy, Jean III Bernoulli et la « république astronomique », 1700–1830." *Revue d'histoire des sciences*, Tome 61 (2008), N° 1: 105–132.
150. Simon, Max. "Über die Entwicklung der Elementar-Geometrie im XIX. Jahrhundert." *Jahresbericht der Deutschen Mathematiker-Vereinigung*, Ergänzungsbände, B. 1. Leipzig: Teubner, 1906.
151. Siukonen, Jyrki. *Mies palavassa hatussa. Professori Johan Welinin maailma* [The Man in the Burning Hat: The World of Professor Johan Welin]. Helsinki: Suomalaisen Kirjallisuuden Seura, 2006.
152. Sjöstrand, Wilhelm. *Grunddragen till den militära undervisningens uppkomst- och utvecklingshistoria i Sverige till året 1792* [Outline of the History of Military Education in Sweden Until 1792]. Uppsala: Lundequistska bokhandeln, 1941.
153. Slotte, Karl Fredrik. *Åbo universitets lärdomshistoria: Matematiken och fysiken* [The History of Learning at the Academy of Åbo: Mathematical and Physical Sciences]. Helsingfors: Svenska Litteratursällskapet i Finland, 1898.
154. Stählin, Karl. *Aus dem Papieren Jacob von Stählins*. Königsberg i. Pr.-Berlin: Ost-Europa-Verlag, 1926
155. Steele, John M. *Ancient Astronomical Observations and the Study of the Moon's Motion (1691–1757)*. New York: Springer, 2012.
156. Sterken, Christiaan and Per Pippin Aspaas (eds.) (2013): *Meeting Venus. A collection of papers presented at the Venus Transit Conference in Tromsø 2012*. Vrije Universiteit Brussel och Universitetet i Tromsø. *The Journal of Astronomical Data*. 19, No. 1. (http://www.vub.ac.be/STER/JAD/jad.htm)
157. Svanberg, Ingvar. "Linnean i österled — Johan Peter Falck i Ryssland" [On Johan Peter Falck's Expeditions in Russia]. *Geografitorget*, N° 4 (2007): 17–29.
158. Tengström, Johan Jacob. *Chronologiska förteckningar och anteckningar öfver finska universitetets fordna procancellerer samt öfver faculteternas medlemmar och adjuncter från universitetets stiftelse inemot dess andra sekularår* [Biographical Data of the Personnel of the Finnish University from Its Two First Centuries]. Helsingfors: Wasenius, 1836.
159. Terrall, Mary. *The Man Who Flattened the Earth: Maupertuis and the Sciences in the Enlightenment*. Chicago: University of Chicago Press, 2002.
160. Trembley, Jean. *Essai de trigonométrie sphérique, contenant diverses applications de cette science à l'Astronomie*. Neuchâtel: Samuel Fauche, 1783.
161. Truesdell, Clifford. *Essays in the History of Mechanics*. Berlin: Springer, 1968.
162. ——. *The Rational Mechanics of Flexible or Elastic Bodies: 1638–1788. Leonhardi Euleri Opera Omnia*, Series II, vol. 11.2. Turici (Zürich): Orell Füssli, 1960.
163. Turner, Gerard L'Estrange. "The Portuguese Agent J. H. de Magellan." *Antiquarian Horology* 9, N° 1 (1974): 74–76.
164. Valipour, Valeska. *La pratique théâtrale dans l'Allemagne de la seconde moitié du dix-huitième siècle (1760–1805)*. Thèse de doctorat en études théâtrales. Université Sorbonne Nouvelle–Paris 3, 2011.
165. Vallinkoski, Jorma. *The History of the University Library at Turku*, Part II. 1722–1772. Helsinki: Publications of the University Library at Helsinki (N° 37), 1975.
166. Valtonen, Mauri, and Hannu Karttunen. *The Three-Body Problem*. Cambridge: Cambridge University Press, 2006.
167. Verdun, Andreas. "The Determination of the Solar Parallax from Transits of Venus in the Eighteenth Century." *Archive des Sciences* 57, Fasc. 1 (2004): 45–68.
168. ——. *Entwicklung, Anwendung und Standardisierung mathematischer Methoden und physikalischer Prinzipien in Leonhard Eulers Arbeiten zur Himmelsmechanik*. Habilitationsschrift. Universität Bern, Philosophisch-Naturwissenschaftliche Fakultät, 2010.
169. ——. "Leonhard Euler's Early Lunar Theories 1725–1752." *Archive for History of Exact Sciences* 67, N° 3 (2013): 235–303.

170. Vucinich, Alexander. *Science in Russian Culture. 1. A History to 1860*. Stanford: Stanford University Press, 1963.
171. Wendland, Folkwart. *Peter Simon Pallas (1741–1811). Materialien einer Biographie* (2 Vols.). Berlin: Walter de Gruyter, 1992.
172. Whittaker, Edmund Taylor. *A Treatise on the Analytical Dynamics of Particles and Rigid Bodies, with an Introduction to the Problem of Three Bodies*, 4th ed. Cambridge: Cambridge University Press, 1952.
173. Williamson, George C., ed. *Bryan's Dictionary of Painters and Engravers*, Vol. III. London: MacMillan, 1903.
174. Wilson, Curtis. "Euler and Applications of Analytical Mathematics to Astronomy." In [18, pp. 121–146].
175. Wolf, Rudolf. *Biographien zur Kulturgeschichte der Schweiz*. Cyclus I–IV. Zürich: Orell, Füssli & Co., 1858–1862.
176. Woolf, Harry. *The Transits of Venus: A Study of Eighteenth-Century Science*. Princeton: Princeton University Press, 1959.
177. Yeo, Richard. "Encyclopaedism and Enlightenment." In [45, pp. 350–365].

Index of Names

J.C.-E. Stén, *A Comet of the Enlightenment*, Vita Mathematica 17,
DOI 10.1007/978-3-319-00618-5,

N

O

P

Q

R

Abbreviations

AASIP	Acta Academiae Scientiarum Imperialis Petropolitanae, St. Petersburg (1777–1782)
AJEB	Astronomisches Jahrbuch oder Ephemeriden, Berlin
Berg. brevs.	"Bergianska brevsamlingen", a collection of letters copied by Bengt Bergius, located at KVAC
E ...	refers to the catalogue[1] of Euler's publications by Gustaf Eneström.
HUB	Helsinki University Library, the National Library of Finland
KB	Kungliga Biblioteket, the National Library of Sweden, Stockholm
KVAC	Centre for history of science at the Royal Swedish Academy of Sciences (Kungliga Vetenskapsakademien, KVA), Stockholm
KVAH	Kongliga Vetenskaps Academiens Handlingar, Stockholm
KVANH	Kongliga Vetenskaps Academiens Nya Handlingar, Stockholm
[Lexell ...]	refers to the list of Lexell's Publications in the index part of this volume (pp. 269–275).
NAASIP	Nova Acta Academiae Scientiarum Imperialis Petropolitanae, St. Petersburg (1783–1802)
NCASIP	Novi Commentarii Academiae Scientiarum Imperialis Petropolitanae, St. Petersburg (1747–1775)
OO	*Leonhardi Euleri Opera Omnia*. Series I–IV. Birkhäuser/Springer, Basel.
(СПФ)АРАН	Санкт-Петербургский Филиал Архива Российской Академии Наук = Saint Petersburg Branch of the Archives of the Russian Academy of Sciences[2]

[1] *Verzeichnis der Schriften Leonhard Eulers*, Jahresbericht der Deutschen Mathematiker-Vereinigung, Ergänzungsband IV, 2. Lieferung. Leipzig: Teubner, 1913.

[2] Archival references used: Фонд = Collection, Опись = Inventory, Дело = Record, Лист = Folio, Книга = Book, об. = *verso*.

J.C.-E. Stén, *A Comet of the Enlightenment*, Vita Mathematica 17,
DOI 10.1007/978-3-319-00618-5,

St. ...	refers to the catalogue of Daniel Bernoulli's publications by Hans Straub in [55].
UBB	Library of the University of Basel
UUB	Uppsala University Library

The manufacturer's authorised representative in the EU is Springer Nature Customer Service Centre GmbH, Europaplatz 3, 69115 Heidelberg, Germany. If you have any concerns regarding our products, please contact ProductSafety@springernature.com

Printed and bound by CPI Group (UK) Ltd, Croydon, CR0 4YY
30/09/2026
02225238-0005